Impact of Environmental Degradation on Ecosystems and Preventive Measures

Melford Mbedzi
Mangosuthu University of Technology, South Africa

Sithembile Innocentia Nkosi
Mangosuthu University of Technology, South Africa

David Olusoji Alabi
Durban University of Technology, South Africa

IGI Global
Scientific Publishing
Publishing Tomorrow's Research Today

Vice President of Editorial	Melissa Wagner
Director of Acquisitions	Mikaela Felty
Director of Book Development	Jocelynn Hessler
Production Manager	Mike Brehm
Cover Design	Jose Rosado

Published in the United States of America by
IGI Global Scientific Publishing
701 East Chocolate Avenue
Hershey, PA, 17033, USA
Tel: 717-533-8845
Fax: 717-533-7115
Website: https://www.igi-global.com E-mail: cust@igi-global.com

Library of Congress Cataloging-in-Publication Data

Names: Mbedzi, Melford, 1988- editor | Nkosi, Sithembile Innocentia, 1984- editor | Slabi, David Olusoji, 1979- editor
Title: Impact of environmental degradation on ecosystems and preventive measures / Melford Mbedzi, Sithembile Innocentia Nkosi, David Olusoji Slabi.
Description: Hershey, PA : IGI Global Scientific Publishing, [2025] | Includes bibliographical references and index. | Summary: "This book aims to raise awareness about the impact of environmental degradation on ecosystems, human health, and the economy in Africa"-- Provided by publisher.
Identifiers: LCCN 2025012310 (print) | LCCN 2025012311 (ebook) | ISBN 9798337315492 h/c | ISBN 9798337315508 s/c | ISBN 9798337315515 ebook
Subjects: LCSH: Environmental protection--Africa | Environmental degradation--Africa | Ecosystem management--Africa
Classification: LCC TD170.3 .I47 2025 (print) | LCC TD170.3 (ebook) | DDC 363.70096--dc23/eng/20250416
LC record available at https://lccn.loc.gov/2025012310
LC ebook record available at https://lccn.loc.gov/2025012311

British Cataloguing in Publication Data
A Cataloguing in Publication record for this book is available from the British Library.

Table of Contents

Detailed Table of Contents

Environmental deterioration poses unprecedented hazards to the natural ecosystem and human health, save for world economy. Environmental damage follows the path of sharp interweaving of human activities including industrialization, deforestation, air pollution, and resource exploitation. This deterioration advances socioeconomic inequality, expounds on the reasons for the reduction in biodiversity, and fuels climate change. This chapter offers a thorough study of environmental damage, with specific attention to its consequences for the African continent including South Africa as well as other sensitive regions all around. The study mostly focuses on doable sustainability projects meant to reduce environmental harm and promote a more beautiful future.

Invasive alien flora can overwhelm riverine network and modify the hydrology. Little is known about riparian invasive alien trees impacts on general ecology although invasive alien trees are by far the biggest threat to universally red-listed dragonflies (Odonata) in South Africa. Twenty-four permanent plots of 10 m2 each were constructed. Before removal, invasive alien species and native species present in the quadrats were counted and recorded. The invasive alien plants were removed by digging and uprooting with no use of chemicals. The difference was not significant when it comes to species diversity and richness before and after removal of Lantana

Türkiye's limited use of renewable energy increases reliance on fossil fuels, widening the trade deficit and heightening energy security risks. This dependency also restricts economic growth, job creation, and worsens environmental damage due to higher carbon dioxide (CO2) emissions. This study examines the impact of renewable and non-renewable energy consumption (REC and NREC) on CO2 emissions in Türkiye from 1965 to 2023, using the asymmetric bootstrap nonlinear cointegration test. Results show that, in the long term, NREC significantly raises CO2 emissions more than REC, while in the short term, only NREC affects emissions. Based on these findings and Türkiye's renewable energy potential, key policy implications are identified. In conclusion, increasing domestic renewable energy production is a vital solution for reducing the trade deficit and energy security risks, aligning with the broader goal of a low-carbon future.

Biomedical Waste is coming from medical institutions and its proper disposal and management is important for public health, environmental protection and occupational safety. International bodies such as the World Health Organization (WHO) and the United Nations Environment Programme (UNEP) and National Regulators have widely provided comprehensive frameworks and guidelines for the standardization of biomedical waste management. The WHO's "Safe Management of Wastes from Healthcare Activities" holds important protocols for segregating, treating and disposing of hazardous medical waste. The Basel Convention sets controls on the trans-boundary movement of hazardous waste to facilitate environmentally sound disposal and disposal methods worldwide. Regulation is further ensured through public awareness, the training of healthcare workers, and stringent monitoring by environmental agencies. In the age of Industry 6.0, international cooperation and strict regulations will ensure that the disposal of biomedical waste complies with environmental and public health requirements.

Urban wetlands serve an important role in cooling cities, sequestering carbon, and offering recreational and educational opportunities for urban residents. They enhance water quality by filtering contaminants and removing surplus nutrients, conserving aquatic habitats. They reduce the emission of GHGs into an urban environment. The current study investigates the diverse significance of the Surajpur wetland. It extends much-needed habitat to a wide range of species in the region. It is significant due to its proximity to metro expansion in Delhi and adjacent districts. However, these benefits are faced with challenges like development pressure and pollution in urban wetlands, which, as a result, require good management and restoration. This concern chapter mainly deals with water quality indices for the year 2024 in the Surajpur wetland. As a constant case, this case study attempts to implement wetlands in city life for a resilient, sustainable city. This chapter discloses that the Surajpur wetland is a fundamental structure that needs to be protected for ecological and human beings.

Increasing soil salinization threatens agriculture and biodiversity by limiting ecosystem productivity and resilience. Carob (Ceratonia siliqua L.), native to Mediterranean regions, is notable for its drought tolerance and adaptability to degraded soils. This study examines the morphological, physiological, and biochemical responses of six Moroccan carob ecotypes to NaCl-induced salt stress under greenhouse conditions. Results reveal significant variability among ecotypes, emphasizing the role of genetic diversity in adaptation. Key parameters, such as proline and total sugar content, highlight promising adaptation mechanisms. Tolerant ecotypes could aid in saline soil restoration and support sustainable agriculture, positioning carob as a strategic resource for combating desertification and promoting resilient agroecosystems.

Today's modern life makes it increasingly difficult for individuals to live in a healthy environment. In this context, sustainability plays a critical role in solving environmental problems. The purpose of this research is to determine the environmental awareness and knowledge levels of university students. The sample consists of 384 students studying at Sivas Cumhuriyet University. The data were analyzed using the independent sample t-test, ANOVA, and Tukey test in the SPSS 23.0 program. The findings showed high environmental awareness of students receiving education on climate and environmental problems, family environment, 2nd-grade and natural students, and vocational school students. The results reveal the effect of education on more sensitivity to the environment and provide an important roadmap for educators and policymakers in developing strategies.

Preface

OVERVIEW OF THE SUBJECT MATTER

This book, Impact of Environmental Degradation on Ecosystems and Preventive Measures, addresses the critical issue of environmental degradation, its impact on ecosystems, and the preventive measures necessary to mitigate its adverse effects. The chapters are structured to provide a comprehensive exploration of how human activities have led to significant ecological imbalances and what strategies can be employed to restore and protect natural environments.

Relevance to the World Today

The importance of environmental sustainability cannot be overstated in today's rapidly changing world. With the intensification of climate change, resource depletion, and biodiversity loss, the themes explored in this book are more relevant than ever. Addressing environmental challenges not only ensures the health of our ecosystems but also safeguards human health, supports economic stability, and promotes social equity.

Target Audience

This book is intended for a diverse audience including university students, educators, policymakers, environmental scientists, and sustainability advocates. Its comprehensive examination of environmental degradation and sustainability measures provides valuable insights for anyone invested in understanding and solving ecological challenges.

Section Descriptions

1. Restoring Ecosystem Balance: Challenges and Opportunities – This section (Chapters 1-3) explores environmental degradation and sustainability trends, the dynamics of invasive species, and ecological control measures. It highlights the critical steps necessary for restoring ecological balance and enhancing ecosystem resilience.
2. Climate Change, Energy Transition, and Social Equity – This section (Chapters 4-5) delves into the role of renewable energy in reducing carbon footprints, addressing climate change, and bridging social inequalities. It focuses on sustainable energy policies and the transformative potential of clean energy solutions for economic and environmental stability.
3. Pollution Control, Waste Management, and Ecosystem Resilience – Focusing on biomedical waste disposal, urban wetland management, and plant resilience to abiotic stresses, this section (Chapters 6-9) emphasizes sustainable waste management practices and the preservation of vital ecosystems against pollution and climate impacts.

CONCLUSIONS

The contents of this book serve as both critical analysis and a call to action. It is a blueprint for understanding the consequences of environmental degradation and the necessary steps toward a sustainable future. By integrating scientific insights with policy recommendations, it aims to inspire collective action and drive meaningful changes in environmental preservation and sustainability.

Acknowledgement

The successful completion of this book would not have been possible without the support and guidance of many individuals and institutions. I would like to express my heartfelt gratitude to the academic and research community whose tireless work and commitment to environmental sustainability inspired much of the content presented here. Special thanks go to my colleagues and mentors who provided valuable insights and feedback during the development of each chapter.

I am also deeply grateful to Durban University of Technology for providing the academic environment that fostered research and exploration of environmental awareness among students. To the students who participated in the surveys and studies, your engagement and insights were invaluable to this work. Finally, I extend my sincerest thanks to my family and friends for their unwavering support and encouragement throughout this journey. Your belief in the importance of this work kept me motivated and inspired to bring it to completion.

Melford Mbedzi
Mangosuthu University of Technology, South Africa

Sithembile Innocentia Nkosi
Mangosuthu University of Technology, South Africa

David Olusoji Alabi
Durban University of Technology, South Africa

Introduction

Environmental degradation is one of the most pressing global issues of our time, threatening the delicate balance of our ecosystems, human health, and the sustainable future of our planet. As climate change accelerates, natural resources are strained, and biodiversity is under unprecedented threat, the need for collective awareness and proactive measures has never been greater. This book, Impact of Environmental Degradation on Ecosystems and Preventive Measures, is a timely contribution to the discourse on sustainable development, conservation, and environmental stewardship.

In a world increasingly dominated by industrialization, urbanization, and technological advancement, the relationship between human activities and the environment has become alarmingly unbalanced. The consequences of this imbalance are evident in rising global temperatures, soil degradation, water pollution, and the loss of species across the globe. This work brings to light not only the challenges posed by environmental degradation but also the potential pathways to mitigation and restoration.

The authors have meticulously explored a range of critical topics, from the dynamics of renewable and non-renewable energy consumption to the challenges of biomedical waste disposal and sustainable urban wetland management. They have further investigated the resilience of ecosystems amidst invasive species, the socioeconomic implications of water scarcity, and the promising role of sustainable agricultural practices.

It is my belief that this book serves as a vital resource for educators, policymakers, environmental scientists, and students alike, offering not just a critique of current environmental practices, but also a vision for sustainable progress. The collective insights provided in these chapters emphasize the urgency of environmental action and the need for innovative solutions to restore ecological balance and secure a sustainable future.

Introduction

Chapter 1
Environmental Degradation and Sustainability:
Trends, Challenges, and Opportunities

Anirudh Dash
https://orcid.org/0009-0004-4036-5036
Christ University, India

ABSTRACT

Environmental deterioration poses unprecedented hazards to the natural ecosystem and human health, save for world economy. Environmental damage follows the path of sharp interweaving of human activities including industrialization, deforestation, air pollution, and resource exploitation. This deterioration advances socioeconomic inequality, expounds on the reasons for the reduction in biodiversity, and fuels climate change. This chapter offers a thorough study of environmental damage, with specific attention to its consequences for the African continent including South Africa as well as other sensitive regions all around. The study mostly focuses on doable sustainability projects meant to reduce environmental harm and promote a more beautiful future.

INTRODUCTION

Environmental deterioration poses unprecedented hazards to the natural ecosystem and human health, save for world economy. Environmental damage follows the path of sharp interweaving of human activities including industrialization, deforestation,

DOI: 10.4018/979-8-3373-1549-2.ch001

air pollution, and resource exploitation. This deterioration advances socioeconomic inequality, expounds on the reasons for the reduction in biodiversity, and fuels climate change. This chapter offers a thorough study of environmental damage, with specific attention to its consequences for the African continent including South Africa as well as other sensitive regions all around. The study mostly focuses on doable sustainability projects meant to reduce environmental harm and promote a more beautiful future. 2013; Rosediana, S.; Anshari, G. The dependence of the African ecosystems and environments-the Columbia basin-and their corresponding shift of rental revenue towards development has coupled climate change and environmental deterioration creating ba-black stories not so pleasing. On the other end of the scale, Africa-afterfurther predisposed is tarnished by uncontrolled exploitation via field conversion, therefore providing another metric to balance such areas in dread of future exploitation. Asnake and Elias (2019) claim that deforestation has far-reaching negative impacts on land degradation from soil erosion and disturbance of natural water cycles, which aggravates the consequences of global climate change even more.

Africa and Asia still face a great challenge from their continuous environmental catastrophe of pollution. In heavily crowded cities like Johannesburg, Lagos, and Jakarta, the major disease risks from air pollution include respiratory diseases, cardiovascular difficulties and decrease human life spans. Industrial waste combined with agricultural runoff and inadequate sanitation systems generates water pollution, which also causes outbreaks of health-threatening waterborne illnesses such cholera and dysentery, hence exacerbating water shortage. While highlighting the overlapping links between environmental stewardship, social equality, and population welfare, environmental degradation exhibits its most severe consequences upon socioeconomic disadvantaged groups.

Since Africa is still especially vulnerable to the consequences of climate change, it is the main reason causing environmental degradation. Along with increasingly frequent severe weather occurrences like droughs and cyclones and floods significantly impacting food production while having populations migrate and overwhelm limited resources, increasing heat levels plus changing rainfall distributions effect. Because environmental forces force people from their homes which necessitated them to leave their natural locations resulting in struggles over access to land and water supplies and economic capacities, climate change migration has developed as a new worldwide concern.

Loss of biodiversity becomes a serious problem when habitat degradation combines with poaching activities together with effects of pollution and climate change. Among the many well-known African species including elephant, lion, rhinoceros, and gorilla, many are under increasing risk of extinction.

Waste management and reduction present major issues for South Africa and the larger African continent that directly affect environmental sustainability and public health. Effective garbage disposal and environmental stewardship are desperately needed in view of fast urbanization, population increase, and growing industrial activity. This chapter investigates many waste management techniques and reduction initiatives influencing sustainability in South Africa and the African continent generally. Poorly handled garbage causes major health hazards, poisoning the environment and destroying ecosystems. Harmful greenhouse gasses like methane are emitted in metropolitan areas with poor landfill management, therefore aggravating problems with climate change. Beyond only ecological damage, environmental degradation has major financial effects. For many African countries to maintain their economic stability, tourism and natural resource management are rather important. Damage to national parks, forests, and water supplies might compromise the tourism industry, therefore affecting employment losses and lower income. Moreover, restricted water availability and diminishing soil quality jeopardize food security and economic development, (Nichanian, D. 2016) which drives many people into additional poverty and raises their sensitivity to environmental threats. This chapter assesses key resilient and ecologically friendly solutions meant to tackle present issues.

What is Environmental Degradation?

In many respects, our environment shapes our health. Many studies have looked at the link between human health and the surroundings; (K. C., & Mohanta, D. K. 2012, May) it has been found that environmental hazards significantly (Mashila, T. 2014) affect human health either directly by means of exposure to harmful substances or indirectly by means of disturbance of ecosystems sustaining life. Environmental degradation includes examples of deterioration of ecosystems, species lost, and depletion of natural resources like soil, water, and air. Environmental degradation is the classification for any change or disturbance of the surroundings seen to be detrimental or unwelcome. Environmental degradation is one of the Ten Threats the UN High Level Threat Panel publicly warned against. The United Nations International Strategy for Disaster Loss defines environmental degradation as "the reduction of the capacity of the environment to meet social and ecological objectives, and needs". Environmental degradation mostly originates from human interference. The habitat, the cause behind which the plants and animals survive, and their presence all determine the degree of effect the surroundings allow. Human activity is one of the key reasons behind environmental degradation. Non-industrial activities generally damage or degrade the environment in Africa; thus, people treat the environment as a result of their jobs in mining, forestry, and agriculture, which are all centered on the production of main commodities. Knowing this abuse of the ecology would help one to spot environmental damage in Africa. Living shapes the surroundings, so there is a link, connection, or interaction between the natural world and the activities people do for a living.

Causes of Environmental Degradation in Africa

Poverty and population increase are causing environmental damage in Africa. Many people involved in environmental activities and conversations about it nowadays believe that population expansion is the key variable or source of environmental problems around the globe which keeps worsening. Though some businesses disagree, generally we accept this all the time since at least six million individuals call this planet home. The population ought to rise from six billion to 9.4 billion in the next fifty years. While population in developing countries would increase by more than 70 percent, populations in developed countries will plateau at about 1.2 billion. Developing nations are supposed to handle the extra 3.4 billion.

African nations generate more basic products; consequently, it is our contention that fast population increase is already testing the natural resource reserves on the continent and will always do so. The environment itself suffers from the very worsening surrounding s in the form of strained or overused resources resulting from some type of exploitation. Average population growth should be slower to raise capital per labour unit.

Much of the debate on poverty at the United Nations, non-governmental organizations, and government agencies during their discussion on climate and development underlines "That process which deprives people, of the basic means for livelihoods, and that undermines his physical, cultural, and spiritual well-being," economists describe poverty. The participatory element associated to the word, which lacked a very clear meaning, had to do with "enhancing the ability of the people to control resources to shape their own life." From what I know, those folks will need to be able to assist themselves from their surroundings utilizing the resources at hand. Abuse of the meager resources nearing poverty is practically the truth nowadays.

Less than $1 a day is lived by about 1.2 billion people; most of them reside in rural areas of the developing world, particularly Africa. The sources assert that the last poor areas are Asia, Africa, and Latin America. Three-quarters of the poorest of the poor worldwide reside in rural regions; Latin America makes 20%, Asia makes 57%, and Africa makes 51%. The only choice left for the rural poor is to squander their already extremely limited resources in general pursuit of a livelihood or particular search for food security. Among the effects is environmental degradation. Consequently, the few means of the people's living produce what has subsequently been called a "downward spiral" or a "vicious circle". Most of the time, the pressing demands for family food—in both consumption and production—are forced decisions against the environmental sustainability. Poverty therefore turns into a monster.

Types of Environmental Degradation in Africa

In majority of the African nations, including Nigeria, Ghana, Togo, Sierra Leone, Gambia, Liberia, etc., environmental damage results largely from nonindustrial activities. But industrial contaminants now present in Africa are significantly more than they used to be. Most of these nations generate main commodities: mining, forestry, fishing, agriculture, timber, and energy. Apart from basic goods, the populace helps in manufacturing; this entails food processing, simple material handling including paper, glass, and textiles, as well as advanced production including pharmaceuticals or machinery assembly. 16. While there is much freedom to alter the content of the produced commodities, the raw ingredients to make simple products, like copper or groundnuts, are, by nature fixed. One should be able to tell simple from manufactured goods. From an environmental perspective, the only option to boost the output of key commodities is to boost local resource exploitation, which also carries the danger of resource degradation including the depletion of fish stocks, loss of soil fertility, and exhaustion of mineral resources.

Conversely, if affordable energy sources and capital are easily accessible, industrial production may be raised using somewhat remote raw material sources. Therefore, the manufacturing of basic products enables overcultivation and land abuse and usage with consequent effects of deforestation, desertification, soil erosion, and pollution. Apart from that, the materials causing pollution and the way energy is disposed of help to determine the environmental effect of manufacturing.

One type of environmental effect resulting from energy generation is contamination of air and water. [2023 Yan, L., Zhou, J., & Li, H.] Loss of rich soil, deterioration of farmlands, extinction of species and biodiversity, and damage of aquatic ecosystems are further elements clearly influencing human health. The social and environmental expenses of oil production add even another level of environmental damage. The Niger Delta of Nigeria is the finest illustration of this deterioration. Among additional contaminants are pipeline leaks, gas flaring, oil waste disposal, and oil spills. Agbu 20 claims, on the other hand, flares account for 75% of the yearly gas output in (Leonard, B., & Regan, S. 2019). Other resources, including land, water, and vegetation, become seriously physically and ecologically damaged.

The Impact of Environmental Degradation in Africa

Humanity is confronted with numerous risks and hazards as a result of environmental degradation. Local resource exploitation results in the depletion of marine stocks, the depletion of mineral resources, and the decline in soil fertility, as previously mentioned. Ultimately, this results in food insecurity or catastrophe, which further exacerbates destitution. In addition to polluting the air and potable water, energy production also destroys animals, farmlands, and aquatic ecosystems. All of these have a significant impact on the health of individuals. Desertification has led to a decrease in the availability of agricultural land, grazing pastures, and rangeland. This frequently leads to conflicts between Arab nomads and African cultivators.

In particular, the pressure of desert encroachment in northeastern Nigeria is closely associated with the confrontations between ranchers and grazers, which results in a greater restriction of both agricultural and grazing areas. Soil erosion is the result of overgrazing, which depletes the soil, and overcultivation, which clears the land of vegetation. This problem is prevalent in the eastern region of Nigeria. Soil degradation also contributes to an increase in the number of river and dam citations. This is the primary cause of deforestation.

What is Environment Sustainability?

Environmental sustainability is the ability to maintain the features appreciated in the physical surroundings. Human life, the natural environment's capacity to sustain human and other species' living conditions (such as clean air and water, a suitable temperature), the elements of the environment that generate renewable resources like water, timber, fish, and solar energy, the ability of society to operate despite the depletion of non-renewable resources, the standard of living for all people, and the environment's liability and beauty.

Some environmental aspects run the risk not being conserved when they are endangered. For example, the great use of non-renewable resources (like coal, oil, and minerals) or damage to the ecosystem runs the danger of major degradation in quality, destruction, or extinction. Environmental managers generally find means to reduce the damage or waste when environmental problems arise. Still, it can be challenging to undo the effects of threats, and specific timing and location of occurrence are not always predictable. Environmental managers are thus using strategies aimed to prevent damage before it starts more and more.

Steps to avert hazards and effects from occurring, protect the environment from damage, and heal past damage must all be part of a thorough sustainability program. Sustainability issues arise whenever there is a possibility that the beloved artifacts or features of the surroundings might be lost in a challenging or irreversible manner.

Furthermore, one has to act right away if such threats arise. Environmental sustainability initiatives include reducing the use of physical resources, adopting a "recycle everything/buy recycled" attitude, using renewable resources instead of depletable ones, redesigning production processes and products to stop the manufacturing of toxic materials, and safeguarding and restoration of natural habitat and environments valued for their liability or beauty.

What Makes an Issue a Sustainability Issue?

Every time a valuable system, object, tool, or process is under danger, a sustainability question emerges. Considerable decrease in the quality of the valued system, item, object, process, or characteristic might risk its existence. Stated differently, a sustainability issue arises whenever anything of values stands the risk of not being sustained. Always a sustainability issue is a great sense of urgency raises. This urgency may have to do with something presently existing or a known potential (Subrahmanyam, K. 2000). For instance, someone's capacity to reach their full potential may be hindered or biodiversity may be at danger of extinction if they live in poverty or if violence or sickness endangers their life.

Starting sustainable development (SD), social, environmental, and financial sustainability should all be integrated into it. Still, the conversation veers drastically and becomes more nebulous the instant the word "development" is spoken. This study does not address sustainable development that is, "development without throughput growth beyond environmental carrying capacity and which is socially sustainable" (27, 28, 30) or development that is socially, economically, and environmentally sustainable.

Similar definition of sustainable development is provided by the World Wildlife Fund (107): "improvement in the quality of human life within the carrying capacity of supporting ecosystems." Though the social components of these definitions call for explanation, they are less vague than Brundtland's (105) definition, "development that meets the needs of the present without compromising the ability of future generations to meet their own needs."

Part of the Brundtland Commission's success may be attributed to its notion of sustainability in a growing environment and its opacity. The paucity of demands for enlargement was startling when the World Commission for Environment and Development (WCED) 106 gathered once more five years later. In their document (106), HRH Prince Charles praised WCED for giving up their 1987 demand for five to ten times the projected pace of economic growth.

REVIEW OF LITERATURE

Climate Change Mitigation

Though Africa contributes for a relatively tiny percentage of the world's greenhouse gas emissions (Nyiwul, L. M. 2019), the World Resources Institute's 2013 shows that Sub-Saharan Africa's contribution is about seven percent but is still growing. Just a small number of nations on the African continent account for a significant fraction of the (Holt, R. M. 1983) world emissions. Out of these countries are Egypt, Algeria, Morocco, South Africa, and Nigeria. Concurrently, it is generally anticipated that the continent would endure an excessive proportion of the projected consequences of climate change. Among the most susceptible industries are those with environmental sensitivity; water and agriculture come first. The availability of enough water resources in many nations all throughout the continent depends much on the total amount of rainfall received. This raises issues about not just water security but also the consequences of drought brought forth by climate change. Many of the agricultural activities on the continent rely on rain-fed systems, hence climate change raises the risk of increasing food insecurity. Calculated as a percentage of GDP, the overall cost of climate change is greater for Africa than for the rest of the world, according calculations done by the African Development Bank (AfDB). Regarding tackling climate change and it generates, the continent is exposed to rather a big range of solutions. Although some of them are not realistic, some are; some will be more helpful than others. Knowing the goals of African countries for activities related to climate change is especially important since they are aiming toward sustainable development aspirations. The fact of the matter is there is evidence suggesting African nations are prioritizing adapting and mitigation of climate change conditions above their sustainable development strategy's linkages. At least in part, this has been made possible by the continent's active involvement in national, regional, and global projects aimed at reduced greenhouse gas emissions (GHG) and other emissions generating climate change. Following the Common But Differentiated Responsibility (CBDR) concepts stated in the United Nations Framework Convention on Climate Change (UNFCCC), there is no expectation of particular emission goals for Africa. This concept is also preserved in Article 2 of the Paris Agreement (PA). On the other hand, they are not excluded from the voluntary, self-identified, national mitigation and adaptation objectives set by the PA to become legally enforced following 2020. One can grasp Africa's priorities for climate change mitigating and adaptation activities undertaken by non-state entities (such as cities, investors, subnational regions, and companies) in their respective countries by means of their individual nationally determined contributions (INDCs) and their commitments in the non-state actor zone for climate action (NAZCA).

Of the 165 nations that have sent in their Internationally Determined Contributions (INDCs), practically all African nations 53 total are engaged. Most of them are still promises rather than particular policy actions that have been really implemented; most of them have been meticulously ready to meet the present development goals. Africa counts 77 of the 517 climate actions currently undertaken by various nations globally and notified to the United Nations Framework Convention on Climate Change (UNFCCC) as of June 2017. The United Nations Framework Convention on Climate Change (UNFCCC) emphasizes in these activities "good practices, approaches, and technologies with significant mitigating potential, which could be tapped into up to the year 2020". Moreover, the NAZCA webpage displays 66 cities all throughout Africa as well as five subnational zones pledging different degrees and scope of climate change efforts. These are in use now.

Mitigation

Given their direct influence on land, agriculture, and forestry, renewable resources have grown to be a main emission response in the framework of African energy options. This is the direction forward, according to conservationists. Actually, policies of global warming among African countries have resulted in some sort of discrimination among Tanzania, Malawi, Somalia, and Zambia. Many of the nations in the region rely mostly on hydropower, in which case renewable energy explains biomass, geothermal, and sometimes solar and wind energies. With roughly 28 million biomass, hydropower, geothermal, and what should be occasionally solar and wind energy, the area is quickly depending heavily on renewable energy sources. Notwithstanding these, per capita renewable energy consumption across the sub-Saharan region is quite much less than the average in other developing economies; overall renewable energy generation by 2012 stood at almost 100 billion kilowatt-hours, with installed capacity of almost 29 million kilowatts. One strategy takes into account mitigating actions in the energy sector ongoing increase of general energy efficiency and replacement of fossil fuels with renewable energy. Since hydropower has seen great and poorly planned intersections between energy use and other infrastructural transformations create barriers to energy efficiency and higher energy losses on account of a colonial era energy transmission system, this later one has quickly gained prominence especially bio-fuels, solar, and very much so efficient cooking technologies. Although official objectives range greatly from 2% in Guinea-Bissau to approximately 80% in Gabon, these features of climate policy are rather universal across Africa. Including South Africa, Ethiopia, Morocco, and The Gambia, certain nations have proposed specific carbon reduction initiatives to show a clear decrease of carbon emissions by 2030. Moreover, Africa boasts the next 17% of the world's forest cover; so, their mitigating actions are perpetuated from past conservation initiatives, especially in the forestry sector. Land use change and forestry are the main causes of GHG emissions in Africa; indeed, the forest sector accounts for between 10% to 30% of world emissions from deforestation and land use changes raising concentrations of atmospheric GHG. Reduced Emissions from Deforestation and Forest Degradation (REDD+), afforestation, and reforestation options are thus the main mitigating strategies underline ones stressing sustainable forest management options that present higher potential for aggregate mitigating at a lower cost per ton of emissions reduced.

Adaptation

Given only the most obvious sectors indicated in the INDCs, their classification of all sectors and subsectors makes it abundantly evident that African countries give adaptation top priority above mitigation in terms of sectors in which some kind of action would be sought. African countries are assigning adaptation top importance in line with predicted sectoral risk profiles for Climate Change, considering the diverse degrees of vulnerabilities in various economic sectors. As expected, more sectors than mitigating ones show signs of adaptation; this serves to highlight the degree of African climate change risk issues. All things considered, followed by water, agriculture frequently rates as the most evident sector where consequences of climate change would be somewhat severe. This ranking corresponds with studies conducted elsewhere on the same topic of adaptation priority (e.g., see Mbeva et al., 2015; Pauw and Pegels, 2013). As shown in Figure 4 in the Appendix, adaptation activities in this sector are aimed toward developing climate smart agriculture, improving the agricultural sector in the areas of irrigation systems, land and soil management, crops, cattle, fisheries and aquaculture, and guaranteeing food security. Climate change is expected to affect Sub-Sahara Africa most especially in sorghum, wheat, corn, and millet; average grain output by 2050 is expected to drop 3.2% and even higher losses in certain areas. These impacts would guarantee food production and reduce the vulnerability of livelihoods for low-income households, therefore enabling the area to fulfill its goal of sustainable development. Given that most of Africa's sustainable development depends on agriculture, there is strong justification to give it priority focus for adaptation. Per AfDB (2017), the percentage of agriculture in Africa ranges from 2.3% of GDP in South Africa to 80% in Liberia, with even higher estimates for employment in the industry.

Data from the INDCS show that African countries are reducing these risks by means of adaption strategies including water management, water supply and quality, increase of water conservation and reuse, (Nyiwul, L. M. 2019) water infrastructure, and enhancement of waste water treatment capacity. These actions are meant to help reduce the risks associated to climate change; among them are extreme weather events like floods and drought that would worsen food scarcity by reducing agricultural production and arable land; consequently, poverty is enhanced and sustainable development is under threat. Regular floods affect many parts of Africa; the Horn of Africa is particularly susceptible to extreme ones. Climate change-driven droughs can reduce water flows from rivers and lakes providing hydropower plants, therefore increasing energy instability. For example, the Nile region is predicted to see a 75% decline in river flow by 2100 as well as biomass resource depletion, which would have negative consequences on sustainable development and energy security (UNEP, 2006). These natural disaster hazards can help to explain why disaster risk and management is another crucial focus of adaptation in Africa.

Sustainable Water Management in South Africa: Challenges and Opportunities

Tasked with meeting the needs of the water users the government opted to assist, the South African Department of Water Affairs and Forestry (DWAF) (Anderson, M. E. 2020) Many of that was used to assist white farmers with agricultural irrigation. Environmental issues were not top of emphasis. This has changed in democratic South Africa, with DWAF now in charge of ensuring all of its citizens have access to safe water and also for guaranteeing an ideal balance between consumptive water use and water conservation via a new Water Act (Republic of South Africa [RSA] 1998), generally regarded as one of the most progressive and enabling pieces of environmental legislation in the world. Progressive new legislation, however, could not necessarily have positive results even with strong intent for change (Van Wyk et al. 2006). Following the guidelines connected with the new regulation has to be done in small increments and is like "climbing an uncharted mountain" (Mackay et al. 2003). Errors are fairly natural and inevitable; the learning process cannot stop them. Periodically having time to evaluate and determine the next action should be a critical but underestimated and neglected part of the "mountain-climbing" process. The challenges in implementing the guidelines aimed to ensure ground water field environmental sustainability are discussed in this study. Although the Water Act does not specifically mention to this term, the environmental concerns of the National Water Act (RSA 1998) are often referred to as "resourcedirected measures," (RDM). Under the "Protection of Water Resources" chapter, the RDM term denotes a set of White Paper (DWAF 1997) policies eventually merged into the Water Act. RDM's principal aim is to set clear targets for the desired degree of protection for every resource—including the ideal balance between "consumptive use" (pulling water out of the system) and "nonconsumptive use"—conservation, leaving water in the system—by means of which the system may be kept intact. These objectives so become legally enforceable and should be taken into account while any other Water

Act application is executed (such as awarding a water allocation license permitting the usage of ground water).

Consumptive use describes things like industrial, irrigation, and household demands including water. Comprising the "silent" goods and services (Mackay and Moloi 2003) for objectives like absorption, breakdown, and transfer of biodegradable wastes, "nonconsumptive use" is part of a considerably greater planetary "life-support system." Heavy metal sequestration; flood attenuation; lower potential flood damage; water storage for later release in the dry season Food supplies including fish, shellfish, and vegetables; material resources including reeds and timber; medicinal plants, Support of sustainable lifestyles includes floodplain farming, preservation of cultural, spiritual, and recreational activities including ecotourism.

Seeing RDM policy, rules, and efforts at execution from a broad viewpoint seems to help one to separate two basic underlying concepts

The second key concept in RDM is that of a "threshold"—that is, if the consumptive use of water keeps growing, it will eventually pass a point whereby the integrity of the resource and/or vital ecosystems reliant on it is compromised. To stress that it is not a precise line that scientists can rapidly determine or that society can typically agree on, the sustainability threshold is shown as a band rather than a line.

Under the National Water Act (RSA 1998), these elements of RDM implementation comprise: 1. Classification: deciding, either totally or partially, the degree of protection every major water supply requires. Two elements constitute "the reserve": (1) volume and quality of water needed to fulfill basic human requirements and; (2) volume and quality of water required to preserve aquatic ecosystems which would guarantee ecologically sustainable development and usage of the water resource in issue. Based on Classification and the Reserve, resource quality goals (RQOs) which aim to set concrete, monitorable criteria for the resource-basing the ideal state of the resource for all relevant stakeholders-are shown. It addresses the whole state of all the components of the water resource, including the aquatic biota, riparian ecosystems, and the water itself in the resource, therefore transcending simple water quality. Though more usually the RDM method completely overlooks ground water ecosystems as such, groundwater might simply be thought of as an underground reservoir providing water either to surface or human settings. It is hypothesised that this attitude towards them results from the "ideal" ground water ecosystems offering little products and services. These presumptions might or could not be accurate.

The Impact of Pollution on Human Health in Africa

Climate change and air pollution are two main global problems with extensive effects on human health. Particularly vulnerable to the consequences of climate change, Africa's issues have become more urgent and gravely compromise the environment as well as the welfare of the people here. Based on its diversified ecosystems and different socio-economic dynamics, Africa experiences a range of climate change-related consequences become more severe in recent years. Rising temperatures have been a main concern; average temperature increases all over the continent of 1.5 to 2 °C (2.7 to 3.6° F). These temperature rise have affected precipitation patterns; (Sullivan, R. C., & Pryor, S. C. 2014). While some places have more variable rainfall and face the risk of floods, others suffer more frequent and severe drenches. Low-lying areas and ecosystems around coasts are particularly vulnerable to sea-level rise—which has been occurring at a rate of around 2 to 5 millimeters each year. Extreme storms, heat waves, floods, and other calamities have become very frequent. This affects several parts of the continent and seriously compromises infrastructure and human lives as well.

Apart from climate change, air pollution is a major environmental concern in Africa, particularly in urban centers and regions of considerable industrial activity. Rising energy consumption, industrialization, and continent-wide rapid urbanization have all contributed to aggravate air quality. The enormous degrees of air pollution approximately 7 out of 10 Africans are exposed to cause cities to suffer most. Particle matter (PM) including fine particles (PM2.5) and coarse particles (PM10) is the principal pollutant of concern in many urban regions with average concentrations above World Health Organization (WHO) guidelines. Other pollutants including nitrogen oxides (NOx), sulfur dioxide (SO2), and volatile organic compounds (VOCs) also contribute generally to the air pollution levels.

Both climate change and air pollution directly and indirectly affect human health in Africa. Direct health implications of climate change include increased heat-associated disorders arising from heat stress, heatstroke, and other related illnesses when average temperatures rise. Higher risk of injuries and deaths from floods, storms, and other natural disasters resulting from the frequency and intensity of extreme weather events has also followed. Indirectly affecting food security, air quality, water availability, and so worsening health risks is climate change. Variations in temperature and rainfall have impacted agricultural productivity, hence generating hunger and food poverty; changes in precipitation patterns and water sources increase the risk of waterborne diseases.

From acute lower respiratory infections to chronic obstructive pulmonary disease (COPD), and asthma, air pollution poses serious health risks all throughout the continent based on significant degrees of exposure. Furthermore linked to air pollution exposure include heart attacks and strokes, among cardiovascular diseases. Long-term exposure to polluted air may especially impact cognitive development and increase the risk of unfavorable birth outcomes including low birth weight and early births, especially in children. In Africa, interactions among human health, air pollution, and climate change are complex and connected. Changing climatic circumstances helps one to influence air quality by thus influencing the lifespan, generation, and dispersion of air contaminants. For instance, altered rainfall patterns and increasing temperatures might hasten wildfires and biomass burning, hence releasing pollutants into the heavens. Conversely, air pollution aggravates global warming by releasing greenhouse gases and aerosols influencing the radiative balance of the Earth, therefore contributing to climate change.

One must understand that interactions among air pollution, climate change, and its consequences are not uniform over Africa. Regional differences in human susceptibility, air pollution levels, and climate create different challenges and health dangers. Already present issues with food security, water scarcity, and the spread of vector-borne diseases in Sub-Saharan Africa are exacerbated by climate change. Particularly air pollution issues exist in cities with large population densities, close urban layouts, and notable industrial activity—Lagos in Nigeria, Nairobi in Kenya, and Johannesburg in South Africa. Bad air quality and negative health effects for the local population are caused in part by limited access to renewable energy sources, traffic congestion, and industrial pollution. Children, the elderly, and those with preexisting respiratory disorders are especially sensitive populations to the detrimental effects of air pollution. Moreover seriously contributing to air pollution in cities are inadequate transportation systems.

Climate change threatens humans. George, N. 2022 & Gangtar, N. N. Global CO2 emissions have changed somewhat dramatically from 14.9 billion metric tons in 1970 to 36.4 billion metric tons in 2021. Thus, throughout the same period, atmospheric CO2 emissions concentration has increased from 325 ppm to 414 ppm. Africa suffers from extreme climatic occurrences and is susceptible, much as other continents. Vulnerability is exacerbated by low continent adaptive capacity and dependency on rain-fed agriculture. There are several accounts of African temperatures rising. North Africa's temperature has been rising between 0.2oC every decade and 0.4oC since the 1970s. West African temperatures have exhibited encouraging 0.28 oC trends. Temperature intensity in the Sahel and West Africa falls between 0.25 and 1.8 oC. South Africa has the highest predicted increase. Regular temperature increases jeopardize arable land and reduce the yield of several African crops. The yearly rainfall of Africa has also varied throughout its several regions. Precipitation in North Africa has seen declining patterns. While diminishing tendencies have also been seen in West Africa, East and Southern Africa get enough precipitation. Generally, African rainfall is declining, which influences food, water, energy security, environment, livelihoods, and economy. About US$ 1.4 billion annually on food crops spread all throughout Africa has been lost. With reported annual production losses of 8.9%, 2.3 million MT of wheat squandered translates to influence on 48.2 million consumers worldwide. About forty percent of African arable land produces less crops, which results in poverty and affects over 57% of the people. East African annual agricultural yields have apparently plummeted around 25%. By 2023, a significant 48% of Africa's GDP—or $1.4 trillion overall—will be vulnerable to climate change. Africa amply illustrates both economic development and quick urbanization. Among the countries whose economic growth has been seen are Rwanda (8.7%), Ethiopia and Côte d'Ivoire (7.4%), Ghana (7.1%), Tanzania (6.8%), and Benin (6.7%). Between 2000 and 2018, Africa's urbanization rate increased from 30.8 to 38.8% with a 2.2% economic progress (Nathaniel and Adeleye, 2021). Among the top 100 fast-growing cities worldwide are seventy-nine African cities, which also face significant risks from climate change. Great demand for energy and greenhouse gas emissions follow from rising urbanization and economic growth. Moreover distinguishing Africa is its quick change of people. Ethiopia, Angola, and Tanzania have reported respectively with yearly population increase rates ranging from 4.8 to 4.5. Population growth has increased CO_2 emission from 399,239Kw in 1990 to 823,424,000 in 2018. South Africa was the most polluting country, having emitted 449 billion metric tons of CO_2 emissions in 2019; Egypt placed second with 247 billion metric tons (Saleh, 2021). Other big CO_2 ($\geq$10 Mt/year) emitters are countries including Nigeria, Algeria, Libya, and Morocco. Many African countries face rapid rise in CO_2 emissions and environmental harm brought on by fast economic expansion and population growth. This draws attention to Africa's role in global climate change above its socioeconomic growth. Policymakers should thus focus more on mitigating and adapting strategies to lower the consequences of climate change on the continent.

National parks are among the most broad classifications of protected areas worldwide and in Africa (Balkwill, K. 2013). They belong to IUCN system of protected areas in class II. Established to protect the ecological integrity of one or more ecosystems for present and future generations, National Parks exclude exploitation or occupation detrimental to the purposes of designation of the area; provide a basis for spiritual, scientific, educational, recreational, and visitor opportunities, all of which must be environmentally and culturally compatible. National parks account for the most of the total territory under protection globally. For instance, Colchester estimates that Africa has more than 1,812 National Parks covering 3,112,027 km2 total on the continent. Of 23 million km2, around 1 million km2 of land has been designated as National Parks alone in SubSaharan Africa alone (about 4%). National Parks employ two main approaches to protect biodiversity: one is the preservation plan, which aims to set aside regions free from human activity save for visitors. This approach prevents direct use of park resources for either business or personal use. Often cited as the "protectionism approach," or "the fines and fences," this type of tactic involves This paper addresses the consistency "preservation approach". The preservation plan aims to exclude human activities deemed incompatible with the objectives of preserving biodiversity in National Parks. The most common technique until the 1980s was the preservation strategy; now, in certain National Parks the second approach known as community-based conservation—which allows people especially those who live nearby benefit socially or economically from parks—has since gained front stage. Suggested to address the problems with the exclusion of the community-based conservation plan was the one based on (McClain, M. E. 2013) PROJECTIONS FOR INCREASED WATER USE IN FOOD AND ENERGY PRODUCTION

Trends for Environmental Sustainability

Among estimates for increasing water demand in Africa, hydropower and agriculture rule. Rising agricultural consumptive consumption has bearing on both irrigated and rainfed crops. Plans for medium-sized and large projects predominate in enhanced non-consumptive use by hydropower; small hydropower is being encouraged but trustworthy information on installed or expected capacity was not acquired during this review. Most hydropower producing and irrigated agriculture rely on infrastructure to store water during wet seasons for usage during dry ones or droughs. Development plans so mostly depend on multipurpose dams. Many recent studies have examined the geographical arrangement and expected rates of increasing water use in the following decades.

WASTE MANAGEMENT

With the (Muzenda, E., Ntuli, F., & Pilusa, T. J. 2012) National Environmental Management Waste Act, 2008 taking effect on July 1, 2009, integrated waste management became required. Under this waste management strategy, South Africa promotes a waste hierarchy by encouraging cleaner manufacturing, waste minimising, reusing, recycling and trash Waste Management, Approach and Context in M South Africa: An All Around View Thomas Edison Muzenda, Freeman Ntuli, Tsietsi Jefrey Piluma 68 2012 149 Treatment World Academy of Science, Engineering and Technology As of 2010, disposal and remediation are seen as the last choices in waste management, Fig. 2. Waste avoidance and minimization is first choice for management of waste. Waste should be retrieved, utilized again, disposed of, processed, if avoidance is not possible. To safeguard human health and ensure the welfare of the environment, the New National Environmental Management Waste Act (2008) and the NWMS offer more actions for the rehabilitation of polluted land.

B. Recovery, Re-Use and Recycling

The second level in the waste hierarchy is recycling, re-use, and recovery. Though they are rather distinct physical processes, recovery, reuse, and recycling have the same goal of recovering materials from the waste stream and thereby lowering the volume of trash produced that climbs the waste hierarchy. The Waste Act's Section 17 lays forth guidelines controlling recovery, reuse, and recycling as well as a variety of other regulatory actions the Minister might have at hand in this regard. Driven mostly by industry-led, voluntary projects with finances administered independently from the government via non-profit groups, which supervise recovery/recycling processes and facilities, South Africa's recycling rates are very well established.

C Storage, Collection and Transportation

The major point of interaction (Sullivan, R. C) between the public and waste service providers is the supply of waste management services comprising the storage, collection and transportation. Directly affecting all phases of the waste hierarchy are also the degree and manner of waste service provided for families and companies. Among other things, the trash Act mandates towns provide access to and sustainability of trash services, offer waste services at reasonable rates, and maintain separate financial records for waste services rendered. While district municipalities in general do not see waste management as part of their duties, all metropolitan municipalities and most local municipalities have a basic responsibility for this service. Based on guidelines created in line with the trash Act, the NWMS outlines a program for universal supply of trash services. To reach this, coordinated work among several areas of government is needed to solve the issues with capacity and finances related to trash service delivery.

D. Treatment and Disposal

This is the lowest and least desirable level in the hierarchy among the chemical and/or physical treatment of trash: landfilling of wastes. As stated in the National Environmental Policy, waste treatment, processing, and disposal in South Africa must abide by environmental justice and equal access to environmental services. This is particularly important considering the inclination of landfills and waste treatment facilities to be located near impoverished regions and informal communities. Right today, most of the trash collected finds its way to landfills. Strong greenhouse gas methane is generated from biodegradable garbage found in landfills. Plastic trash in particular is a challenge as it takes a long time to break down and occupies valuable space in landfills. The number of legally allowed landfill sites demonstrates the commitment of a country to effectively manage the complete range of produced waste. Although more than half of the population does not get a continuous municipal trash collecting service, municipal rubbish collecting has improved as the 1996 and 2001 census demonstrate. NEMA, Management Act 107 of 1998.

Although in certain circumstances the smaller towns offer no services at all, the metropolitan municipalities supply around 100% of them. General landfills facilities accept household, business, and industrial non-hazardous waste in addition to construction and garden trash. Most of the sites spread around South Africa are owned and administered by local governments. Well-planned landfill sites are essential for effective rubbish disposal fulfilling regulatory requirements. Table 2 shows, as of 2010, the legal landfill sites per province in South Africa [10]. Given the massive volumes of waste generated, 20% of the 817 landfills in South Africa are in Gauteng [11]. Of the 1336 waste facilities included in the DEAT (2005) census, just 553 (43%) are visible.

With an emphasis on socioeconomic and environmental concerns in the framework of Africa over the past ten years, all examined publications particularly address land-use change and climate change. While only a small number of research were normative, most of the participatory futures thinking projects used experimental techniques. Themes covered a spectrum of sustainability challenges related to intersections between environmental and socioeconomic change, e.g., food security and biodiversity ecosystem services, biodiversity or deforestation, and livelihoods agriculture and water or climate change wildlife conservation and land use and major systemic transformation to combat challenges of poverty, inequality, and food security.

Chanlanges in Environmental Sustainability and Opportunities in Environmental Sustainability

The several obstacles preventing South Africa's groundwater consumption to be environmentally sustainable. (Seward, P. 2010) A major problem is the unequal sharing of water resources; 60% of the water is found in just 20% of the land, which causes overuse in some areas. The 1998 Water Act sought to correct these inequalities by encouraging fair water usage; yet, its execution has encountered challenges including insufficient monitoring and enforcement systems. Further more challenging for parties to negotiate is the Act's convoluted regulatory structure, therefore impeding good management. By changing precipitation patterns, (Seward, P. 2010) climate change increases these difficulties and strains groundwater supplies even further. Moreover, it is still challenging to balance theoretically strong but sophisticated techniques with basic, readily understood management notions. These elements taken together hamper the country's efforts at sustainable groundwater use.

(Johansson, E.L. 2021) Involvement has always been sought for many various reasons by many different parties in Africa. Not always have involvement helped to boost the voices of marginalized groups. Claiming that all "sides" were considered has also helped strong performers keep the status quo (Arnstein 1969, White 1996). Although participation has significant potential to bring about social change, its main negative consequences might be related to power relations between those who lead communities and those who are led and its limitations to give ways for those with limited agency to deal with these interactions. Although involvement is said to have been the norm of development in the 1970s, its use for development goes back over a century (Cornwall 2006, Kyamusugulwa 2013). Popular participation affected the Colonial Development Act of 1929 for the United Kingdom, which thereafter organized and standardized the expansion of British colonial territory. Popular involvement came from a "moral" requirement of colonial powers reacting to people under their authority who were demanding protection against exploitation, preservation of their tribal areas, and some role in the making of their own future (Cornwall 2006). Though participation took many varied forms in different colonial territories, it was primarily based on the concept of "civilization," wherein certain tasks and powers of institutions were allocated to traditional leaders, who were then under colonial authorities' oversight. The capacity of the colonial powers to continue extracting under local authorities' cooperation rested heavily on "civilization". Participation in the 1930s stayed anchored in such ideas of development; aspects of "modernization" and "governmentality" became more clear in the 1940s and 1950s through participatory policy processes. Understood in the framework of democratic institutions in the 1960s, foreign aid to post-colonial nations was committed to enhance education and training for effective citizen engagement in political processes essential for self governance (Cornwall 2006). In the 1970s, participation was redefined as something that each person should freely and democratically support development initiatives and fairly share the outcomes. The United Nations recommended countries all over to inspire public participation in national development programs and decision-making (Cohen and Uphoff 1980). Thus, involvement becomes even more important for defining goals, formulating policies and strategies, and implementing social and economic development projects (Cornwall 2006). As participation became an even more basic feature of development, the U.S. Foreign Assistance Act of 1973 stressed the requirement of including the "beneficiaries" in various American Aid initiatives to share the fruits of development (Cohen and Uphoff 1980). In the 1980s, neo-liberalism transformed participation so that participants were seen as "beneficiaries" expected to be active in implementing and in incurring the costs of development rather than as passive consumers of development aid (Cornwall

2006). Highly influenced by marketizing regimes such the UK and the U.S., cost-sharing and co-production of services became mainstream in global development, and the discourse of self-reliance came to promote more individualistic public participation not only to do it by yourself, but also for yourself (Cornwall 2006). To meet rural demands for little investment on basic facilities, governments in former British colonies instituted self-help projects such village development committees (Vengroff 1974). Self-reliance and local development activity also inspired rural people, so helping the public to be more aware of national development projects and so increase support for the government (Vengroff 1974). To offset grassroots resistance to reforms, international agencies such as the World Bank actively pushed involvement all through the 1980s.

Opportunity

Normative scenarios include thinking about desired goals and then looking at the paths leading to them via visioning or imaginaries. As they seek futures outside of current socioeconomic and cultural structures and limits, visions can seem more "radical" than other alternatives. Visionaries can reveal challenges in present socioeconomic and cultural systems and help to uncover opportunities for grassroots activities to flourish by relating to current programs for sustainable development. Moreover defined as generally agreed upon desired or accessible futures, imaginations enable one to envision the possibilities and motivate effort toward fresh directions of development. Often used in both exploratory scenario creation and normative visioning, backcasting helps evaluate the relative feasibility of future scenarios and produces strategic action plans for how to accomplish them (van der Voorn et al. 2012, Wiek and Iwaniec 2014).

In the 1990s, participation in development was steadily pushed as a new, pragmatic strategy to handle a range of development challenges. Many existing mainstream development agencies promised a new approach that would empower the impoverished more in terms of development (Cornwall 2006). The Human Development Report of 1993 by the United Nations Development Program included remarks on people's willingness to participate and positioned their engagement within an overall development strategy that enabled them to acquire access to a wide range of options. Emphasizing "civil society" involvement during this time, civil society groups were regarded as autonomous from the state and therefore able to champion the interests of the impoverished (Cornwall 2006). Another paradigm for participation that became well-known during this period was the creation of "stakeholder language" and "stakeholder participation," realizing that stakeholders had varied degrees of ability to influence outcomes.

This paper seeks to identify opportunities for management of air pollution and prevention of pollution-related sickness as well as to quantify the consequences of home and ambient air pollution on health, human capital, and the economy in a quickly changing Africa. Though we examine statistics for the whole continent, (P. J. 2021) we especially pay great attention to three quickly rising sub-Saharan countries: Rwanda, Ghana, and Ethiopia.

Many African countries are at really early phases of development. They so have opportunity to be free from being ensnared in systems depending on fossil fuels and to pick non-polluting paths to growth. Through wise, forward-looking investments in renewable energy and clean technologies, African countries may avoid ambient air pollution, build human capital, and expedite their own development.

There is increasing pollution of the ambient air all throughout Africa. Without deliberate action, it will increase disease and mortality, reduce economic output, slow down the creation of human capital, and thus hinder development. Since most African countries are still in early development and avoid depending on economies based on fossil fuels and therefore lower pollution, most of them have potential to rapidly convert to wind and solar energy.

The authors assess how these guidelines assist relevant project participants in being involved. Jimenez and Perez-Foguet look at how chores are distributed to village councils and regional governments in order to provide enough water supply for rural communities. The biggest challenges to effectively reaching the vulnerable populations, the authors found, include lack of accountability at the village level, inadequate top-down help, and political interference even if local officials are seen as more responsive and have a better grasp of needs. They assert that the dispersed strategy may be quite successful with greater central government engagement. As also observed by Ishmael Kosamu, analyzing the quality grading of recently published EIA reports and a field survey indicates numerous major limitations to conducting environmental impact assessments (EIs) for development projects in Malawi. Among these limitations include political will to appropriately link the assessments to the process of development, inadequate human capacity to execute EIs, and too high expenses.

The land change patterns of Central Africa highlight the benefits of forest preservation for the slowing down of climate change. They found that in some cases small holder agroforestry was crucial for preserving carbon reserves in natural forest stands; habitat management reduced impact logging to support biodiversity. They end with a synopsis of the existing projects under the Clean Development Mechanism and Reduction of Emissions through Deforestation and Degradation (REDD) that would compensate communities for maintaining vegetation biomass.

The countries have applied numerous tools in order to accomplish their goals of sustainable development and environmental protection. Policy for innovation, regulations, environmental subsidies, information sharing and awareness campaigns and environmental levies are among the several economic tools governments have at hand. Braathren and Greene claim that " Taxes are in particular a key part of this tool kit". Though other measures are at their disposal in terms of climate change and environmental preservation, environmental taxes have become the most crucial and successful economic instrument accessible to governments. 2022: Mpofu, F. Y. Although the terms green taxes and environmental taxes are occasionally used synonymously in the literature, they will be used synonymously throughout this work. Others refer to the green taxes as either climate taxes, eco taxes, or ecological taxes. These phrases convey the need of maintaining the surroundings, temperature, and biodiversity.

Based on the theory that despite reducing other taxes and maintaining the same level of government income mobilized, environmental tax policies enhancing these taxes produce a two-fold advantage, research and policy ideas promoting environmental taxes also exist. According to the double dividend (DD) idea, one might maybe enhance the environmental as well as the economic ones by imposing green or environmental taxes and utilizing the mobilized money to reduce the tax rates for other taxes.

Recommendation

To handle environmental damage issues and promote sustainability, one must apply comprehensive and forceful strategies. Strict environmental policies and norms must be enforced by governments to ensure that resource exploitation and industrial activities meet sustainable criteria. Shari, B. E., Moumouni, Y., Ohunakin 2024 Among other renewable energy sources, solar and wind power help to reduce dependency on fossil fuels, therefore cutting greenhouse gas emissions. Reforestation projects and afforestation programs should also be given first importance in order to aid to restore degraded regions and boost carbon sequestration. Among other waste management practices, reversing plastic usage and starting recycling programs would assist to significantly reduce pollution and conserve ecosystems. Public awareness efforts and educational projects need to be intensified in order to encourage sustainable consumption patterns and ecologically friendly behavior. To maintain soil fertility and offer food security in rural areas, sustainable agricultural practices like crop rotation, organic farming, and efficient water usage should be urged to be promoted. Moreover, implementing conservation initiatives and protecting biodiversity depends on government, non-governmental organization, and local population participation. By means of air quality monitoring and water purification technology, technological advancements can help to lower health concerns related to pollution. Dealing with environmental harm demands for a varied approach combining technical developments, community engagement, and legislative interventions to generate a sustainable and robust future for people and the world.

CONCLUSION

The chapter has comprehensively summarized environmental degradation together with its origins, effects, and urgent need for sustainable practices. Emphasizing the significant ecological damage resulting from resource exploitation, urbanization, and industrial expansion, it has underlined the unstable equilibrium between human activity and environmental health. Among the issues that the emphasis on Africa—especially South Africa has highlighted are water shortages, waste management, and (Agan, B., & Balcilar, M. 2023) the consequences of climate change on various industries. Talks on the need of integrated waste management, the preservation role of national parks, and the ratio of consumptive to non-consumptive water use have helped to clarify alternative strategies for lowering environmental damage.

The chapter has also underlined the need of a complete plan to handle these complex issues by clarifying the link between environmental sustainability and social, economic, and governmental elements.

REFERENCES

Adukaite, A., & Cantoni, L. (2016). Raising awareness and promoting informal learning on World Heritage in Southern Africa: The case of WHACY, a gamified ICT-enhanced tool. *International Journal of Education and Development using ICT, 12*(2). https://www.learntechlib.org/p/173447/

Agena, T. (2007). The relationship between economic activities and environmental degradation in Africa. *Journal of the Historical Society of Nigeria, 17*, 28–40. https://www.jstor.org/stable/41857144

Alam, A., Upadhyay, S., Murthy, C. H., Reddy, M. J. B., Jana, K. C., & Mohanta, D. K. (2012, May). *Reliability evaluation of solar photovoltaic microgrid. In 2012 11th International Conference on Environment and Electrical Engineering*. IEEE., https://ieeexplore.ieee.org/document/6221427

Anderson, M. E. (2020). From 'exorcism' to engagement: The private sphere of perpetratorship in twenty-first century South African documentary film. *Continuum (Perth), 34*(2), 210–225. DOI: 10.1080/10304312.2020.1737433

Andrews, N. (2014). Driving factors for national competitiveness in Africa as measured by GDP per capita. https://repository.up.ac.za/bitstream/handle/2263/43965/Andrews_Driving_2014.pdf?sequence=1&isAllowed=y

Baloyi, E. (2015). The Christian church's role in the escalating mob justice system in our black townships-An African pastoral view. *HTS Theological Studies, 71*(2), 01-07. https://www.tandfonline.com/doi/abs/10.1300/J118v04n04_12

Baninla, Y., Sharifi, A., Allam, Z., Tume, S. J. P., Gangtar, N. N., & George, N. (2022). An overview of climate change adaptation and mitigation research in Africa. *Frontiers in Climate, 4*, 976427. https://www.frontiersin.org/journals/climate/articles/10.3389/fclim.2022.976427/full. DOI: 10.3389/fclim.2022.976427

Devall, B. (2006). Conservation of biodiversity: Opportunities and challenges. *Human Ecology Review*, •••, 60–75. https://www.jstor.org/stable/24707557

Fisher, S., Bellinger, D. C., Cropper, M. L., Kumar, P., Binagwaho, A., Koudenoukpo, J. B., Park, Y., Taghian, G., & Landrigan, P. J. (2021). Air pollution and development in Africa: Impacts on health, the economy, and human capital. *The Lancet. Planetary Health, 5*(10), e681–e688. https://www.thelancet.com/journals/lanplh/article/PIIS2542-5196(21)00201-1/fulltext. DOI: 10.1016/S2542-5196(21)00201-1 PMID: 34627472

Goodland, R. (1995). The concept of environmental sustainability. *Annual Review of Ecology and Systematics*, *26*(1), 1–24. https://www.jstor.org/stable/2097196?seq=1. DOI: 10.1146/annurev.es.26.110195.000245

Holt, R. M. (1983). Dynamic Dallas: A Building Critique. *Public Library Quarterly*, *4*(4), 67–84. DOI: 10.1300/J118v04n04_12

Leonard, B., & Regan, S. (2019). Legal and institutional barriers to establishing non-use rights to natural resources. *Natural Resources Journal*, *59*(1), 135–180. https://digitalrepository.unm.edu/cgi/viewcontent.cgi?article=4037&context=nrj

Mashila, T. (2014). Spatial planning for climate change adaptation: Developing a climate change local area adaptation plan for Khayelitsha. https://open.uct.ac.za/server/api/core/bitstreams/4ab3b5fe-2f63-4d05-bbc2-aefa27671370/content

McClain, M. E. (2013). Balancing water resources development and environmental sustainability in Africa: A review of recent research findings and applications. *Ambio*, *42*(5), 549–565. https://link.springer.com/article/10.1007/s13280-012-0359-1. DOI: 10.1007/s13280-012-0359-1 PMID: 23238866

Mpofu, F. Y. (2022). Green Taxes in Africa: Opportunities and challenges for environmental protection, sustainability, and the attainment of sustainable development goals. *Sustainability (Basel)*, *14*(16), 10239. https://www.mdpi.com/2071-1050/14/16/10239. DOI: 10.3390/su141610239

Muhumuza, M., & Balkwill, K. (2013). Factors affecting the success of conserving biodiversity in national parks: A review of case studies from Africa. *International Journal of Biodiversity*, *2013*(1), 798101. https://onlinelibrary.wiley.com/doi/full/10.1155/2013/798101. DOI: 10.1155/2013/798101

Muzenda, E., Ntuli, F., & Pilusa, T. J. (2012). Waste management, strategies and situation in South Africa: An overview. Johansson, E. L. (2021). Participatory futures thinking in the African context of sustainability challenges and socio-environmental change. *Ecology and Society*, *26*(4). https://pdfs.semanticscholar.org/6970/c6a60b8a7c2236f6c36290e8eb6b0ac06f9e.pdf

Nakawuka, P., Langan, S., Schmitter, P., & Barron, J. (2018). A review of trends, constraints and opportunities of smallholder irrigation in East Africa. *Global Food Security*, *17*, 196–212. https://oaktrust.library.tamu.edu/server/api/core/bitstreams/4ab0ea7f-027e-428e-a3d4-6b6c8481b16f/content. DOI: 10.1016/j.gfs.2017.10.003

Nathaniel, S. P., & Iheonu, C. O. (2019). Carbon dioxide abatement in Africa: The role of renewable and non-renewable energy consumption. *The Science of the Total Environment, 679*, 337–345. https://www.sciencedirect.com/science/article/abs/pii/S0048969719320224. DOI: 10.1016/j.scitotenv.2019.05.011 PMID: 31085413

Nichanian, D. (2016). Seizing a Seat at the Table: Participatory Politics in the Face of Disqualification. https://knowledge.uchicago.edu/record/1733?v=pdf

Nyiwul, L. M. (2019). *Climate change mitigation and adaptation in Africa: Strategies, synergies, and constraints*. Springer International Publishing., https://link.springer.com/chapter/10.1007/978-3-030-02662-2_11

O'Donoghue, R., & Rosenberg, E. A review of the staging and enactment of environment and sustainability education in South Africa: An illustrative case study. *World Review*, 22-39. https://www.taylorfrancis.com/chapters/edit/10.1201/9781003145202-3/review-staging-enactment-environment-sustainability-education-south-africa-rob-donoghue-eureta-rosenberg

Ogeto, G. M. (2017). *Role of Media in Political Conflict: A Comparative Study of Kenya and Zimbabwe*. https://erepository.uonbi.ac.ke/bitstream/handle/11295/108075/Ogeto_Role%20Of%20Media%20In%20Political%20Conflict.pdf?sequence=1&isAllowed=y

Omokhua, A. G., Madikizela, B., Aro, A., Uyi, O. O., Van Staden, J., & McGaw, L. J. (2018). Noxious to ecosystems, but relevant to pharmacology: Four South African alien invasive plants with pharmacological potential. *South African Journal of Botany, 117*, 41–49. https://repository.up.ac.za/bitstream/handle/2263/69205/Omokhua_Noxious_2018.pdf?sequence=1&isAllowed=y. DOI: 10.1016/j.sajb.2018.04.015

Schrier-Uijl, A. P., Silvius, M., Parish, F., Lim, K. H., Rosediana, S., & Anshari, G. (2013). Environmental and social impacts of oil palm cultivation on tropical peat. *Reports from the Technical Panels of the 2nd Greenhouse Gas Working Group of the Roundtable on Sustainable Palm Oil Kuala Lumpur*. https://d2ouvy59p0dg6k.cloudfront.net/downloads/the_environemental_and_social_impacts_of_oil_palm_on_peat.pdf

Seward, P. (2010). Challenges facing environmentally sustainable ground water use in South Africa. *Groundwater, 48*(2), 239–245. https://ngwa.onlinelibrary.wiley.com/doi/full/10.1111/j.1745-6584.2008.00518. DOI: 10.1111/j.1745-6584.2008.00518.x PMID: 19016897

Shari, B. E., Moumouni, Y., Ohunakin, O. S., Blechinger, P., Madougou, S., & Rabani, A. (2024). Exploring the role of green hydrogen for distributed energy access planning towards net-zero emissions in Nigeria. *Sustainable Energy Research*, *11*(1), 16. DOI: 10.1186/s40807-024-00107-1

Subrahmanyam, K. (2000). Self-reliant defence and Indian industry. *Strategic Analysis, 24*(7), 1221–1234. https://www.tandfonline.com/doi/abs/10.1080/09700160008455283DOI: 10.1080/09700160008455283

Sullivan, R. C., & Pryor, S. C. (2014). Quantifying spatiotemporal variability of fine particles in an urban environment using combined fixed and mobile measurements. *Atmospheric Environment*, *89*, 664–671. https://www.sciencedirect.com/science/article/abs/pii/S135223101400168X?via%3Dihub. DOI: 10.1016/j.atmosenv.2014.03.007

Sutton, P. (2004). A perspective on environmental sustainability. *Paper on the Victorian Commissioner for Environmental Sustainability*, *1*, 32. https://www.donboscogozo.org/images/pdfs/energy/A-Perspective-on-Environmental-Sustainability.pdf

Takeuchi, K., & Aginam, O. (2011). Sustainability challenges and opportunities in Africa. *Sustainability Science*, *6*(1), 3–5. https://link.springer.com/article/10.1007/s11625-010-0124-y. DOI: 10.1007/s11625-010-0124-y

Tyagi, S., Garg, N., & Paudel, R. (2014). Environmental degradation: Causes and consequences. *European researcher, 81*(8-2), 1491. file:///C:/Users/dasha/Downloads/1409224435.pdf

Yan, L., Zhou, J., Li, H., Zhong, R., Zhuang, J., Xu, X., Wang, Y., Liao, X., & Shi, B. (2023). Wearable synthetic leather-based high-performance X-ray shielding materials enabled by the plant polyphenol-and hierarchical structure-facilitated dispersion. *Collagen and Leather*, *5*(1), 12. https://jlse.springeropen.com/counter/pdf/10.1186/s42825-023-00119-5.pdf. DOI: 10.1186/s42825-023-00119-5

Chapter 2
The Rate of Recolonization by Native Plant Species After the Eradication of the Invasive Alien Plant Species (Lantana camara L.) at Valdezia, Limpopo

Melford Mbedzi
https://orcid.org/0000-0002-8330-1356
Mangosuthu University of Technology, South Africa

Milingoni Peter Tshisikhawe
University of Venda, South Africa

Ndidzulafhi Innocent Sinthumule
University of Johannesburg, South Africa

Sebataolo Johannes Rahlao
https://orcid.org/0000-0003-4451-0988
The South African Association for Marine Biological Research, South Africa

ABSTRACT

Invasive alien flora can overwhelm riverine network and modify the hydrology. Little is known about riparian invasive alien trees impacts on general ecology although

DOI: 10.4018/979-8-3373-1549-2.ch002

invasive alien trees are by far the biggest threat to universally red-listed dragonflies (Odonata) in South Africa. Twenty-four permanent plots of 10 m2 each were constructed. Before removal, invasive alien species and native species present in the quadrats were counted and recorded. The invasive alien plants were removed by digging and uprooting with no use of chemicals. The difference was not significant when it comes to species diversity and richness before and after removal of Lantana camara in the study area. L. camara re-established in large numbers even after the treatment. The emergence of L. camara after the clearing is thought to be due to its ability to reproduce through the stem cuttings which might be lying on the ground, thereby reproducing via stem layering and from the seeds in the ground as some of the seedlings were seen being produced from the seeds.

INTRODUCTION

The introduction of an invasive alien species is often responsible for an increase in predation and competition, habitat reduction, an assortment of illnesses in animals and hereditary change in population (the indigenous species influenced by these invasive alien species may change their hereditary makeup). Once these species are well established, it is often very difficult to remove them. At the point when the removal is possible, it comes at a staggering expense financially and ecologically. In any case, ecosystems are forever in the process of trying to control invasive alien plants (PA Department of Conservation and Natural Resources, Invasive species Management Plan 2005). Invasive species cause major ecological and related financial misfortunes which can add up to billions of dollars each year.

Invasive alien species have been distinguished as the second main driver of biological diversity depletion following habitat annihilation (Alonso et al., 2001) and the primary driver of species depletion in Island landscapes (Clout and Veitch, 2002). Invasive alien trees can overwhelm riverine ecosystem and modify the hydrology (Castro-Diez and Fernandez, 2017). Little is known about riparian invasive alien trees' impact on general ecology (Richardson and Van Wilgen, 2004) although invasive alien trees are by far the biggest threat to universally red-listed dragonflies (Odonata) in South Africa (Samways and Taylor, 2004). South Africa is imperative for biological diversity since two of the world's hotspots are within it (Myers et al., 2000).

It has been demonstrated that invasive species (including plants, animals and microbes) can impact the distribution, abundance and recovery of indigenous species (Aravind et al., 2010). Exotic plant species can also divert pollinators and dispersers of indigenous plant species towards themselves, therefore obstructing the reproductive achievement of indigenous species. Non-indigenous species are imperative operators of worldwide ecological change (Aravind et al., 2010) and are perceived as the second most noteworthy risk on biodiversity after anthropogenic habitat loss and ecosystem destruction (Gooden et al., 2009). The aim of this study was to investigate the rate at which the native plant species would recolonize the area which was previously infested with invasive alien plant species (*Lantana camara*) and to assess the factors to consider for successful recolonization or restoration of the invaded site.

1.1 Occurrence of the Species Under Study

1.1.1 Lantana Camara

Lantana camara is found in various climatic and geographic ranges. It is found in a range of ecosystems including riparian zones, open grazing lands, forest regions, national parks, rainforest edges, and along roads, creeks and fence lines, and are additionally cultivated as ornamentals in parks and gardens. Pervasions go from single shrubberies to thick monocultures, displacing local vegetation or pasture species (Day et al., 2003).

METHODOLOGY

2.1 Description of the Study Area

The study was conducted in the Makhado District Municipality, in a farm at a village known as Valdezia (Figure 1). The study site is along the road (R524) towards Louis Trichardt and was highly infested with *L. camara*. It lies at 23.8'17''S and 30.8'17''E. The soil type in this region is mainly clay red to gravel type. The species composition found here is mainly *Vachellia karroo, Sclerocarya birrea* subsp *caffra,* and *L. camara*. The area receives summer rainfall and is dominated by Tshivenda and Xitsonga speaking people.

Figure 1. The Makhado District Municipality including the study area (The Green Empire (Pty) Ltd).

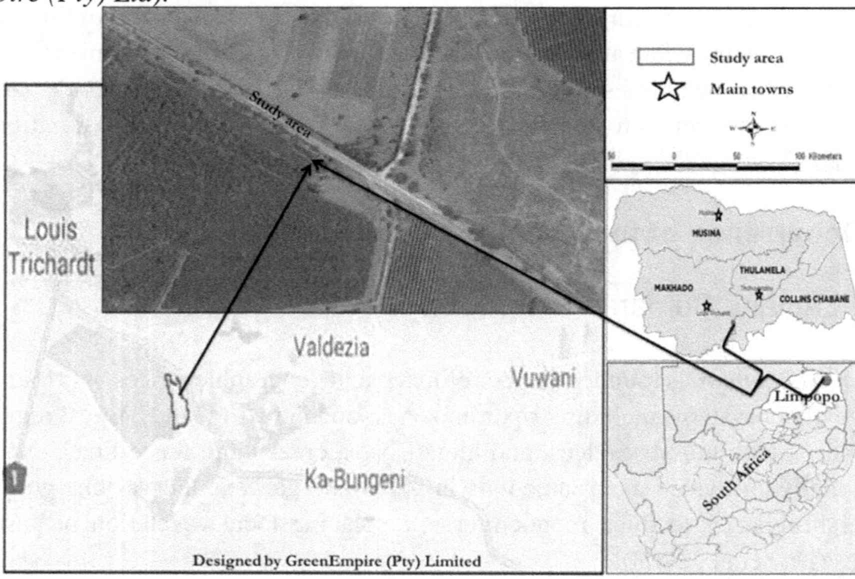

2.2 Description of Species Under Study

2.2.1 Lantana Camara

Lantana camara L. is an indigenous woody shrub of tropical America (Holm et al. 1997) which is found in 47 nations and has been portrayed as one of the world's ten most exceedingly awful weeds (Cronk and Fuller 1995). *L. camara* is a low upright, vigorous, climbing shrub with stout bent prickles and a black current repulsive smell; it reaches heights of 2.4 meters (or 7 meters in some cases). It has a very strong root system and produces a new flush of shoots even after repeated cutting (Rosacia et al., 2004).

Lantana camara is mostly indigenous to tropical and subtropical America, yet limited taxa are indigenous to Africa and tropical Asia. It currently exists in approximately 50 nations where under a few hundred species are vascular plants (Pyšek et al., 2020). The documented quantities of *Lantana* species vary from 50 to 270 specific and sub-specific units, but it seems that 150 species are a better approximation (Mushi, 2019). The genus is a troublesome one to classify systematically since species are not steady and hybridization is broad, with age, the shape of inflorescence changes, and the colour of the flowers fluctuate with maturity and age (Munir and Adelaide, 1996).

Dutch travellers brought *Lantana camara* from Brazil into the Netherlands in the late 1600s and later pioneers from different nations conveyed seeds to North America, Europe and Great Britain (Mohamed et al., 2016). It soon outspread to the islands of Pacific, Australia and Southern Asia, following its introduction into Hawaii for ornamental purpose. In comparative route from Natal (South Africa), it was quickly dispersed by birds to hotter regions of South Africa (Mohamed et al., 2016). In the eighteenth and nineteenth century, nurserymen marketed, and advanced numerous vivid forms and it is currently cultivated worldwide as an ornamental plant (Mohamed et al., 2016). Of the 650 vascular plant names in the genus, the numerous parts are related with the *L. camara* complex (Mishra, 2015). The plant is an aggressive and constrain out-breeder invader that has infested huge spans of fields, plantations and forest areas in numerous tropical and subtropical territories. It has been estimated that 4 million ha in Australia and 160 000 ha in Hawaii are infested with *L. camara* (Ross, 1999). It has been viewed as one of the 10 most noxious weeds on the planet and much has been written on its encroaching habit, techniques for eradication and control (Parson and Cuthberson, 1992; Munir and Adelaide, 1996). Besides its notoriety as an orchard plant, *L. camara* is said to form an important windbreak and gives a decent readiness to crops, making the progress with fine leaf with a protective cover. The plant is not voluntarily ingested by cattle except if fodder is lacking. In tropical nations, the mature blue-dark berries are ingested, but ingestion of the green berry has ended up to human fatalities (Morton, 1994; Ross, 1999).

The ecological limitations to distribution of *Lantana* are obscure. Inside the limits of its present occurrence, it is proceeding to invade new regions and increases in number. The way that *Lantana* is absent in a few regions does not mean it is unsuited to them; it has been portrayed as a 'sleeper weed', anticipating conditions to spread. *Lantana camara* was put at category one on the CARA legislation in the country. Records demonstrate that *Lantana* was brought into South Africa in 1858 in Cape Town, Western Cape Province (WC), where, under Mediterranean climatic conditions it spread very little, and in 1883 at Durban, KwaZulu-Natal Province (KZN) (Stirton, 1977), where it prospered under subtropical situations. It was pronounced a poisonous weed in 1946 in KZN, which incorporated South African invasion of 80%, every decade doubling in area (Marr, 1964; Kwembeya et al., 2013). Measured by ongoing requests from people in general for exhortation on the control of *Lantana*, the weed is currently spreading and expanding in population size mainly in the province of Mpumalanga (MP) and Limpopo (LP), and in addition in the North West (NW), Eastern Cape (EC), Gauteng (GP) and the southern part of Western Cape (WC), that is, essentially in the warmer and external parts of its present occurrence. *Lantana* was evaluated by experienced researchers to have invaded more than 2.2 million ha in South Africa, which, if condensed, would totally cover a zone in excess of 69 000 ha (Versfeld et al., 1998). An ongoing, statistically valid, national, invasive alien plant study found that *Lantana* covers 560 000 ha of the landscape, including riparian zones (Kotze et al., 2010).

Flowers are small, generally orange, occasionally changing from white to red in different shades and having a yellow throat in axillary heads. Fruits are small, greenish-blue black, blackish, drupaceous, and sparkling with two nullets, nearly all year round, dispersed by birds. Seeds sprout easily. *Lantana camara* berries draw in frugivorous birds and mammals that assist to scatter its seeds widely. The species forms dense, impenetrable thickets that smother indigenous vegetation and pasture (Zachariades et al., 2017). The outspread of *L. camara* in forests is an exceptional reason for worry as it might change forest structure and composition affecting indigenous species assemblages (Aravind et al., 2006). This species can out-compete indigenous plants since it can develop on nutritiously poor soils (Platt et al., 2016). It flowers all year-round and produces a lot of nectar which attracts potential pollinators (Aravind et al., 2010). It changes forest structure by replacing the native understorey species, and this, in turn, can influence the distribution and behaviour patterns of animal populations.

South African specimens of the weed *Lantana* that Sanders has investigated are generally crossbreeds between *L. nivea* subsp. *mutabilis* and *L. camara* var. *aculeata* (Zachariades et al., 2017). Others seem to have been contributed from both above parent varietal plus *L. scabrida, L. horrida* Kunth subsp. *Tilifolia* (cham.) R.W. Sanders from South Brazil, *L.* (X) *strigocamara, L. hirsute*, conceivably *L. depressa* var. *depressa* from Florida and the Callowiana Hybrid group (*Lantana* (X) *strigocamara* X *L. depressa var. depressa*), and there are also some unblended *L. nivea* subsp. *mutabilis* exhibit (Zachariades et al., 2017).

Noxious triterpines and lantadenes A and B that lead to fatalities to, cattle, horses, goats, sheep, and rabbits by the miss-performance of the liver and different organs are found in the leaves of Lantana (Munyua et al., 1990; Morton, 1994). Nevertheless, many animals would cautiously abstain from ingesting this plant when they have a choice. The green fruits also contain the toxic substances and have led to sickness and fatalities in kids (Morton, 1994). *Lantana camara* leaves and their leachates exert allelopathic impacts *in vitro* and to a lesser degree in soil on the germination of the seeds, root elongation, and plant development of numerous species (Casado, 1995).

2.2.2 Habitat Description

The various and wide geographic dissemination of *L. camara* reflects its broad environmental resistance. It is distributed in varying habitats and soil types. It normally establishes best in open sunny wastelands, for example, rainforest boundaries, coastal areas, and in forests recuperating from fire (Zachariades et al., 2017). *Lantana camara* does not invade solid rainforests, but rather, is located on their edges (Zachariades et al., 2017). Where wet sclerophyll forests and rainforests have been disturbed by logging, spaces are constructed; this enables *Lantana* to invade on the forests.

Lantana camara develops under extensive variety of climatic constraints. In Australia, the inland limits of its geographical range coincide with the 750 mm isohyet in southern Queensland and the 1250 mm isohyet in the north, with invasion being limited to creek lines in drier zones. It does not seem to have a higher temperature or precipitation limitation and is frequently found in tropical regions getting 3000 mm of precipitation annually, if soils are adequately well drained. *L. camara* rarely occurs where temperatures fall beneath 5°C most of the times, and in South Africa, it occurs in zones with a mean yearly surface temperature greater than 12.5°C (Zachariades et al., 2017). Extended chilling temperatures damage aerial woody parts and cause defoliation. *Lantana camara* dominates low laying vegetation in disturbed zones, regularly producing thick mono-specific bushes, and is compromised in widespread depletion of indigenous plant species (Gooden et al., 2009).

2.2.3 Reproduction

If adequate moisture and light are available, *L. camara* flowers all year round in most places, in wet summer months, the flowering is at its peak. Flowering takes place only in the wetter or warmer months in cooler or drier regions due to drought or frost damage (Mello et al., 2005). As early as the second growing season, plants can start to flower. The variations in the length of corolla, number of flowers per inflorescence and its diameter lead to some butterfly species visiting certain taxa of *Lantana* regularly than others. According to this view, different varieties of *Lantana* would have different pollinator species. Therefore, there may be little cross-pollination between species or varieties of *Lantana* both in naturalized and native ranges of the section Camara (Zachariades et al., 2017).

2.2.4 Benefits

In numerous agricultural diversities, *L. camara* is sown worldwide as an ornament. In temperate zones, it is planted as a yearly bedding species. *Lantana camara* oil, an aromatic blend which differs from native plant heterogeneity, is conveyed from Brazil (Weyerstahl et al., 1999). Steeps of the leaves and other parts of the plant are utilized as healing remedies because of their anti-inflammatory properties (Oyedapo et al., 1999), they also serve as an expectorant and tonic, and are also added to baths as antirheumatics. *Lantana camara* extracts have also been shown to be strong medicinal drugs (Liogier, 1990). Since the leaves and some different parts of *L. camara* are toxic, caution must be observed when they are utilized therapeutically. The mature fruit is tender and eaten in large quantities by birds and regularly eaten by people in some nations (Herzog et al., 1994). In some agricultural settings, invasions of *Lantana* are believed to prevent soil compaction, and are valued as source of organic matter for pasture field remodel or improvement. It is also valuable in steep zones and stream banks for stabilizing soil and preventing erosion. At times, *L. camara* smothers weeds perceived to be more threatening. The sale of ornamental *Lantana* in a few states provides income to the nursery business. Concentrates of *L. camara* leaves have demonstrated solid pesticide and antimicrobic action in various tests. Preserving potatoes with *Lantana* leaves almost eliminates harm by *Phthorimaea operculella* Zeller, the potato tuber moth (Lal, 1987). Stems and leaves are utilized as mulch. However, of lesser standard on account of magnitude and morphology, *Lantana* stems are broadly utilized as fuel wood in less developed nations.

2.3 Data Sampling and Analysis

Twenty-four permanent plots of 10 m x 10 m each were constructed using iron rods. The number of alien invasive species and native species present in the quadrats were recorded before they were removed. The invasive alien plant species were removed mechanically. No chemicals were used, since there are concerns about their adverse environmental implications. The areas were monitored for the reestablishment of alien invasive species; if they occurred, they were removed. Only the native plant species were left alone, if they happened to be seen establishing in the study site.

All data collected were entered and stored in Microsoft Excel 2010, and were also used in the descriptive statistical analysis of the results. They were then analized using Primer V6 and PERMANOVA.

RESULTS

3.1 Species Richness

Species richness was not significantly different between the sampling periods. Observations of the eradicated sampling plots showed that species richness in the dug area of *L. camara* was found to be 08 (Sep-2016) before eradication and 08 (Jun-2017), 10 (Sep-2017), 09 (Dec-2017) and 11 (Mar-2018) after eradication **(Table 1).** Only Mar-2018 was richer with 11 species, with the least rich (Sep-2016 and Jun-2017) having 08 species respectively.

Table 1. Floral species richness during the sampling periods following the eradication of Lantana camara.

Response variable	Treatment	Sep-2016	June-2017	Sep-2017	Dec-2017	Mar-2018
Species richness	Digging	08	08	10	09	11

Individuals of *L. camara* slightly increased after eradication (2016-Sep to 2017-Dec), decreased thereafter (2017-Dec to 2018-Mar). On the other hand, the number of the rest of the species slightly decreased after eradication (2016-Sep to 2017-Jun), and the highest number of *L. camara* species were observed when the rest of the species were at their lowest (2017-Dec) (Table 5.1, Figure 5.2)

Figure 2. Line graph showing the number of Lantana camara and other plant species pre (Sep-2016) and post eradication.

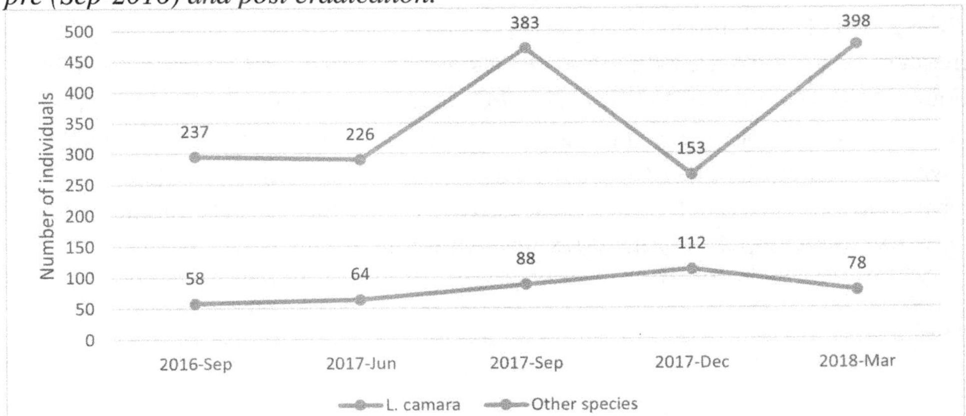

Figure 3. Multidimensional scaling ordination plots of species composition in different sampling periods between Sep-2016 and Dec-2017 following the eradication of Lantana camara.

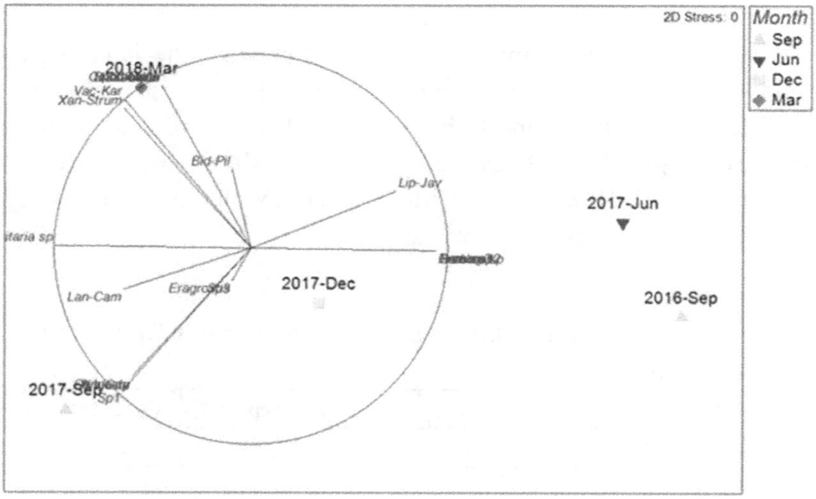

The MDS plot of species composition clearly indicated that species were shared almost similarly between the sampling periods (Figure 5.2). More species are observed in 2018-Mar, which is similar to what is presented on species richness table (Table 5.1). The similarity in all the sampling periods was attributed to species such as *L. camara, V, karoo, B. pilosa* and *L.javanica*, which are the only species that were observed throught the sampling periods.

Different plant species were observed during the five sampling periods, however, 2016-Sep and 2017-Jun had completely similar plant species differing only in the frequencies (Table 5.2), with other species observed from 2017-Sep to 2018-Mar. Four species (*L. camara, V. karoo, B. pilosa* and *L. javanica*) were the only species observed throughout the sampling periods. These four species are attributed to have contributed the most to the similarity between 2016-Sep and 2017-Jun and have contributed also to the similarity across the sampling periods. Frequency was observed to be high during the 2018-Mar sampling period.

Table 2. Frequency of native plant species following eradication of Lantana camara

Species Name	2016-Sep	2017-June	2017-Sep	2017-Dec	2018-Dec
Lantana camara L.	83.333	100	100	100	83.333
Vachellia Karoo (Hayne) Banfi & Galasso.	66.666	100	66.666	66.666	66.666
Bidens Pilosa L.	83.333	66.666	100	83.333	100
Lippia Javanica L.	83.333	83.333	33.333	66.666	33.333
Herb spp1	33.333	33.333	0	0	0
Combretum spp.	16.666	16.666	0	0	0
Grass spp2	16.666	16.666	0	0	0
Tree spp3	16.666	16.666	0	0	0
Hibiscus cannabis L.	0	0	83.333	0	0
Spp1	0	0	83.333	50	0
Digitaria spp.	0	0	83.333	66.666	33.333
Cynodon dactylon (L.) Pers	0	0	33.333	0	0
Congesta	0	0	16.666	0	0
Xanthium Strumarium L.	0	0	16.666	16.666	66.666
Eragrostis spp.	0	0	0	100	0
Spp3	0	0	0	100	0
Sida cordifolia L.	0	0	0	0	16.666
Combretum molle R.Br. ex G.Don.	0	0	0	0	16.666
Cleome	0	0	0	0	16.666
Sporobolus spp.	0	0	0	0	16.666

DISCUSSION

Preceding clearing by 'Working for Water' in 2016, transects had invasive alien species density of up to 97%. These high IAP densities seemed to negatively affect community diversity which is congruent to various other researcher's investigations (Morris et al., 2008; Holmes et al., 2000; Pyšek and Pyšek, 1995). After the lowering of invasive alien species density, there was an eminent corresponding increment in the native diversity of vegetation in the already densely-infested regions. The significant contribution to the high invasive alien species densities was from annuals or short-lived perennial species, for example, *X. strumarium* (L.).

Despite being uprooted several times during the observation and monitoring of the area since 2016, *L. camara* kept on emerging in the same area. This is probably because of the soil seed bank which had not yet been exhausted by then. In other words, *L. camara,* because of its presumably large seed bank continues to regenerate for several seasons until the seed bank is depleted. It is believed that with further monitoring, the soil seed bank might be exhausted, and the *Lantana* population might start disappearing for good in the area. In regions where it is prevailing, *L. camara* is likely to reduce indigenous species composition by hindering recolonization after some environmental pressures. The influence of *L. camara* invasion on indigenous plant biodiversity was prevalent in such a way that indigenous herbs, shrubs and trees were compromised seriously in such vegetation. Attempting to eradicate *L. camara,* led to its stimulation as several individuals re-established after trying to eliminate them.

There was an increase in the newly established individuals compared to what existed before disturbance. This may have been initiated by the increase in propagule supply during follow-up observation events by invader species control officers, as well as soil disturbance during manual eradication. Mason and French (2007) reported on a large increase in the number of *L. camara* following disturbances such as soil disturbance and others. Species richness in each growth form declined significantly with increasing *Lantana* cover, indicating that the threat of Lantana is pervasive across all life forms in the recipient community. Similar pervasive losses of plant species across multiple growth forms have been detected elsewhere for *Lantana* (Gooden et al., 2009), possibly representing a general effect of the invader on native communities.

Some researchers announced an increase in both abundance and richness of indigenous species recovery after *Lantana* eradication in north Queensland (Gooden et al., 2009), and this was not the situation with this investigation. Just other invasive herb species occupied the site in large numbers than indigenous plant species. Trees, relative to alternate life forms, largely drove the abundance in indigenous species recovery. This compares with others' discoveries of contrasts among plant species considering *Lantana* eradication, however they revealed a noteworthy increase in herbs and shrubs with respect to woody trees (Gooden et al., 2009).

There is exact proof that Lantana can allelopathically suppress indigenous species regeneration (Gentle and Duggin, 1998); however, there is no genuine quantifiable proof for this from field evaluations (Gooden et al., 2009). Expansion in species richness in controlled areas proposes that institute *L. camara* is an obstruction to recuperation of species variety after disturbance of the vegetation. Then again, expansion in species richness may correlate to control rather than the original *L.*

camara invasion, for example, soil disturbance and amplification of nutrient availability (Davis et al., 2000).

Figure 4. (a) A dense population of Lantana camara before removal, (b) the cutting and digging of L. camara in progress, (c) cleared site just three months after clearance, and (d) L. camara recolonizing the site after a few months

The emergence of *L. camara* after clearing is thought to be due to its ability to reproduce through the stems which might be lying on the ground. The species can produce roots on the stem if it comes into contact with the ground, thereby reproducing via stem layering, and also from the seeds.

CONCLUSIONS AND RECOMMENDATIONS

Lantana camara is a difficult species to deal with because it spreads in three ways. Firstly, through vegetative reproduction where a stem sends roots into the soil (layering) which allows for dense stands to form and spread to short distances very quickly. Secondly, birds and other animals such as foxes consume the fruits, and the droppings they pass would contain seeds, in this way they get dispersed over a large distance. Digestion of seeds generally increases the rate of their germination. Lastly, it is also capable of re-seeding and re-sprouting. The site previously infested with *L. camara* needs more monitoring after eradication and possibly active restoration in the form of planting the native plant species which previously occupied the area before the infestation by the invasive species. It is therefore more important to study the effect the seed bank might have on the recovery of this problematic species and find out better ways of control.

REFERENCES

Alonso, A., Dal Lmeier, F., Granek, E., & Raven, P. (2001). Biodiversity connecting with the tapestry of life, Smithsonian Institution / Monitoring and assessment of biodiversity Program and President's committee of Advisors on Science and Technology. Washington DC, United States of America.

Aravind, N. A., Dinesh, R. A. O., Ganeshaiah, K. N., Uma Shaanker, R., & Poulsen, J. G. (2010). Impact of the invasive plant, *Lantana camara*, on bird assemblages at Male Mahadeshwara Reserve Forest, South India. *Tropical Ecology, 51*(2S), 325–338.

Aravind, N. A., Rao, D., Vanaraj, G., Ganeshaiah, K. N., Shaanker, R. U., & Poulsen, J. G. (2006). *Impact of Lantana camara on plant communities at Malé Mahadeshwara reserve forest, South India. Invasive Alien Species and Biodiversity in India*. Banaras Hindu University.

Casado, C. M. (1995). Allelopathic effects of *Lantana camara* (Verbenaceae) on morning glory (*Imomoea tricolor*). *Rhodora, 97*, 264–274.

Castro-Díez, M. P., & Fernández, Á. (2017). Effects of non-native riparian plants in riparian and fluvial ecosystems: A review for the Iberian Peninsula. *Limnetica, 36*, 525–541.

Clout, M. N., & Veitch, C. R. (2002). *Turning the tide of biological invasion: the potential for eradicating invasive species. Turning the tide: the eradication of invasive species*. IUCN SSC Invasive Species Specialist Group.

Cronk, Q. C. B., & Fuller, J. L. (1995). *Plant Invaders*. Chapman and Hall.

Davis, M. A., Grime, J. P., & Thompson, K. (2000). Fluctuating resources in plant communities: A general theory of invasibility. *Journal of Ecology, 88*(3), 528–534. DOI: 10.1046/j.1365-2745.2000.00473.x

Day, P. K., LeDuc, H. G., Mazin, B. A., Vayonakis, A., & Zmuidzinas, J. (2003). A broadband superconducting detector suitable for use in large arrays. *Nature, 425*(6960), 817–821. DOI: 10.1038/nature02037 PMID: 14574407

Gentle, C. B., & Duggin, J. A. (1998). Interference of shape *Choricarpia leptopetala* by shape *Lantana camara* with nutrient enrichment in mesic forests on the Central Coast of NSW. *Plant Ecology, 136*(2), 205–211. DOI: 10.1023/A:1009784706714

Gooden, B., French, K., & Turner, P. J. (2009). Invasion and management of woody plant, *Lantana camara* L., alters vegetation diversity within wet sclerophyll forest in south eastern Australia. *Forest Ecology and Management*, *257*(3), 960–967. DOI: 10.1016/j.foreco.2008.10.040

Herzog, F., Farah, Z., & Amado, R. (1994). Composition and consumption of gathered wild fruits in the V-Baoule, Cote d'Ivoire. *Ecology of Food and Nutrition*, *32*(3–4), 181–196. DOI: 10.1080/03670244.1994.9991399

Holm, L., Doll, J., Holm, E., Pancho, J., & Herberger, J. (1997). *Word weeds: Natural Histories and Distribution*. John Wiley & Sons, Inc.

Holmes, P. M., Richardson, D. M., Van Wilgen, B. W., & Gelderblom, C. (2000). Recovery of South African fynbos vegetation following alien woody plant clearing and fire: Implications for restoration. *Austral Ecology*, *25*(6), 631–939. DOI: 10.1111/j.1442-9993.2000.tb00069.x

Kotze, I., Beukes, H., Van Den Berg, E., & Newby, T. (2010). National invasive alien plant survey. Report No. GW/A/2010/21 to the Department of Water and Environmental Affairs, Working for Water Programme. Agricultural Research Council-Institute for Soil, Climate and Water. Pretoria, South Africa.

Kwembeya, A., Rugare, J. T., & Mabasa, S. (2013). Allelopathic effects of lantana (lantana camara) on blackjack (bidens pilosa) and pearl millet (pennisetum glaucum). *Asian Journal of Agriculture and Rural Development*, *3*(8), 543–553.

Lal, L. (1987). Studies on natural repellents against potato tuber moth (*Phthorimaea operculella* Zeller) in country stores. *Potato Research*, *30*(2), 329–334. DOI: 10.1007/BF02357672

Liogier, H. A. 1990. Plantas medicinales de Puerto Rico y del Caribe. Ibero americana de Ediciones, Inc., San Juan, PR 563p.

Marr, R. S. (1964). Get rid of *lantana.Farming in South Africa*, (February), 42–45.

Mason, T. J., & French, K. (2007). Management regimes for a plant invader differentially impact resident communities. Institute for Conservation Biology, University of Wollonong, NSW 2522, Australia. DOI: 10.1016/j.biocon.2006.11.023

Mello, F. B., Jacobus, D., Carvalho, K., & Mello, J. R. (2005). Effects of *Lantana camara* (Verbenaceae) on general reproductive performance and teratology in rats. *Toxicon*, *45*(4), 459–466. DOI: 10.1016/j.toxicon.2004.12.004 PMID: 15733567

Mishra, A. (2015). Allelopathic properties of Lantana camara. *International Research Journal of Basic and Clinical Studies*, *3*(1), 13–28.

Mohamed, N. M., Makboul, M. A., Farag, S. F., Jain, S., Jacob, M. R., Tekwani, B. L., & Ross, S. A. (2016). Triterpenes from the roots of *Lantana montevidensis* with antiprotozoal activity. *Phytochemistry Letters*, *15*, 30–36. DOI: 10.1016/j.phytol.2015.11.008

Morris, T. L., Witkowski, E. T. F., & Coetzee, J. A. (2008). Initial response of riparian plant community structure to clearing of invasive alien plants in Kruger National Park, South Africa. *South African Journal of Botany*, *74*(3), 485–494. DOI: 10.1016/j.sajb.2008.01.177

Morton, J. F. (1994). *Lantana*, or red sage (*Lantana camara* L. Verbenaceae), notorious weed and popular garden flower; some cases of poisoning in Florida. *Economic Botany*, *48*(3), 259–270. DOI: 10.1007/BF02862327

Munir, A. A., & Adelaide, J. 1996. A taxonomic review of *Lantana camara* L. and *Lantana montevidensis* (Spreng) Briq. (Verbenaceae) in Australia. Board of the Botanic Gardens and State Herbarium, Adelaide, South Australia. 1976–2015 (Vol. 1, 29).

Munyua, S. J. M., Nienga, M. J., Karitu, T. P., Kimoro, T. P., Kiptoon, J. E., & Buoro, I. B. J. (1990). A note on clinical-pathological findings and serum enzyme activity in sheep, goats and Friesian calves with acute *Lantana camara* poisoning. *Bulletin of Animal Health and Production in Africa*, *38*(3), 275–279.

Mushi, G. V. (2019). *Local communities' perceptions on lantana camara and management responses in East Usambara, Tanzania* (Doctoral dissertation, Sokoine University of Agriculture).

Myers, N., Mittermeier, R. A., Mittermeier, C. G., Da Fonseca, G. A. B., & Kent, J. (2000). Biodiversity hotspots for conservation priorities. *Nature*, *403*(6772), 853–858. DOI: 10.1038/35002501 PMID: 10706275

Oyedapo, O. O., Sab, F. C., & Olagunju, J. A. (1999). Bioactivity of fresh leaves of *Lantana camara.Biometrical Letters*, *59*, 179–183.

Parsons, W. T., Parsons, W. T., & Cuthbertson, E. G. (2001). *Noxious weeds of Australia*. CSIRO publishing.

Pennsylvania Department of Conservation and Natural Resources, Invasive species Management Plan., 2005.

Platt, R. V., Ogra, M. V., Badola, R., & Hussain, S. A. (2016). Conservation-induced resettlement as a driver of land cover change in India: An object-based trend analysis. *Applied Geography (Sevenoaks, England)*, *69*, 75–86. DOI: 10.1016/j.apgeog.2016.02.006

Pyšek, P., Hulme, P. E., Simberloff, D., Bacher, S., Blackburn, T. M., Carlton, J. T., Dawson, W., Essl, F., Foxcroft, L. C., Genovesi, P., Jeschke, J. M., Kühn, I., Liebhold, A. M., Mandrak, N. E., Meyerson, L. A., Pauchard, A., Pergl, J., Roy, H. E., Seebens, H., & Richardson, D. M. (2020). Scientists' warning on invasive alien species. *Biological Reviews of the Cambridge Philosophical Society*, *95*(6), 1511–1534. DOI: 10.1111/brv.12627 PMID: 32588508

Pyšek, P., & Pyšek, A. (1995). Invasion by *Heracleum mantegazzianum* in different habitats in the Czech Republic. *Journal of Vegetation Science*, *6*(5), 711–718. DOI: 10.2307/3236442

Richardson, D. M., & Van Wilgen, B. W. (2004). Invasive alien plants in South Africa: How well do we understand the ecological impacts? *South African Journal of Science*, *100*, 45–52.

Rosacia, W.Z., Achivar, A.N. and Avanzado, M.B., (2004). *Lantana* and Hagonoy: Poisonous weeds prominent in rangeland and grassland areas. *Research Information Series on Ecosystems (Department of the Environment and Natural Resources, Republic of the Philippines),* 16(2).

Ross, I. A. (1999). *Medicinal plants of the world. Chemical constituents, traditional and modern medical uses*. Human Press.

Samways, M. J., & Taylor, S. T. (2004). Impacts of invasive plants on red-listed South African dragonflies (Odonata). *South African Journal of Science*, *100*, 78–80.

Stirton, C. H. 1977. Some thoughts on the polyploid complex *Lantana camara* L. (Verbenaceae). In *Proceedings of the Second National Weeds Conference of South Africa.* pp. 321–340.

Versfeld, D. B., Le Maitre, D. C., & Chapman, R. A. (1998). Alien invading plants and Water Resources in South Africa: A Preliminary Assessment. Water Research Commission Report No. TT99/98.

Weyerstahl, P., Marschall, H., Eckhardt, A., & Christiansen, C. (1999). Constituents of commercial Brazilian *Lantana* oil. *Flavour and Fragrance Journal*, *14*(1), 15–28. DOI: 10.1002/(SICI)1099-1026(199901/02)14:1<15::AID-FFJ777>3.0.CO;2-M

Zachariades, C., Paterson, I. D., Strathie, L. W., Hill, M. P., & Van Wilgen, B. W. (2017). Assessing the status of biological control as a management tool for suppression of invasive alien plants in South Africa. *Bothalia –. African Biodiversity and Conservation*, *47*(2), 1–19.

Chapter 3
Abiotic and Biotic Stresses and Their Interactions in Plants:
Concepts and Mechanisms

Melford Mbedzi
https://orcid.org/0000-0002-8330-1356
Mangosuthu University of Technology, South Africa

Kuben Naidoo
Mangosuthu University of Technology, South Africa

ABSTRACT

Abiotic stress refers to the negative effects on plant growth and development caused by abiotic factors such as extreme temperatures, drought, salinity, floods and pollutants. On the other hand, biotic stress is caused by living organisms such as pathogens (disease-causing microorganisms) and pests (insects, mites, and rodents) that can harm plants. The interaction between abiotic stress and biotic stress can have a significant impact on plant health and productivity. Drought stress weakens plants, making them more susceptible to pathogen attack. Lack of water can hinder a plant's defense response, making it harder for it to fight disease. Pathogens can quickly colonize and spread in drought-stressed trees, causing more serious damage. Increased soil salinity can interfere with normal physiological processes in plants, leading to reduced growth and yield.

DOI: 10.4018/979-8-3373-1549-2.ch003

INTRODUCTION

Nature is constantly changing and plants must maintain their inherent ability to survive in the ecological environment through natural genetic variation and adaptive mechanisms. In addition to natural variations, induced variations have led to the domestication of plants and the development of agriculture, which in turn transformed human civilization and are now feeding a growing global population. However, plants grown in agricultural fields are constantly exposed to various biotic and abiotic stresses (Mohanta et al., 2017).

The impact of various abiotic stresses on plant growth and development is evident in the increasing ecological challenges due to climate change (Bellard et al., 2012). Physiological constraints on agricultural production are increasingly exacerbated by growing populations and competition for environmental resources, creating new problems for agricultural production (Mohanta et al., 2017). In natural and agricultural ecosystems, plants respond rapidly to abiotic and biotic stressors, individually or in combination, through complex and long-term adaptive processes through evolutionary adaptive strategies.

Future climate change scenarios are expected to increase the impact of abiotic stresses on plants and further increase the virulence and prevalence of existing pathogens and pests, thereby negatively affects crop productivity and plant survival rate. Indeed, global warming produces multiple and sometimes contradictory impacts on global and regional climate, subjecting plants to higher temperatures, higher vapor pressure deficits, and often droughts. more frequent and intense, but also unusual frosts, floods, etc. More importantly, plants will face a complex series of simultaneous stress factors that will challenge their survival in certain habitats and sometimes promote their expansion into new geographical areas. With this in mind, understanding the resistance and recovery mechanisms of plants to the complex and diverse impacts of climate change is an urgent task (Balestrini et al., 2022).

During plant growth and development, plants face many abiotic stresses, including drought, flood, salinity, freezing, cold temperatures, hot temperatures, intense light high and low altitude, heavy metal toxicity, ozone pollution, carbon dioxide, sulfur dioxide, soil pollution, pH, UV radiation and more. Among these stress factors, drought, heat, salinity, and metal toxicity are the main causes of crop loss. Plants can resist the effects of these stresses to some extent, but it is difficult for plants to overcome combined stress conditions. Understanding and identifying the key regulatory factors involved in many stress responses in plants is a complex process. It is essential to identify traits that may confer tolerance to such environmental stresses, and genomic studies may answer these questions (Mohanta et al., 2017).

METHODOLOGY

This paper summarizes previous research (2010-2025) on the biotic and abiotic stresses that affect plants in different ways, positive of negative. All the information on plant stresses was gathered through the use of internet databases (Google Scholar, Science Direct, and Pub Med) and library documents. With no date or language restrictions, a simple keyword/phrase search (such as, biotic stresses, abiotic stresses in plant, agricultural crop yield) on the plant of interest was conducted. Articles relevant to plant stresses affecting plants were identified by screening all titles and abstracts.

Review of Literature

Plant responses to abiotic stresses throughout their evolutionary history, plants have developed a variety of very complex and efficient mechanisms to sense, respond and adapt to many environmental changes. When found under unfavorable or restrictive growing conditions, plants respond by activating resistance mechanisms at multiple levels of organization (molecular, tissue, anatomical and morphological), by regulating regulate the membrane system and cell wall structure, regulate the cell cycle and cell growth ratio division and metabolic regulation (Xie et al., 2019).

At the molecular level, many genes are induced or repressed by abiotic stress, implying the precise regulation of large stress gene networks (Yamasaki, 2005; Hancock and Veal., 2021). The products of these genes may function in stress response and tolerance at the cellular level. Proteins involved in the biosynthesis of osmoprotective compounds, detoxification enzyme systems, proteases, transporters, and chaperones are among the many protein functions activated as the first line of direct defense against stress. Additionally, activation of regulatory proteins (e.g., transcription factors, protein phosphatases, and kinases) and signaling molecules is essential in the simultaneous regulation of signal transduction and stress responsive genes (Kharbech et al., 2020; Manafi et al., 2021). Early plant response mechanisms prevent or minimize stress-induced cell damage, restore homeostatic conditions, and allow continued growth (Gupta et al., 2022).

Restoring the energy balance, osmotic and redox imbalance caused by stress factors is the first target of the plant's immediate response. Observed tolerance responses to abiotic stresses in plants often include stress-specific response mechanisms and more general adaptive responses that provide strategic benefits under adverse conditions. General response mechanisms involve central pathways that are involved in energy maintenance and include calcium signaling cascades (Nabaei and Amooaghaie., 2019; Kumar and Ohri, 2023), signaling factors signaling/identification of reactive oxygen species (Zhou et al., 2021) and energy deficiency, (energy-sensing protein kinase, SnRK1). Induction of these central pathways is observed during plant adaptation to different types of stress. For example, the protein kinase SnRK1 is a central metabolic regulator of gene expression associated with energy depletion, but this kinase also becomes active when plants are exposed to various abiotic stresses, such as drought, salt, flood, or nutrient degradation (Astier et al.., 2018; Kumar and Ohri, 2023). Kinase SnRK1 modifies the expression of more than 1000 stress-responsive genes, allowing homeostasis to be re-established by inhibiting energy-consuming processes, thereby promoting stress tolerance (Gupta et al., 2011). Optimizing cellular energy resources during stress is essential for plant adaptation; Energy-intensive processes are partially shut down, such as reproductive activities, translation, and some biosynthetic pathways. For example, nitrogen and carbon assimilation is impaired in maize under salt stress and potassium deficiency; the synthesis of free amino acids, chlorophyll and proteins is also affected (Prochazkova et al., 2014).

Systems Biology Approach in Abiotic Stresses

Due to the sessile nature of plants, they are exposed to multiple abiotic stressors at the same time. Plants activate stress signaling cascades to overcome the negative effects of stress. However, it is difficult to understand whether plants simultaneously activate defense signaling pathways or whether a single signaling cascade controls a multitude of stress responses. The biological system of an organism is complex, and stress signaling mechanisms begin with the first reactions (within minutes) to stressful events. During this process, countless signaling molecules, including genes, proteins, metabolites, and hormones, play a vital role in overcoming stress. However, it is quite difficult to understand the interaction network and reaction rates of all these molecules at the same time. To understand all these events requires a comprehensive knowledge of morphological, physiological, genetic, molecular, and biochemical events in plants. Metabolic signaling networks are central to physiological responses to fluctuating environmental stresses. Advances in genomics, computational tools, and biotechnology have dramatically changed the understanding of stress signaling in plants, and functional genomics studies can be applied holistically to the functional code of the gene in a very short time. Systems biology approaches enable in-depth understanding of complex physiological and cellular processes. The principles of the systems biology approach are based on the advancement of omics sciences (genomics, transcriptomics, metabolomics, proteomics, etc.) and their integration to establish meaningful relationships between genotype, phenotype, and stress tolerance (Mahanta et al., 2017).

Plant Metabolomics and Drought Stress

Rainfall is variable and often insufficient in large areas of rainfed agriculture, unsustainable use of groundwater for irrigated agriculture worldwide and growing urban water demand are putting pressure enormous growth in global food crop production. The need for water to support agricultural systems in many countries will continue to increase due to population growth (FAO, 2007). This worsening water shortage causes water stress for rainfed and irrigation crops. Water stress induces a series of physiological and biochemical changes in plants; The cessation of cell growth and photosynthesis as well as improved respiration are among the first effects. Genomic expression is extensively remodelled, activating and repressing a variety of genes with diverse functions (Shinozaki and Yamaguchi-Shinozaki, 2007; Liu et al., 2004).

Detection of dehydration and activation of defense mechanisms require chemical signals in which abscisic acid (ABA) plays a central role. ABA accumulates in plant tissues under water stress and promotes transpiration reduction through stomatal closure. Through this mechanism, plants minimize water loss and reduce damage caused by stress. ABA regulates the expression of many stress-responsive genes, including late embryonic abundant (LEA) proteins, leading to enhanced drought tolerance in plants (Aroca et al., 2008). Osmotic regulation is one of the main mechanisms that help plants cope with water shortage. These adjustments maintain positive cellular tension through the active accumulation of compatible solutes. Traditionally, the analysis of metabolic responses to drought stress has been limited to the analysis of one or two classes of compounds considered "factors" in the development of tolerance. The application of metabolomic methods provides a less biased perspective on metabolic reaction profiles and contributes to the discovery of new metabolic phenotypes.

Plant Metabolomics and Oxidative Stress

An increase in intracellular ROS levels is a common consequence of unfavourable growth conditions. The imbalance between ROS synthesis and scavenging is induced in a manner that is independent of the nature of the stress; it is caused by both biotic and abiotic stresses. Toxic ROS concentrations severely damage protein structures, inhibit the activity of certain enzymes in important metabolic processes, and lead to the oxidation of macromolecules, including lipids and DNA. All of these side effects compromise cellular integrity and can lead to cell death (Gill and Tuteja

2010; Kar, 2011). Normal cellular metabolic activity also leads to the generation of ROS under normal growth conditions.

Therefore, cells detect uncontrolled increases in ROS levels and use them as a signaling mechanism to trigger protective responses (Møller and Sweetlove, 2010; Vahdati and Leslie, 2013). In this context, plants have developed effective mechanisms to eliminate toxic ROS concentrations. The antioxidant system includes protective enzymes (e.g., superoxide dismutase, catalase, peroxidase, reductase and redoxin) and free radical scavenging metabolites (mainly GSH and ascorbate). GSH is an essential component of the antioxidant system, donating an electron to unstable molecules such as ROS to make them less reactive and can act as a redox buffer during recycling ascorbic acid from the oxidized form to the reduced form thanks to the enzyme dehydroascorbate reductase (Jozefczak et al., 2012). Organized restructuring of metabolic networks is a critical response that gives cells the best chance of survival against oxidative challenge.

It is known that drought stress in the early stages of plant life, immediately after germination, can have devastating effects, on the one hand due to the root system not being fully established and on the other hand, controlling Stomatal control has not yet been perfected. However, drought stress at this early stage of life has not attracted much research attention because it can be easily overcome by farmers through careful selection of sowing dates (Vahdati and Leslie, 2013). Drought stress in later phenological stages has received the most attention, especially comparisons between the effects of drought on vegetative stages and reproductive stages on seed production. It is now well established that the effects of stress can vary significantly depending on the morphological stage of the plant. Reproductive stages are often more sensitive to stress than vegetative stages, but there may also be differences between different stages of the reproductive stage.

Mouhouche et al. (1998) found in Phaseolus vulgaris that the flowering stage is more sensitive than the fruit elongation and seed production stages. Casanovas et al. [83] reported a decrease in both leaf physiology and grain yield of drought-stressed maize during flowering. Boonjung and Fukai (1996) reported that when drought occurs during the vegetative stage, it has little effect on later seed development and yield. The effects of water stress on yield are more severe when drought occurs during cotton development. Grapevines provide an interesting example of the complexity of the relationship between water stress and plant phenology. Traditionally, grapevines are non-irrigated crops, occupying large areas in arid and semi-arid regions (Cifre et al., 2005).

Recently, in the Mediterranean region, irrigation systems have been applied to increase lowland productivity. However, the quality of wine depends largely on the organoleptic properties of the grapes, which in turn, especially in terms of soluble sugar content, depend on moderate water pressure during fruit blooming (i.e. in the stage from installing fruit to veraison). Therefore, irrigation strategies must maximize vineyard yields without sacrificing berry quality, a goal consistent with irrigation deficit (IRD) programs. Furthermore, an in-depth understanding of the mechanisms of plant carbon assimilation and distribution under different water regimes will be necessary in the context of precision agriculture, because in reality, these mechanisms play a role. key role in adjusting the balance between yield and quality of berries. This will hopefully lead to the adoption of irrigation planning criteria based on vine physiology (Cifre et al., 2005).

Biotechnology for Biotic Stress Tolerance

Damage to plants by various living organisms including viruses, fungi, bacteria, nematodes, insects, and weeds has caused biotic stress. Other than that abiotic stress has been shown to have severe and detrimental effects due to environmental changes. Different pesticides, fungicides, and herbicides are commonly used to control crop loss and become the cause of damage to crop yield and yield (Isman and Grieneisen, 2014; Law et al., 2017; Yoon et al., 2013). The use of chemicals has caused harmful effects to crop as well as environmental quality due to chemical pollution of the environment and soil (Ahanger et al., 2017; Mattah et al., 2015). The use of pesticides, fungicides, and herbicides must undergo basic analysis before being applied to crops (Mabe et al., 2017; Mahmood et al., 2014). Pesticide and fungicide residues commonly found on fruits and vegetables have become a cause of immediate harm to human health. Crop yields fluctuate depending on how susceptible they are to a particular pesticide, herbicide and fungicide, the use of which also causes problems in plant metabolism pathways (Aktar et al., 2009; Kim et al., 2017).

Figure 1. An overview diagram showing major environmental stresses and mechanisms of plant–microbe interactions that influence plant fitness during these stresses. Tolerance against these stresses is mainly mediated by enhancement in antioxidant defense enzymes, osmotic adjustmentand production of phytohormones. PGPR: plant growth-promoting rhizobacteria (Kamran et al., 2022).

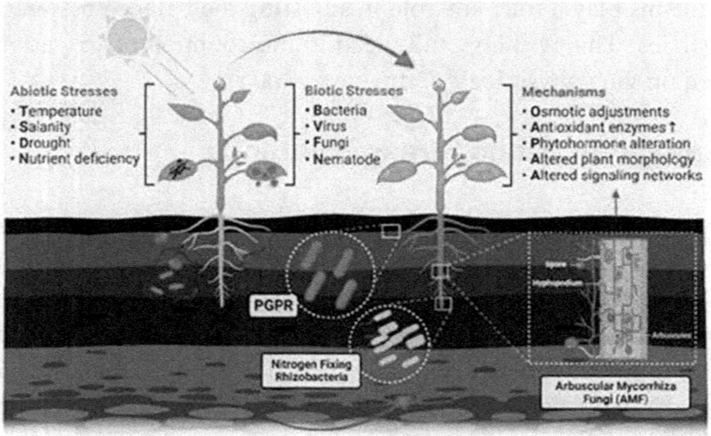

Biotechnology for Abiotic Stress

Tolerance Along with abiotic stresses, drought and high temperature are the two main stresses that damage crop potential and productivity. Such abiotic stresses reduce agricultural income and agricultural profits. Reducing water by up to 40% resulted in up to 40% lower corn yield while wheat yield loss was 21% (Daryanto et al., 2016; Ronald et al., 1992). In Africa, agricultural crops such as cowpea are currently experiencing dry season stress, reducing yields by 34% to 68% (Farooq et al., 2017). Under abiotic stress, generation of reactive oxygen species (ROS) occurs, leading to harmful effects on carbohydrates, nucleic acids, lipids, and proteins. This oxidative stress negatively affects plant growth (Zhu-Salzman et al., 2004). Additionally, water deficit and heat stress can reduce transpiration, stomatal conductance, and photosynthesis in plants (Varsani et al., 2014).

Physiological Effects

Plant physiological processes are largely influenced by temperature changes in the surrounding environment. Plant resistance to extreme temperatures is a complex process determined by environmental factors and the plant's genetic capabilities. In general, the stability of life processes in most plants is relatively wide, ranging from a few degrees above freezing to about 35°C (Ahmad and Prasad, 2011).

Figure 2. Induced oxidative tress caused by climate change in plants.

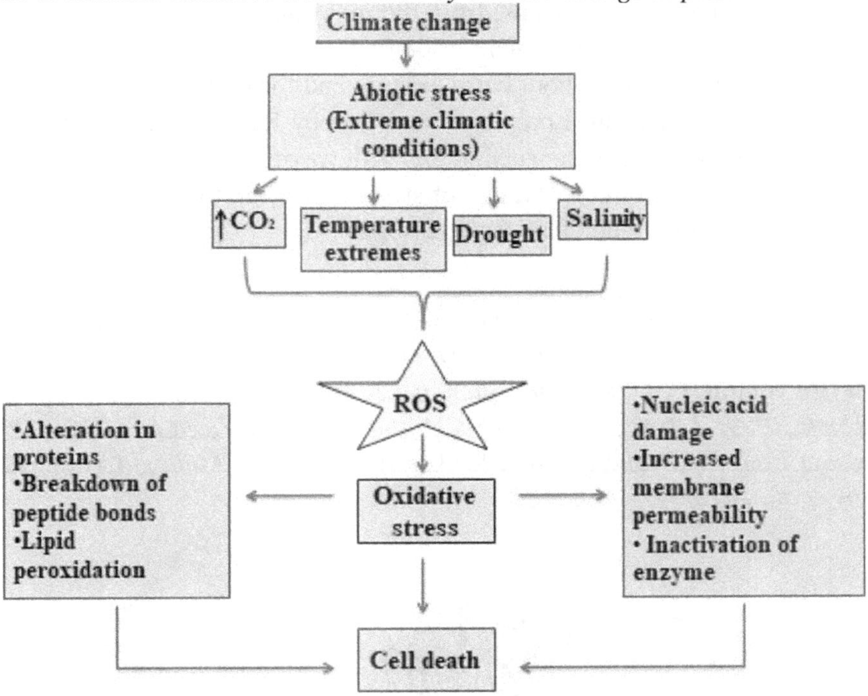

Increasing temperature to a certain level increases growth, photosynthesis, respiration, and enzyme activity, after which these parameters tend to decrease. Respiration increases rapidly with temperature and decreases significantly once the temperature reaches an acceptable level. Photosynthesis is a relatively less sensitive process than respiration, but its pattern of decline is similar to that of respiration. The average rate of enzymatic reactions doubles with every 10°C increase in temperature for approx. The optimal temperature for structural integrity and activity of most enzymes is between 30 and 45°C; and enzymes are irreversibly denatured and inactivated at temperatures above 60°C, with the exception of thermophilic organisms. Therefore, each life process has its own critical or lethal temperature, after which it cannot continue and causes permanent damage to the cellular structure and ultimately death of the cell and the plant (Ahmad and Prasad, 2011).

Eco-Physiological Perspective on Woody Plant Multi-Stress Tolerance

The ability of plants to survive in stressful environments is closely related to a number of tolerance mechanisms, from the molecular to the physiological and ecological levels. Understanding plant responses to stress and associated tolerance mechanisms is important for both basic biology and for improving predictions of the impacts of climate change on plants, and thereby Raise social awareness and plan better to minimize crop impacts. In this VSI, we organize five ecophysiology-oriented reviews that examine a range of studies covering the main responses of woody plants to a multitude of abiotic stressors.

Figure 3. Effects of stresses and some plant ecophysiological responses, ET: Ethylene, ROS: Reactive oxygen species, NO: Nitric oxide, JA: Jasmonic acid, GLVs: Green leaf volatile(s), O3: Ozone (Blum and Jordan, 1985; Bohnert et al., 1995; Gutschick, 1999; Ackerly et al., 2000; Mittler, 2006; McSteen and Zhao, 2008; Wang and Frei, 2011; Shanker and Venkateswarlu, 2011; Kadioglu et al., 2012, Gürsoy et al., 2012).

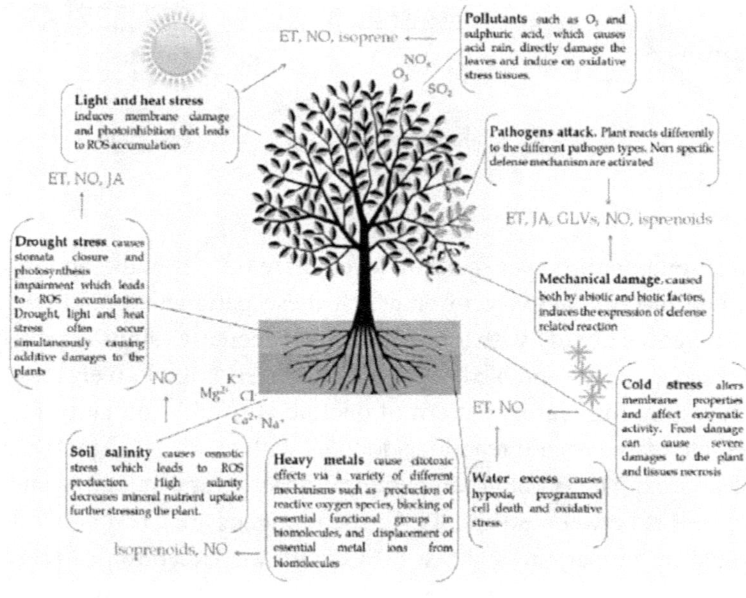

In summary, these reviews focus on:

(i) The main stressors for two related woody species in their natural environment, namely *Quercus ilex* L. (an evergreen oak that dominates the landscapes of some Geological forests) and *Polylepis tarapacana* Phil. (a typical tree of the Altiplano, one of the tallest forests on the planet) (Martín-Sanchez et al., 2022; Rojas et al., 2022).
(ii) Modulate anatomical and physiological morphology against combined abiotic stressors (Lopez' et al., 2022; Nardini, 2022).
(iii) The importance of a global understanding of plant responses to stress to improve agroecosystem functioning in the context of climate change (Pinheiro et al., 2022).

Figure 4. The interaction between soil dryness and high air temperature, and thus vapor pressure deficiency (VPD), aggravates plant stress. During dry soil periods, closed stomata reduce transpiration and thus cool the leaves, which can lead to extremely high foliage temperatures, especially in warmer atmospheres. At the same time, high air temperature and VPD increase leaf water loss through stomata and epidermis (represented by minimum leaf conductivity) as well as soil evaporation, increasing drought can cause plants to fall into a state of deadly dehydration (Puglielli et al., 2023).

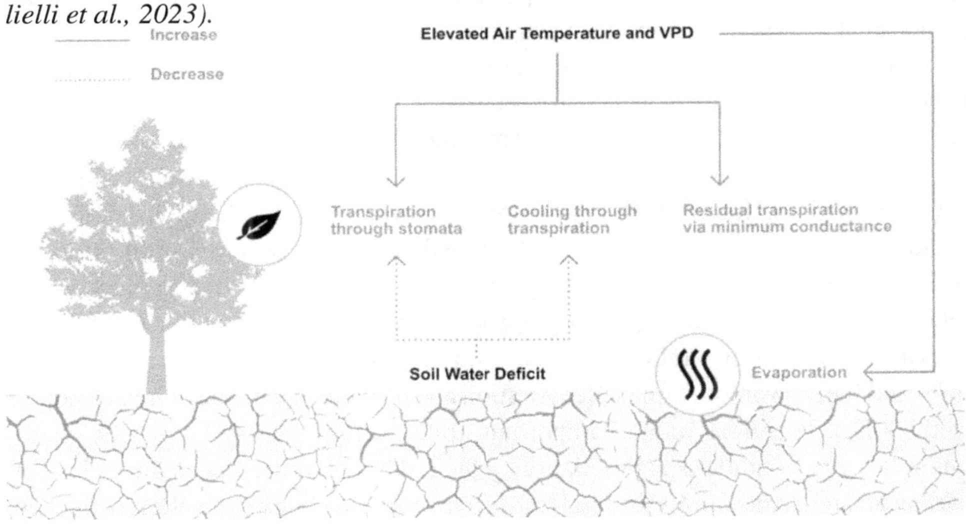

Interestingly, all the eco-physiological assessments in this issue share a common concern about the impact of water scarcity on the species studied and describe recent advances in research on drought stress, including the most common responses and tolerance mechanisms. This concern can be explained by the fact that drought poses a major threat to both natural ecosystems and modern agriculture (Gupta et al., 2020), which is intrinsically linked to increasing mortality of woody plant species around the world over the past century (Allen et al., 2010; Choat et al., 2018), and is expected to increase in frequency and intensity in future levels due to climate change (Trenberth et al., 2014).

It is worth mentioning that increased air temperature and high-water vapor pressure deficit (VPD) often accompany droughts and pose additional challenges to plant survival and growth (Grossiord and al., 2020; Lopez ` et al., 2022; McDowell et al., 2022) (Figure 1 above). One of the most interesting facts about the combination of drought and atmospheric warming (and thus increased VPD) is how each stress can exacerbate the other. On the one hand, high temperatures increase water evaporation from soil, thereby exacerbating soil water shortage (McDowell et al., 2018). On the other hand, drought-induced stomatal closure reduces evapotranspirative cooling, thereby increasing leaf temperature and potentially threatening leaf integrity (Shirley, 1936; Lopez' et al., 2022). In this context, studies such as those presented by Martín-Sanchez ` et al. (2022) and Rojas et al. (2022), it is extremely important to evaluate the physiology of woody plant species living in drought environments that are often accompanied by extreme temperatures.

From a physiological perspective, stress responses often lead to changes in carbon-water relationships, which are particularly sensitive to both drought and heat (Choat et al., 2018). For example, during drought, stomatal closure is one of the first mechanisms to reduce water loss and dehydration in plants (Cardoso et al., 2018; Creek et al., 2020), which occurs around time of loss of leaf turgor (Cardoso et al., 2020). However, this defense mechanism directly leads to reduced photosynthetic capacity, which if prolonged can lead to carbon deficiency and leaf dysfunction (McDowell et al., 2008). In all cases, stomatal closure is important for the plant to avoid drying out quickly and to maintain healthy hydration levels until the rains resume. Even after stomatal closure, plants continue to lose water at a much lower rate through the cuticle and partially opened stomata (Duursma et al., 2019).

This minimal transpiration rate can be rapidly exacerbated by high air temperatures and VPD, promoting plant dehydration and ultimately causing damage (Lopez et al., 2022). Therefore, another important strategy often observed in woody plants under hot, dry conditions is defoliation, which although represents a direct loss of carbon and nutrients, limits the total transpiration of the plants more effectively than simple stomatal closure (Ruehr et al., 2019; Lopez et al., 2022). Under more severe plant dehydration conditions, the water state may fall below the threshold at which air bubbles are drawn into the xylem, expanding, and completely blocking water transport into this conduit (i.e.is embolism) (Tyree and Sperry, 1989). In contrast, widespread xylem embolism leads to complete failure of the tree's hydraulic system, thus being a major cause of drought-related mortality (Choat et al., 2018). Therefore, the lower the water state threshold for the induction and propagation of embolism, the higher the resistance of xylem to embolism. This defense mechanism has been observed in several drought-tolerant species (McAdam and Cardoso, 2019), including *Q.ilex* subsp. balla occurs in drier regions of the Mediterranean region (Martín-S'anchez et al., 2022). Lumpy embolism can also result from freeze-thaw cycles, and *Q.ilex* plants grow under a typical sub-continental type of Mediterranean-like climate, where cold winters are common, may also be able to tolerate freeze-induced embolism (Martín-S'anchez et al., 2022).

Plants occurring in water-limited environments often exhibit hard leaves with high mass per unit area (LMA), which leads to a compelling discussion about the coordination between leaf mechanical resistance and drought tolerance (Nardini, 2022). Combined tolerance to these stressors arises because some anatomical modifications observed in hard leaves (i.e., leaves with high LMA) – including thick cell walls, high major vein density, and narrow xylem conduits – are mechanistically correlated to physiological traits conferring tolerance to dehydration – such as low turgor loss point and high xylem resistance to embolism (Nardini, 2022).

In fact, increases in drought tolerance (lower turgor loss point and higher xylem resistance to embolism), paralleled by increases in LMA (indicating higher leaf mechanical resistance), have been observed for sunflower plants exposed to drought conditions during growth (Cardoso et al., 2018). Further studies assessing variations in both drought tolerance and leaf mechanical resistance within woody species (da Silva Brito et al., 2022) and genotypes are thus necessary for a better understanding of whether trees can exhibit the same adaptive patterns as observed in herbaceous plants. Uncertainties about the vulnerability of woody species in different ecosystems to multiple abiotic stressors are still large. It is therefore essential to evaluate how plants in different ecosystems respond to extreme abiotic stressors, as adaptation and adaptation mechanisms can then be mining to support agroecosystems and biodiversity conservation in the context of climate change (Pinheiro et al., 2022).

CONCLUSIONS

The most important elements that have an impact on a plant's growth and development are stresses, either biotic or abiotic. The interaction between various abiotic and biotic stresses and their combination can have a positive or negative effect on plants. In order to comprehend the overall effect of stresses on plants, it is necessary to study how these stresses interact with one another. The need to develop comprehensive models that can accurately forecast the effects of stress and evaluate the efficacy of mitigation techniques cannot be overstated. The effective mitigation of plant biotic and abiotic stresses necessitates a collaborative endeavor involving researchers, farmers, and policymakers. Collaboration is crucial for creating robust agricultural systems that can endure the challenges posed by both living and non-living factors.Understanding how plants react to combined drought and pathogen stress will help in better analyzing the performance of superior or tolerant genotypes. Crop modeling experts, agronomists, field pathologists, breeders, physiologists, and molecular biologists can work together to lead the development of combined stress tolerant crops with good performance.

REFERENCES

Ahanger, M. A., Akram, N. A., Ashraf, M., Alyemeni, M. N., Wijaya, L., & Ahmad, P. (2017). Plant responses to environmental stresses—From gene to biotechnology. *AoB Plants*, *9*(4), 9. DOI: 10.1093/aobpla/plx025 PMID: 28775828

Ahmad, P., & Prasad, M. N. V. (Eds.). (2011). *Environmental adaptations and stress tolerance of plants in the era of climate change*. Springer Science & Business Media.

Aktar, W., Sengupta, D., & Chowdhury, A. (2009). Impact of pesticides use in agriculture: Their benefits and hazards. *Interdisciplinary Toxicology*, *2*(1), 1–12. DOI: 10.2478/v10102-009-0001-7 PMID: 21217838

Allen, C. D., Macalady, A. K., Chenchouni, H., Bachelet, D., McDowell, N., Vennetier, M., & Cobb, N. (2010). A global overview of drought and heat-induced tree mortality reveals emerging climate change risks for forests. *Forest Ecology and Management*, *259*(4), 660–684. DOI: 10.1016/j.foreco.2009.09.001

Aroca, R., Vernieri, P., & Ruiz-Lozano, J. M. (2008). Mycorrhizal and non-mycorrhizal Lactuca sativa plants exhibit contrasting responses to exogenous ABA during drought stress and recovery. *Journal of Experimental Botany*, *59*(8), 2029–2041. DOI: 10.1093/jxb/ern057 PMID: 18469324

Atkinson, N. J., & Urwin, P. E. (2012). The interaction of plant biotic and abiotic stresses: From genes to the field. *Journal of Experimental Botany*, *63*(10), 3523–3543. DOI: 10.1093/jxb/ers100 PMID: 22467407

Balestrini, R., Chitarra, W., Ghirardo, A., Nardini, A., & Nerva, L. (2022). A stressful life: How plants cope with multiple biotic and abiotic adverse factors. *Plant Stress*, *5*, 5. DOI: 10.1016/j.stress.2022.100095

Bellard, C., Bertelsmeier, C., Leadley, P., Thuiller, W., & Courchamp, F. (2012). Impacts of climate change on the future of biodiversity. *Ecology Letters*, *15*(4), 365–377. DOI: 10.1111/j.1461-0248.2011.01736.x PMID: 22257223

Boonjung, H., & Fukai, S. (1996). Effects of soil water deficit at different growth stages on rice growth and yield under upland conditions. 2. Phenology, biomass production and yield. *Field Crops Research*, *48*(1), 47–55. DOI: 10.1016/0378-4290(96)00039-1

Cardoso, A. A., Brodribb, T. J., Kane, C. N., DaMatta, F. M., & McAdam, S. A. (2020). Osmotic adjustment and hormonal regulation of stomatal responses to vapour pressure deficit in sunflower. *AoB Plants*, *12*(4), 12. DOI: 10.1093/aobpla/plaa025 PMID: 32665827

Cardoso, A. A., Brodribb, T. J., Lucani, C. J., DaMatta, F. M., & McAdam, S. A. (2018). Coordinated plasticity maintains hydraulic safety in sunflower leaves. *Plant, Cell & Environment, 41*(11), 2567–2576. DOI: 10.1111/pce.13335 PMID: 29748980

Choat, B., Brodribb, T. J., Brodersen, C. R., Duursma, R. A., Lopez, ´. R., & Medlyn, B. E. (2018). Triggers of tree mortality under drought. *Nature, 558*(7711), 531–539. DOI: 10.1038/s41586-018-0240-x PMID: 29950621

Cifre, J., Bota, J., Escalona, J. M., Medrano, H., & Flexas, J. (2005). Physiological tools for irrigation scheduling in grapevine (Vitis vinifera L.): An open gate to improve water-use efficiency? *Agriculture, Ecosystems & Environment, 106*(2-3), 159–170. DOI: 10.1016/j.agee.2004.10.005

Creek, D., Lamarque, L. J., Torres-Ruiz, J. M., Parise, C., Burlett, R., Tissue, D. T., & Delzon, S. (2020). Xylem embolism in leaves does not occur with open stomata: Evidence from direct observations using the optical visualization technique. *Journal of Experimental Botany, 71*(3), 1151–1159. DOI: 10.1093/jxb/erz474 PMID: 31641746

da Silva Brito, N. D., dos Santos Medeiros, M. J., de Souza, E. S., & de Lima, A. L. A. (2022). Drought response strategies for deciduous species in the semiarid Caatinga derived from the interdependence of anatomical, phenological and bio-hydraulic attributes. *Flora (Jena), 288*, 152009. DOI: 10.1016/j.flora.2022.152009

Daryanto, S., Wang, L., & Jacinthe, P.-A. (2016). Global synthesis of drought effects on maize and wheat production. *PLoS One, 11*(5), e0156362. DOI: 10.1371/journal.pone.0156362 PMID: 27223810

Duursma, R. A., Blackman, C. J., Lop'ez, R., Martin-StPaul, N. K., Cochard, H., & Medlyn, B. E. (2019). On the minimum leaf conductance: Its role in models of plant water use, and ecological and environmental controls. *The New Phytologist, 221*(2), 693–705. DOI: 10.1111/nph.15395 PMID: 30144393

Farooq, M., Gogoi, N., Barthakur, S., Baroowa, B., Bharadwaj, N., Alghamdi, S. S., & Siddique, K. (2017). Drought stress in grain legumes during reproduction and grain filling. *Journal Agronomy & Crop Science, 203*(2), 81–102. DOI: 10.1111/jac.12169

Food and Agriculture Organization. (2007). *Coping with water scarcity*. FAO Rome.

Gill, S. S., & Tuteja, N. (2010). Reactive oxygen species and antioxidant machinery in abiotic stress tolerance in crop plants. *Plant Physiology and Biochemistry, 48*(12), 909–930. DOI: 10.1016/j.plaphy.2010.08.016 PMID: 20870416

Grossiord, C., Buckley, T. N., Cernusak, L. A., Novick, K. A., Poulter, B., Siegwolf, R. T., & McDowell, N. G. (2020). Plant responses to rising vapor pressure deficit. *The New Phytologist, 226*(6), 1550–1566. DOI: 10.1111/nph.16485 PMID: 32064613

Gupta, A., Rico-Medina, A., & Cano-Delgado, ˜. A. I. (2020). The physiology of plant responses to drought. *Science, 368*(6488), 266–269. DOI: 10.1126/science.aaz7614 PMID: 32299946

Gupta, K. J., Igamberdiev, A. U., Manjunatha, G., Segu, S., Moran, J. F., Neelawarne, B., Bauwe, H., & Kaiser, W. M. (2011). The emerging roles of nitric oxide (NO) in plant mitochondria. *Plant Science, 181*(5), 520–526. DOI: 10.1016/j.plantsci.2011.03.018 PMID: 21893247

Gupta, K. J., Kaladhar, V. C., Fitzpatrick, T. B., Fernie, A. R., Møller, I. M., & Loake, G. J. (2022). Nitric oxide regulation of plant metabolism. *Molecular Plant, 15*(2), 228–242. DOI: 10.1016/j.molp.2021.12.012 PMID: 34971792

Gürsoy, M., Balkan, A., & Ulukan, H. (2012). Ecophysiological responses to stresses in plants: A general approach. *Pakistan Journal of Biological Sciences, 15*(11), 506–516. DOI: 10.3923/pjbs.2012.506.516 PMID: 24191624

Hancock, J. T., & Veal, D. (2021). Nitric oxide, other reactive signalling compounds, redox, and reductive stress. *Journal of Experimental Botany, 72*(3), 819–829. DOI: 10.1093/jxb/eraa331 PMID: 32687173

Isman, M. B., & Grieneisen, M. L. (2014). Botanical insecticide research: Many publications, limited useful data. *Trends in Plant Science, 19*(3), 140–145. DOI: 10.1016/j.tplants.2013.11.005 PMID: 24332226

Jozefczak, M., Remans, T., Vangronsveld, J., & Cuypers, A. (2012). Glutathione is a key player in metal-induced oxidative stress defenses. *International Journal of Molecular Sciences, 13*(3), 3145–3175. DOI: 10.3390/ijms13033145 PMID: 22489146

Kamran, M., Imran, Q. M., Ahmed, M. B., Falak, N., Khatoon, A., & Yun, B. W. (2022). Endophyte-mediated stress tolerance in plants: A sustainable strategy to enhance resilience and assist crop improvement. *Cells, 11*(20), 3292. DOI: 10.3390/cells11203292 PMID: 36291157

Kar, R. K. (2011). Plant responses to water stress: Role of reactive oxygen species. *Plant Signaling & Behavior, 6*(11), 1741–1745. DOI: 10.4161/psb.6.11.17729 PMID: 22057331

Kharbech, O., Sakouhi, L., Massoud, M. B., Mur, L. A. J., Corpas, F. J., Djebali, W., & Chaoui, A. (2020). Nitric oxide and hydrogen sulfide protect plasma membrane integrity and mitigate chromium-induced methylglyoxal toxicity in maize seedlings. *Plant Physiology and Biochemistry*, *157*, 244–255. DOI: 10.1016/j.plaphy.2020.10.017 PMID: 33152643

Kim, K.-H., Kabir, E., & Jahan, S. A. (2017). Exposure to pesticides and the associated human health effects. *The Science of the Total Environment*, *575*, 525–535. DOI: 10.1016/j.scitotenv.2016.09.009 PMID: 27614863

Law, J. W.-F., Ser, H.-L., Khan, T. M., Chuah, L.-H., Pusparajah, P., Chan, K.-G., Goh, B.-H., & Lee, L.-H. (2017). The potential of Streptomyces as biocontrol agents against the rice blast fungus, Magnaporthe oryzae (Pyricularia oryzae). *Frontiers in Microbiology*, *8*, 3. DOI: 10.3389/fmicb.2017.00003 PMID: 28144236

Liu, H. S., Li, F. M., & Xu, H. (2004). Deficiency of water can enhance root respiration rate of drought-sensitive but not drought-tolerant spring wheat. *Agricultural Water Management*, *64*(1), 41–48. DOI: 10.1016/S0378-3774(03)00143-4

Lopez, ´R., Ramírez-Valiente, J.A., Pita, P., (2022). How plants cope with heatwaves in a drier environment. Flora 295, 152148. https://doi.org/. flora.2022.152148.DOI: 10.1016/j

Mabe, F. N., Talabi, K., & Danso-Abbeam, G. (2017). Awareness of health implications of agrochemical use: Effects on maize production in Ejura-Sekyedumase municipality, Ghana. *Advances in Agriculture*, *2017*, 2017. DOI: 10.1155/2017/7960964

Mahmood, Q., Bilal, M., & Jan, S. (2014). Herbicides, pesticides, and plant tolerance: an overview. In *Emerging technologies and management of crop stress tolerance* (pp. 423–448). Elsevier. DOI: 10.1016/B978-0-12-800876-8.00017-5

Manafi, H., Baninasab, B., Gholami, M., & Talebi, M. (2021). Nitric oxide induced thermotolerance in strawberry plants by activation of antioxidant systems and transcriptional regulation of heat shock proteins. *The Journal of Horticultural Science & Biotechnology*, *96*(6), 783–796. DOI: 10.1080/14620316.2021.1927206

Mattah, M. M., Mattah, P. A., & Futagbi, G. (2015). Pesticide application among farmers in the catchment of Ashaiman irrigation scheme of Ghana: Health implications. *Journal of Environmental and Public Health*, *2015*, 2015. DOI: 10.1155/2015/547272 PMID: 26798369

McAdam, S. A., & Cardoso, A. A. (2019). The recurrent evolution of extremely resistant xylem. *Annals of Forest Science*, *76*(1), 1–4. DOI: 10.1007/s13595-018-0786-7

McDowell, N., Pockman, W. T., Allen, C. D., Breshears, D. D., Cobb, N., Kolb, T., & Yepez, E. A. (2008). Mechanisms of plant survival and mortality during drought: Why do some plants survive while others succumb to drought? *The New Phytologist, 178*(4), 719–739. DOI: 10.1111/j.1469-8137.2008.02436.x PMID: 18422905

McDowell, N. G., Michaletz, S. T., Bennett, K. E., Solander, K. C., Xu, C., Maxwell, R. M., & Middleton, R. S. (2018). Predicting chronic climate-driven disturbances and their mitigation. *Trends in Ecology & Evolution, 33*(1), 15–27. DOI: 10.1016/j.tree.2017.10.002 PMID: 29146414

McDowell, N. G., Sapes, G., Pivovaroff, A., Adams, H. D., Allen, C. D., Anderegg, W. R., & Xu, C. (2022). Mechanisms of woody-plant mortality under rising drought, CO2 and vapour pressure deficit. *Nature Reviews. Earth & Environment, 3*(5), 294–308. DOI: 10.1038/s43017-022-00272-1

Mohanta, T.K., Bashir, T., & Hashem, A. and Abd_Allah, E.F. (. (2017). Systems biology approach in plant abiotic stresses. *Plant Physiology and Biochemistry, 121*, 58–73. DOI: 10.1016/j.plaphy.2017.10.019 PMID: 29096174

Møller, I. M., & Sweetlove, L. J. (2010). ROS signalling–specificity is required. *Trends in Plant Science, 15*(7), 370–374. DOI: 10.1016/j.tplants.2010.04.008 PMID: 20605736

Mouhouche, B., Ruget, F., & Delécolle, R. (1998). Effects of water stress applied at different phenological phases on yield components of dwarf bean (Phaseolus vulgaris L.). *Agronomie, 18*(3), 197–205. DOI: 10.1051/agro:19980303

Nardini, A. (2022). Hard and tough: The coordination between leaf mechanical resistance and drought tolerance. *Flora (Jena), 288*, 152023. DOI: 10.1016/j.flora.2022.152023

Pinheiro, C., David, T. S., Baptista, P., & Guerra-Guimarães, L. (2022). Mediterranean woody agroecosystems in a warming and drier climate: The importance of sustainable knowledge-based management. *Flora (Jena), 291*, 152070. DOI: 10.1016/j.flora.2022.152070

Puglielli, G., Laanisto, L., Gori, A., & Cardoso, A. A. (2023). Woody plant adaptations to multiple abiotic stressors: Where are we? *Flora (Jena), 299*, 152221. DOI: 10.1016/j.flora.2023.152221

Rojas, R., Flexas, J., & Coopman, R. E. (2022). The highest elevation treeline in the world: Polylepis tarapacana Phil. as a unique opportunity to study ecophysiological adaptations to extreme conditions. *Flora : the Journal of Infectious Diseases and Clinical Microbiology = Infeksiyon Hastaliklari ve Klinik Mikrobiyoloji Dergisi*, *292*, 152076. DOI: 10.1016/j.flora.2022.152076

Ronald, P. C., Albano, B., Tabien, R., Abenes, L., Wu, K., McCouch, S., & Tanksley, S. D. (1992). Genetic and physical analysis of the rice bacterial blight disease resistance locus, Xa21. *Molecular & General Genetics*, *236*(1), 113–1. DOI: 10.1007/BF00279649 PMID: 1362973

Ruehr, N. K., Grote, R., Mayr, S., & Arneth, A. (2019). Beyond the extreme: Recovery of carbon and water relations in woody plants following heat and drought stress. *Tree Physiology*, *39*(8), 1285–1299. DOI: 10.1093/treephys/tpz032 PMID: 30924906

Shirley, H. L. (1936). Lethal high temperatures for conifers, and the cooling effect of transpiration. *Journal of Agricultural Research*, *53*, 239–258.

Trenberth, K. E., Dai, A., Van Der Schrier, G., Jones, P. D., Barichivich, J., Briffa, K. R., & Sheffield, J. (2014). Global warming and changes in drought. *Nature Climate Change*, *4*(1), 17–22. DOI: 10.1038/nclimate2067

Tyree, M. T., & Sperry, J. S. (1989). Vulnerability of xylem to cavitation and embolism. *Annual Review of Plant Physiology and Plant Molecular Biology*, *40*(1), 19–38. DOI: 10.1146/annurev.pp.40.060189.000315

Vahdati, K., & Leslie, C. (Eds.). (2013). *Abiotic stress: plant responses and applications in agriculture*. BoD–Books on Demand. DOI: 10.5772/45842

Varsani, A., Navas-Castillo, J., Moriones, E., Hernández-Zepeda, C., Idris, A., Brown, J. K., Zerbini, F. M., & Martin, D. P. (2014). Establishment of three new genera in the family Geminiviridae: Becurtovirus, Eragrovirus and Turncurtovirus. *Archives of Virology*, *159*(8), 2193–2203. DOI: 10.1007/s00705-014-2050-2 PMID: 24658781

Xie, X., He, Z., Chen, N., Tang, Z., Wang, Q., & Cai, Y. (2019). The roles of environmental factors in regulation of oxidative stress in plant. *BioMed Research International*, *2019*, 2019. DOI: 10.1155/2019/9732325 PMID: 31205950

Yamasaki, H. (2005). The NO world for plants: Achieving balance in an open system. *Plant, Cell & Environment*, *28*(1), 78–84. DOI: 10.1111/j.1365-3040.2005.01297.x

Yoon, M.-Y., Cha, B., and Kim, J.-C. (2013). Recent trends in studies on botanical fungicides in agriculture. The plant pathology journal 29, 1.20.

Zhou, X., Joshi, S., Khare, T., Patil, S., Shang, J., & Kumar, V. (2021). Nitric oxide, crosstalk with stress regulators and plant abiotic stress tolerance. *Plant Cell Reports*, *40*(8), 1395–1414. DOI: 10.1007/s00299-021-02705-5 PMID: 33974111

Zhu-Salzman, K., Salzman, R. A., Ahn, J.-E., & Koiwa, H. (2004). Transcriptional regulation of sorghum defense determinants against a phloem-feeding aphid. *Plant Physiology*, *134*(1), 420–431. DOI: 10.1104/pp.103.028324 PMID: 14701914

Chapter 4
The Role of Renewable Energy in Reducing Africa's Carbon Footprint

Mohammad Saad Azam
https://orcid.org/0009-0006-9736-176X
Woxsen University, India

Bysani Sona Sree
Woxsen University, India

Tenneti AppalaNarsimha P. Bhargav Sastry
Woxsen University, India

Mohd Azhar
https://orcid.org/0000-0003-3222-5565
Woxsen University, India

ABSTRACT

Africa's energy crisis leaves millions without electricity while driving carbon emissions through fossil fuel dependence. Yet, the continent's vast renewable resources—solar, wind, hydropower, and geothermal—offer a sustainable solution. Landmark projects like Morocco's Noor Ouarzazate and Kenya's Lake Turkana Wind Power highlight this potential. However, financial, infrastructural, and policy barriers hinder progress. Overcoming these challenges through strategic reforms and investment is vital for reducing emissions, expanding energy access, and ensuring sustainable economic growth.

DOI: 10.4018/979-8-3373-1549-2.ch004

1. INTRODUCTION

1.1. Background

The primary power suppliers for electricity generation in Africa consist of coal together with oil and gas because fossil fuels maintain their prominent status. The complete process of electricity production from fossil fuels results in large GHG emissions that intensify environmental destruction and escalate global warming (IPCC, 2022). The energy sector of Africa depends mainly on non-renewable power sources even though the region has excellent renewable alternatives including solar, wind, hydropower and geothermal capacity because of insufficient power networks and limited financing and weak government policy measures (IEA, 2023).

African countries must address two fundamental energy issues which encompass expanding electricity access for their expanding population numbers and minimizing carbon emissions for environmental protection (UNFCCC, 2022). Sub-Saharan Africa needs additional energy resources because of the fast-rising rates of urbanization and industrialization even though 600 million people do not have access to stable electricity (World Bank, 2023). The quick shift to renewable power leads sustainability goals to the front position (African Development Bank, 2022).

African nation states require immediate solutions for their two energy challenges which consist of increasing power capabilities to serve population growth and decreasing carbon emissions for environmental defense (UNFCCC, 2022). Sub-Saharan Africa needs extra energy resources because its populations keep growing faster while 600 million people remain without steady access to electricity (World Bank, 2023). The quick adoption of renewable energy stands as the central requirement to attain sustainability according to the African Development Bank (African Development Bank, 2022).

Nations need to select greener energy solutions to honor both the Paris Agreement along with worldwide climate agreements (UNFCCC, 2022). AREI serves as a vital program under African governments to fight carbon emissions through their partnership with international climate finance programs (AREI, 2023). Renewable energy stands as a top initiative for African nations because it enables economic expansion through environmentally friendly sustainable practices (World Economic Forum, 2023).

1.2. Purpose of the Chapter

The chapter demonstrates that renewable energy systems can effectively reduce Africa's carbon emission levels. Renewable energy technology exploration includes a discussion about both its technological possibilities and its emission cutting potential.

The analysis considers potential opportunities together with strategic approaches along with main challenges that block Africa's renewable energy transition (IEA, 2023).

Key objectives include:

- The current emission levels throughout Africa require analysis.
- An analysis of how renewable energy technologies decrease emission levels.
- The process requires identifying important barriers that discourage renewable energy implementation.
- The analysis examines both policies and strategies to boost the transition process.

1.3. Key Questions Addressed

a) How Do Carbon Emissions Stand in Africa at Present?

The continent produces 4% of worldwide carbon dioxide emissions which remains lower because of its decreased rate compared to North America, Europe and Asian sectors. The upcoming years will likely see this figure increase as growing economies continue to boost their energy consumption levels.

- **Electricity generation:** The heavy reliance on coal and diesel power plants leads to substantial CO_2 emissions.
- **Deforestation:** Biomass burning for cooking and heating contributes to high carbon output.
- **Industrialization:** Manufacturing, mining, and oil extraction release greenhouse gases into the atmosphere.
- **Transportation:** The growing use of fossil-fuel-powered vehicles adds to Africa's carbon footprint.

While Africa's historical emissions are low, the continent is vulnerable to climate change effects such as extreme weather patterns, droughts, and rising temperatures. The need to shift towards clean energy solutions is, therefore, critical for both environmental and economic resilience.

b) Renewable Energy Technology Deployment Results in What Level of Emission Cut?

Renewable energy serves as a lasting method to decrease Africa's fossil fuel consumption. Different technological solutions exist which demonstrate success in reducing emissions.

Africa stands out for its extensive solar resources because high sunlight intensity exists in both the Sahara and Sahel regions. Two major examples showing solar energy's ability to replace traditional power sources include Morocco's Noor Ouarzazate Solar Complex and South Africa's Renewable Energy Independent Power Producer Procurement Programme (REIPPPP).

Wind Energy provides emissions reduction opportunities to Kenya and Ethiopia because of their robust wind conditions. The Lake Turkana Wind Power Project in Kenya, along with the Ashegoda Wind Farm in Ethiopia, show how wind energy can operate effectively.

The hydroelectric sector stands as the most significant renewable energy source across Africa to power Ethiopia as well as the Democratic Republic of the Congo and Zambia. The instability of water levels caused by changing climate requires African nations to seek power generation techniques beyond hydropower.

A successful low-emission power facility demonstrates the great potential for geothermal energy within East Africa's Rift Valley region through the existence of the Kenyan Olkaria Geothermal Power Station.

The production and use of sustainable biomass and biogas for generating electricity alongside cooking functions provides a substitute for wood and charcoal usage which minimizes forest destruction emissions.

Africa's energy transition towards sustainable solutions will serve two purposes of reduced carbon emissions and expanded power network access for underprivileged areas.

c) What Challenges Prevent Renewable Energy Systems From Being Accepted and Which Methods Would Help Address These Problems?

The development of clean energy solutions in Africa confronts several obstacles when striving to harness its enormous renewable energy resources.

Two main obstacles to renewable energy development include the high barriers that deter investment from both private and public organizations. The implementation of tax incentives and subsidies coupled with international organizations' climate finance from World Bank and AfDB should serve to stimulate investment.

A lack of stable policies alongside inadequate regulations stands as a barrier to the expansion of renewable energy systems. Governments need to create unambiguous and constant policies which draw investors and reduce the time required for project approvals.

Several African nations face barriers in adopting major renewable energy projects because they do not have sufficient electrical grid systems. Nationwide deployment of renewable energy requires significant investment in transmission and storage capabilities to connect renewable power sources with the power grid.

Implementation speed of renewable energy projects slows down because there are few experts available to work in this sector. Revenue generation programs combined with knowledge sharing initiatives function to eliminate this deficiency.

Renewable technology adoption is slowed down by widespread social ignorance of these technologies and social refusal to accept change.

Community involvement along with public education efforts should be used to show citizens how clean energy provides clear advantages.

2. AFRICA'S CARBON FOOTPRINT: CURRENT STATUS

2.1. Sources of Carbon Emissions

Various African sectors including industrial and transportation systems together with fossil fuel production along with deforestation create carbon emissions. Fossil fuel consumption continues to be the dominant energy source throughout Africa due to its provision of most of the continent's energy requirements (IPCC, 2022).

- **Fossil Fuel-Based Energy Production**
- The generation of electricity across Africa depends mainly on coal together with oil and natural gas resources. Heating sources that use fossil fuel combustion disperse significant volumes of CO_2 and additional greenhouse gases into the air which intensifies climate change (Carbon Brief, 2022)
- The African country of South Africa emits the highest amount of carbon dioxide on the continent because it generates power using coal power plants that produce 80% of the electricity (IEA, 2023).
- The oil and gas production activities of Nigeria and Angola as leading African oil producers result in substantial greenhouse gas emission amounts because they refine crude oil and flare natural gas (IPCC, 2022).
- The natural gas power plants using gas as fuel in North Africa emit substantial amounts of emissions since Egypt and Algeria maintain their energy systems primarily based on gas power (Noor Ouarzazate Solar Complex, 2023).

- The renewable energy resources across Africa remain underutilized due to financial limitations combined with inadequate infrastructure along with policy restrictions (World Bank, 2023).
- **Deforestation and Land-Use Changes**

Forest removal stands as one of the primary sources that produces Africa's carbon emission levels. Deforestation of forests causes carbon emissions by releasing captured atmospheric CO_2 into the air according to UNEP (UNEP, 2022).

Farm expansion through agriculture coupled with commercial logging operations leads to the greatest deforestation in Africa. The Congo Basin rainforest uniting the world's second-largest rainforest continues to lose excessive forest regions every year thus speeding up carbon emission levels (UNEP, 2022).

Biomass stands as the main fuel source for residential cooking among thousands of African citizens. The emission of CO_2 occurs when wood and charcoal are burned for cooking purposes while such behavior produces hazardous indoor air pollution which leads to respiratory health problems (World Bank, 2023).

Faster urban growth has led to reckless deforestation activities for constructing houses along with roads and infrastructure which minimizes Africa's capability to capture carbon emissions (IEA, 2023).

- **Industrial Activities, Transportation, and Urbanization**

Industrial operations together with transport solutions represent expanding emission sources throughout African economic growth.

Various manufacturing and mining industries use intense energy sources to create cement products while mining and steel production significantly increase carbon emission output. The industrial sectors of South Africa together with Nigeria and Egypt maintain their operations through reliance on fossil fuels according to IPCC (IPCC, 2022).

The oil extraction activities in Nigeria and Angola result in carbon pollution from methane and CO_2 emissions which occur when operators burn off excess natural gas released during oil drilling (Carbon Brief, 2022).

Modern transportation requires a continuous increase in the use of diesel and petrol which powers automobiles and other heavy vehicles continuously. The ongoing increase of carbon emissions is linked to deteriorating air quality due to urban traffic congestion in major cities like Lagos and Nairobi and Cairo (World Bank, 2023).

The carbon footprint of Africa continues to rise because of these combined environmental factors which necessitate immediate adoption of cleaner energy alternatives.

2.2. Regional Variations

The quantity of carbon emissions in Africa's regions demands attention because it varies greatly based on energy production methods as well as industrial output and local economic advancement levels.

North Africa

The North African states of Egypt together with Libya as well as Algeria present the highest carbon emissions per person numbers because they depend heavily on fossil fuel resources for power.

Natural gas remains essential for electricity generation in Algeria and Egypt hence producing elevated carbon emission levels (Noor Ouarzazate Solar Complex, 2023).

The cities in these countries face elevated emissions levels because they produce petrochemicals together with steel products and cement through their industrial operations (IEA, 2023).

The Noor Ouarzazate Solar Complex stands as one of the biggest concentrated solar power facilities globally among Morocco's renewable energy advancement initiatives (Noor Ouarzazate Solar Complex, 2023).

Sub-Saharan Africa

The carbon emissions rate in Sub-Saharan Africa remains lower than North Africa while the overall emissions numbers are increasing because of growing energy needs and urban development as well as tree removal (IEA, 2023).

The excessive use of coal power plants has made South Africa the largest CO_2 polluting country in Sub-Saharan Africa based on Carbon Brief (Carbon Brief, 2022) data.

Kenya and Ethiopia have directed their efforts toward wind and geothermal energy development to decrease their dependence on fossil fuels according to the World Bank (World Bank, 2023).

The region of Central Africa particularly the Democratic Republic of the Congo faces intense deforestation which causes atmospheric emissions to increase (UNEP, 2022).

Oil-Producing Regions (Nigeria & Angola)

Nigeria together with Angola produce oil at the highest levels across Africa while producing elevated emissions during the processing of oil and through gas flaring activities (IPCC, 2022).

According to Carbon Brief (2022) Nigeria participates in the list of top gas flare nations while it discharges substantial quantities of CO_2 and methane emissions to air.

These economies heavily depend on oil as their primary source while trying to develop alternative energy sources because oil remains their essential foundation (IEA, 2023).

2.3 Impact of Carbon Emissions

Africa experiences the adverse effects of climate change despite emitting only small amounts of planet-warming gases globally. The existing rise in temperature along with irregular climatic events and environmental deterioration leads to major economic issues in affected communities.

- **Environmental Degradation:**

Computing projections show that Africa faces multiple climate change impacts which include desertification together with droughts and severe weather patterns (IPCC 2022).

Deserts continue to spread throughout the Sahel region thus posing a threat to local food security along with water supply resources. Droughts occur regularly in Mali and Niger and Chad because of the current climate situation (IPCC, 2022).

The increase in sea levels poses a flooding danger to coastal cities like Lagos and Dar es Salaam according to the World Bank (World Bank, 2023).

The decrease of biological diversity progresses because deforestation together with climate change causes habitats of wildlife to shrink (UNEP, 2022).

- **Health Impacts:**

Air pollution emerges from carbon emissions causing severe negative impacts on human health (World Economic Forum, 2023).

Air pollution caused by traffic and industrial activity in Cairo and Lagos creates respiratory diseases that result in more asthma cases and lung infections and heart disease according to the World Bank (World Bank, 2023).

Millions of humans and primarily women and children suffer from dangerous air contaminants when they use firewood and charcoal for cooking (UNEP, 2022).

- **Economic Costs:**

African economies face substantial economic expenses because of climate change (AfDB, 2022).

Extremely adverse weather conditions create expense-heavy damage to roads alongside bridges and residential housing structures requiring repairs that cost a substantial amount (World Bank, 2023).

Healthcare Costs Grow Because pollution-related illnesses create extra healthcare expenses according to the World Economic Forum (World Economic Forum, 2023).

- **Africa's Vulnerability to Climate Change:**

The minimal emissions from Africa do not protect the continent against the escalating impact of climate change. People in numerous communities lack sufficient capability to fight environmental adjustments, thereby creating a combination of food instability and forced population movements which generate economic challenges. Rising seas endanger Lagos and Dar es Salaam alongside drought conditions in the Horn of Africa, which has worsened human emergencies.

3. RENEWABLE ENERGY POTENTIAL IN AFRICA

3.1. Abundant Renewable Resources

The continent possesses extensive renewable energy potentials that elites have poorly developed (IEA, 2022). The physical landscape along with atmospheric conditions in Africa make it an ideal spot for establishing big-scale renewable energy projects (IRENA, 2021).

Solar Energy

The high solar irradiation levels recorded throughout Africa create strong opportunities for solar power exploitation across this area (World Bank, 2020). The joint Sahara-Sahel area constantly experiences uninterrupted high solar radiation levels. The power generation facilities Noor Ouarzazate Solar Complex in Morocco together with Renewable Energy Independent Power Producer Procurement Programme (REIPPPP) in South Africa show that large-scale solar projects reduce fossil fuel usage (REN21, 2021).

Wind Energy

The main opportunities from wind energy occur in locations which present consistent rapid wind flows at either coastal or inland sites (IRENA, 2022). The rapid development of wind power arises from two notable achievements in Kenya and Ethiopia. National power grids utilize supplementary power from wind-generated energy through projects like the Lake Turkana Wind Power Project in Kenya and Ashegoda Wind Farm in Ethiopia thus reducing dependence on fossil fuels (African Development Bank, 2020).

Hydropower, Geothermal, and Biomass

Hydropower stands as the largest renewable energy source throughout the African continent. Ethiopia together with the Democratic Republic of the Congo along with Zambia lead major hydropower production throughout Africa since these countries contain abundant water-supplied river valleys (World Bank, 2021). Water levels fluctuating due to climate change create significant obstacles for using hydropower as the sole power generation source. Other alternative energy solutions must be implemented immediately.

Around the East African Rift Valley region geothermal energy proves effective for power generation. Olkaria Geothermal Power Station in Kenya proves that geothermal power generates sustainable and enduring electricity supply (IRENA, 2021).

Biomass along with biogas provides alternative ways to produce power and develop cooking systems. The energy source supports deforestation prevention through its work to eliminate the need for wood and charcoal (UNEP, 2022).

3.2.Current Renewable Energy Deployment

Such investments in renewable power systems in the area have increased yet overall large-scale deployment continues to be minimal (REN21, 2022). Two renewable energy projects achieve multiple objectives because they reduce both energy deficits and minimize harmful environmental impacts.

Case Studies

Noor Solar Plant functions as a giant concentrated solar power installation worldwide through solar energy production to lower Moroccan coal usage and advance nationwide renewable energy targets (World Bank, 2022). The eastern part of Africa depends on the Olkaria Geothermal Plant as its main geothermal facility to shift away from fossil fuels while achieving better energy stability (IRENA, 2021).

3.3.Comparative Advantages of Renewable Energy

Environmental Benefits

The implementation of renewable energy systems creates dual benefits by minimizing both dangerous releases to the atmosphere and environmental air pollution (UNEP, 2021). Unfossilised power generation systems serve as a valuable tool to decrease carbon footprint which helps in achieving worldwide climate change reduction goals.

Economic Benefits

Employment positions and established secure energy systems and power distribution opportunities appear from renewable energy investments which primarily benefit isolated and underserved population areas (African Development Bank, 2021). Renewable energy systems will determine African economic expansion toward sustainable development as it moves toward urbanization and industrialization.

4. THE ROLE OF RENEWABLE ENERGY IN REDUCING CARBON EMISSIONS

Renewable energy provides the solution for sustainable energy transformation since Africa requires it to cut its carbon pollution. African carbon emissions make up only 4% of the total worldwide amount despite their usage of coal oil and gas as power sources that cause severe environmental damage (IEA, 2022). The adoption of renewable energy solutions allows both emission reductions and it enables sustainable power systems for growing communities (REN21, 2021).

4.1. Direct Emission Reductions

The main African electricity generation sources consist of coal and oil and gas while both fuel types act as significant greenhouse gas (GHG) emission sources (World Bank, 2020). When fossil fuels burn they release substantial amounts of CO_2 and CH_4 along with various pollutants that enhance climate warming patterns.

How Renewable Energy Lowers Emissions

Multiple emission reduction mechanisms arise from standardizing energy production by using solar power alongside wind power and hydropower facilities and geothermal systems according to the IEA. (IEA, 2021).

Solar power systems alongside wind power systems produce electricity which does not produce CO_2 emissions during the process. Morocco operates the Noor Solar Complex as well as Kenya operates Lake Turkana Wind Power Project which replaces fossil fuel generation with renewable power sources to decrease carbon emissions (IRENA, 2022).

Stable power output characterizes the electricity production from hydropower systems and geothermal stations that also inflict minimal harm to the environment. Renewable energy stands as an effective power source because of both Ethiopian hydroelectric plants and Kenyan geothermal units at Olkaria (African Development Bank, 2020).

Creating electricity and fuels from biomass and biogas gives countries an opportunity to stop their citizens from cooking with wood and charcoal thus reducing deforestation while reducing resulting pollution (UNEP, 2022).

The conversion to clean energy systems offers African nations the chance to decrease carbon outputs as they construct stable renewable power networks (REN21, 2021).

4.2. Indirect Benefits of Renewable Energy

The environmental benefits of renewable energy produce further advantages that help achieve a low-carbon future and emissions reduction (World Bank, 2021).

Reducing Deforestation and Land Degradation

Firewood together with charcoal serve as the primary energy products derived from biomass. Wood functions as the primary energy source that African people use for their heating and cooking requirements. Argumentative statements targeting the environment show that the demand for wood leads to intense forest destruction that not only harms carbon storage sites but also creates emissions from burning wood (UNEP, 2022). Solar stove systems together with biogas power plants and small hydroelectric plants combine to provide environment-friendly alternatives that protect wildlife populations and defend forests (IEA, 2021).

Promoting Sustainable Development

New renewable power infrastructure delivers countrywide power security by providing economic progress through the combination of numerous advantages (IRENA, 2022). Rural regions lack the benefit of stable electric power systems because a power crisis continues to affect these communities. Outlying areas beyond electric grid control gain electricity from autonomous renewable power systems which unite solar microgrids and wind turbines to create power for economic advancement and enhanced living criteria (REN21, 2021).

4.3. Integration into National Energy Systems

African states encounter various barriers when they try to implement renewable energy technologies into their national main electricity networks (World Bank, 2021). The lack of modern power distribution systems together with insufficient infrastructure prevents clean energy transitions because of policy challenges. Off-grid distributed power solutions prove as dependable solutions for reaching electricity distribution.

Challenges in Grid Integration

Two major problems face African countries since their outdated and restricted electric grid cannot accept renewable power from wind and solar resources (IRENA, 2021). Building transmission infrastructure and storage systems requires essential investment which enhances standard power supply networks.

Decentralized and Off-Grid Renewable Solutions

The implementation of decentralized independent renewable energy systems offers African countries their preferred method to address electricity generation problems (African Development Bank, 2020).

5. CHALLENGES TO RENEWABLE ENERGY ADOPTION

Africa has very rich renewable energy resources, yet different financial, technical, policy, and social constraints impede their adoption. To create a clean energy supply for the continent, these constraints need to be tackled through policy measures, infrastructure investment, and mobilization at the level of the community-wide population for long-term energy security and economic growth (African Development Bank [AfDB], 2022; International Energy Agency [IEA], 2023).

5.1. Financial Barriers

Some of the large financial barriers confronting renewable energy on the continent include the high initial investments and limited financial backing. While fossil fuel-based electricity generation is typically subsidized by the government and receives large capital support, in contrast, a renewable energy project requires enormous upfront investment, something economically challenged African nations can hardly afford (World Bank, 2023).

- **Key Financial Challenges:**

1. **High initial capital cost:** The establishment of solar farms, wind farms, geothermal power plants, and hydropower plants entails enormous capital expenditures on setting them up, integrating them to the grid, and maintaining them thereafter (Noor Ouarzazate Solar Complex, 2023).
2. **Limited access to financing:** Governments in Africa, along with private financiers, cannot access the needed financing due to their limited borrowing capacity, high interest rates, and the unavailability of green funding institutions (African Development Bank [AfDB], 2022).
3. **Fossil fuel subsidies:** Most governments subsidizing fossil fuels reduce the price of traditional energy in the short run, thus rendering alternative sources of energy financially uncompetitive and discouraging private sector investments in alternative technologies (International Energy Agency [IEA], 2023).
4. **Currency and economic instability:** Most African states are characterized by instability and currency devaluation, which reduces investor confidence and makes those long-term financing options for alternative energy projects very difficult to find (World Economic Forum, 2023).

- **Potential Solutions**

1. **Domestic Climate Investment Funds:** So, financing still comes from international climate investment funds; for instance, World Bank Clean Technology Fund (CTF); African Development Bank Sustainable Energy Fund for Africa (SEFA) offers financing options for renewable projects (World Bank, 2023).
2. **Government incentives:** Governments should motivate investments in renewables through financial incentives, such tax breaks, financial subsidies, and favorable loan schemes to encourage them (African Development Bank [AfDB], 2022).

3. **Public-Private Partnerships:** Public-Private Partnerships may bridge finance by marrying government support with private investments. This reduces financial risk and improves project feasibility (International Energy Agency [IEA], 2023).
4. **Green Bonds and Microfinance Options:** Expanding access to green bonds and microfinance options will enable smaller-scale renewable energy projects and local entrepreneurs to participate in the clean energy transition (African Development Bank [AfDB], 2022).

5.2. Technological and Infrastructure Challenges

As such, the continent suffers from technological and infrastructure bottlenecks, slowing down the implementation of renewable energy projects. The update of current infrastructure, low energy storage capacities, and capacity constrictions imply constraints for integrating renewables into national power grids, continuously weakened by aging installations and, in most cases, the low levels of technical skill (Intergovernmental Panel on Climate Change[IPCC], 2022).

- **Key Technological Challenges**

1. **Shortage of skilled personnel:** The limited skill set and few qualified personnel knowledgeable in solar panel installations, wind turbine maintenance, power grid management, and energy storage solutions plague smooth set-and-forget implementation and inevitable monitor-and-maintain practices with renewable energy systems terribly (African Renewable Energy Initiative [AREI], 2023).
2. **Aging infrastructure and inadequate grid supply:** Most of the African power grids primarily can handle base-load energy sources. In reference to the variable nature of solar and wind energy and the concerns of grid stability, they are stuck with insufficient capacity to inject large-scale renewable energy projects (International Energy Agency [IEA], 2023).
3. **Intermittence of renewable energy:** The generation of solar and wind energy varies, depending quite highly on the weather. Because of inadequate energy storage configurations, this variable generation would not allow for continuous and reliable electricity supply (World Economic Forum, 2023).
4. **Low investments in R&D:** Lower investments into innovations and research in the renewable sector make countries largely dependent on foreign technologies, thus rising operational costs and constraining the technologies in terms of local needs (United Nations Framework Convention on Climate Change [UNFCCC], 2022).

- **Potential Solutions**

1. **Constant Capacity-Building On-hands training:** Governments and educational institutions should put in place investment programs to build capacity in a workforce proficient in renewable energy technology (African Renewable Energy Initiative [AREI], 2023).
2. **Smart Program Retrofitting in Existing Lines/Grids:** The need for more stability would be catered for through national power grid upgrades with smart grid technologies and energy storage mechanisms to enable vast opportunities for the integration of variable renewable energy sources (Intergovernmental Panel on Climate Change[IPCC], 2022).
3. **Hybrid Systems:** A mix of energy solutions from wind, solar, hydropower, and battery-type systems can instead help improve electricity reliability and lessen dependence on any one kind of energy source (International Energy Agency [IEA], 2023).
4. **Providing Sufficient Financial Support for R&D and Local Manufacturing:** Through R&D and the manufacture of indigenous technologies funded from the Government budget, they will be less dependent on expensive imports and have sustainable energy solutions specific to African needs (United Nations Framework Convention on Climate Change [UNFCCC], 2022).

5.3. Policy and Regulatory Barriers

Weak policy frameworks coupled with inconsistent regulatory environments substantiate various challenges in the development of renewable energy in Africa. Investors, among other stakeholders, require clear long-term policy framework directions to ensure stability and thus mitigate risks attached to renewable energy investments (International Energy Agency [IEA], 2023).

- **Key Policy Challenges**

1. **Unclear and Inconsistent Policies:** An unclear and inconsistent policy environment in many African countries towards renewable sources of energy has created a situation of uncertainty which puts off many would-be investors (World Bank, 2023).
2. **Cumbersome Regulatory Approvals:** Poor regulatory frameworks, which often become muddled and unclear, prove to be marred by unnecessary red tape and administratively sluggish, thus halting field development in renew-

able energy (United Nations Framework Convention on Climate Change [UNFCCC], 2022).

3. **Unstable Tariff Structures:** Unstable energy pricing structures and inconsistent feed-in tariffs serve to inhibit private participation due to unpredictable financial returns on renewable energy projects (African Renewable Energy Initiative [AREI], 2023).
4. **Little Government Support for Decentralized Solutions:** Most policies focus on large-scale generation yet provide little support to decentralized systems such as mini-grids and home solar systems, which are critical elements of rural electrification (Intergovernmental Panel on Climate Change[IPCC], 2022).

- **Potential Solutions**

1. **Develop Clear and Consistent Policy Frameworks:** Governments must introduce clear and uniform renewable energy policies along with long-term goals for investments, development, and grid integration (International Energy Agency [IEA], 2023).
2. **Establish Special Regulatory Bodies:** The regulator should be created specifically for renewable energy, which can speed up the approval process, reduce bureaucratic delays, and boost investor confidence (United Nations Framework Convention on Climate Change [UNFCCC], 2022).
3. **Ensure Attractive Financial Mechanisms for Private Investors:** Fixed feed-in tariffs and tax incentives for private investments would guarantee predictable financial returns for renewable energy producers (African Renewable Energy Initiative [AREI], 2023).
4. **Support for Decentralized Energy Systems:** Policies must encourage expansion of energy access to nonurban regions through support for micro-grids, off-grid solar solutions, and community-led renewable energy efforts (World Economic Forum, 2023).

5.4. Social and Cultural Barriers

Public attitude and resistance from the community to renewable energy, combined with historical energy usage habits, have partly been reasoned as some of the reasons impacting the slow deployment of renewable energy across Africa (International Renewable Energy Agency [IRENA], 2021). Problems such as misinformation, land conflicts, and the cultural attachment to fossil fuels might impede some communities from embracing clean alternatives (World Bank, 2020).

- **Key Social Challenges**

1. **Lack of public awareness:** The lack of awareness on the benefits of renewable energy further leads to misconceptions whereby many communities see solar and wind power as unreliable and expensive (REN21, 2022).
2. **Use conflicts over land:** Large-scale renewable energy projects normally demand vast tracts of land, which eventually creates conflicts with the local communities that often fear displacement or environmental damage or loss of agricultural land (UNEP, 2019).
3. **Cultural resistance to change:** Many African populations have long existed without access to modern forms of energy; because of this, they often have a strong preference for traditional biomass and fossil fuels (IEA, 2020).

- **Potential Solutions**

1. **Community Engagement and Education Programs:** Community engagement and education programs will not only be established to use awareness of the benefits of renewable energy but also to let the local populations participate in planning the projects (IRENA, 2021).
2. **Fair Land Acquisition and Compensation Policies:** Enact Fair and transparent land acquisition policies and compensation for those displaced (World Bank, 2020).
3. **Promotion of decentralized renewable solutions:** Encourage household, solar systems, biogas digesters, and community-based microgrids (REN21, 2022).

6. STRATEGIES FOR SCALING RENEWABLE ENERGY IN AFRICA

The construction of renewable energy in Africa is, however, dependent upon investment, technology, and policy support (IRENA, 2021). Regions abundant with sunshine and strong winds can allow for the setting up of multiple powers, such as solar and wind, which certainly can provide viable and cheap electricity (IEA, 2020). Weather conditions drive the use of these energy resources, and further development in battery storage systems becomes key toward the supply of uninterrupted power (World Bank, 2021). The infrastructure development is also very important, since there is little provided in many areas by way of strong power grids needed to transmit clean energy more effectively (REN21, 2022). In certain situations, such as in the remote parts of Africa, where solutions may sometimes require extensive power lines to reach them, small solar plants or mini-grids would be impossible to set up without establishing large power lines, while off-grid solutions can quickly and easily take electricity to communities (AfDB, 2020). Accordingly, government support can do a lot to encourage investments into renewable energy using investment incentives, tax breaks, or through ensuring low-interest loans that can attract even more businesses into the quickly developing industry of renewable energy (UNEP, 2019). Such incentives also further foster local manufacturing of solar panels, wind turbines, and batteries which will in turn lower their cost and enhance job creation (IRENA, 2021). Public awareness campaigns assist communities and businesses in appreciating the advantages of switching to renewable energy (World Bank, 2020). Lastly, strong policies and regulations will guarantee that these efforts last, creating a stable ground in which renewable energy projects can thrive (REN21, 2022). These combinations will provide Africa with a great chance to rely lesser on fossil fuels, make electricity affordable, and consequently provide millions of people with access to clean energy (IEA, 2020). The promotion of resourceless energy in Africa cannot be unbundled from various series of strategies that include, but are not limited to, support policies, financing mechanisms, skill developments, and community engagement (AfDB, 2020). In this manner, renewable energy generation, including solar, wind, and hydro power, will become equally accessible, affordable, and sustainable.

6.1. Policy and Regulatory Reforms

The development of renewable energy will need the support of policies by governments aimed at encouraging investments and development in the sector (IRENA, 2021). One such significant policy is the Feed-in Tariff (FiT), which stipulates that renewable energy producers must be guaranteed a fixed subsidy for the electricity they supply to the grid (World Bank, 2021). This will encourage ventures by industries and individuals into clean electricity as they know they will get value for their investment (REN21, 2022).

Another important step is setting renewable energy targets. Governments can put targets on how much renewable energy should be accounted for in the total generation of electricity by a certain year (IEA, 2020). Such targets are useful as they give investors and developers an idea of the right direction to take (AfDB, 2020).

Furthermore, there will need to be stronger institutional frameworks and governance for these policies to be operationally effective (UNEP, 2019). By this, it is meant that government institutions must have well-defined roles and responsibilities, there should be a considerable reduction of corruption, and laws must be consistently enforced (IRENA, 2021). Without strong governance, even the best policies would not work due to implementation failures (World Bank, 2021).

6.2. Financing Mechanisms

Renewable energy projects require a huge amount of money to fund, and many African countries struggle to secure that funding (REN21, 2022). Several methods have been identified to attract greater capital into the sector:

- **International climate finance:** Organizations such as the Green Climate Fund and the World Bank provide funding for renewable energy projects in developing countries (IEA, 2020). African governments can apply for this funding, which is generally aimed at supporting large-scale clean energy projects (AfDB, 2020).
- **Green Bonds:** These are special types of loans taken to finance ecological projects (IRENA, 2021). They offer governments and commercial companies a possibility of being able to raise money, specifically for the renewable energy infrastructure (World Bank, 2021).
- **Public-Private Partnerships (PPP):** Governments could work with private companies to develop renewable energy projects (REN21, 2022). For example, a government could give the company land and basic infrastructure, while the company would invest in building and maintaining solar or wind farms (IEA, 2020).
- **Foreign Direct Investment (FDI):** By attracting international companies to invest in Africa's renewable energy sector, they can bring capital, expertise, and technology critical to making such solutions possible (AfDB, 2020). Governments should provide investment-friendly policies that include taxation relaxation or incentives for renewable energy companies in order to entice international companies (IRENA, 2021). This will enable African states to realize large-scale renewable energy projects while not solely depending on budgets (World Bank, 2021).

6.3. Capacity Building and Technology Transfer

One of Africa's major challenges is a lack of skilled individuals and advanced technologies in renewable energy projects, and for this reason, there is a need to provide training and information-exchange initiatives (REN21, 2022). In light of these, the training programs would encompass teaching local engineers, technicians, and entrepreneurs in the design, installation, and maintenance of renewable energy systems (IRENA, 2021).

Collaboration with developed countries and international organizations can enable such advanced technologies to be transferred to Africa (IEA, 2020). An example of such a collaboration can be with countries that have been able to successfully develop renewable energy, such as Germany or China, which would provide cutting-edge solar panel and wind turbine technology (AfDB, 2020); moreover, universities and research institutions may collaborate to develop new solutions that suit the peculiar energy challenges of Africa (World Bank, 2021).

By investing in human capital and technology transfer, Africa can build its own renewable energy industry and no longer depend on foreign firms (REN21, 2022).

6.4. Community Engagement and Inclusivity

The African continent must be endowed with renewable energy, which will be useful for all communities, including the most distant and therefore, underserved African people (IRENA, 2021). Such decentralized energy solutions, such as mini-grids and stand-alone solar systems, might well deliver electricity to very remote rural communities while providing an alternative to large, very capital-intensive power plants with sometimes great distances to cover for transmission lines (World Bank, 2021).

Moreover, both governments and companies need to guarantee equitable clean energy access (IEA, 2020). Following this principle, this entails that low-income households must be guaranteed inexpensive electricity while renewable energy project development should not merely focus on urban centres at the total neglect of rural areas (AfDB, 2020).

Furthermore, decisions on the implementation of such projects should always involve direct engagement with local communities (REN21, 2022). Such an approach would ensure that people support and maintain all manner of projects, as otherwise they would feel that they played no helpful role in them (UNEP, 2019). Further investing in programs offering opportunities to women and youth would not only create good jobs but help sustain the sector in the checks and balances of the future (IRENA, 2021).

7. CONCLUSION AND RECOMMENDATION

7.1. Summary of the Key Insights

The collection and use of renewable energy stands at the forefront of transforming African energy markets in two distinct ways: emission reduction and sustainable progress. The utilization of solar power and wind power and hydroelectric energy will strengthen energy reliability along with economic survival. Large-scale implementation requires both the resolution of financial limitations and policy improvement as well as improvements to existing infrastructure.

7.2. Policy Recommendation

Governments must create specific national policies which will draw investment into renewable power sectors. To harness renewable energy opportunities organizations should establish integrated financing partnerships that combine technological development. The successful transition of Africa needs financial aid from international partners who will also provide technical expertise along with policy guidance. Improvements to renewable energy research and development efforts will generate innovative solutions as well as optimized productivity.

7.3. Call to Action

New efforts in renewable energy operations require immediate expansion to fight climate-change effects effectively. Renewable energy speed-up efforts require shared activity between the government sector along with business components and civil society stakeholders. Building both the education infrastructure and the technical expertise of workers will help establish the qualified workforce renewable energy needs. The national energy grids need integration with this system to immediately supply power to underserved rural areas.

REFERENCES

African Development Bank (AfDB). (2022). *Renewable Energy in Africa: Trends and Opportunities.* Retrieved from https://www.afdb.org/en/topics-and-sectors/initiatives-partnerships/africa-renewable-energy-initiative

African Renewable Energy Initiative (AREI). (2023). *African Energy Transition Report.* Retrieved from https://www.arei.org/energy-transition

Carbon Brief. (2022). *Africa's Carbon Emissions and Climate Impact.* Retrieved from https://www.carbonbrief.org/

Deloitte. (2023). *Healthcare in India: Trends and Opportunities.* Retrieved from https://www2.deloitte.com/in/en/pages/life-sciences-and-healthcare/articles/shaping-healthcare-for-all-in-india.html

Goldratt, E. M. (1990). *The Goal: A Process of Ongoing Improvement.* North River Press.

Intergovernmental Panel on Climate Change (IPCC). (2022). *Climate Change and Africa: Impacts, Adaptation, and Vulnerabilities.* Retrieved from https://www.ipcc.ch/report/ar6/wg2/

International Energy Agency (IEA). (2023). *Africa Energy Outlook.* Retrieved from https://www.iea.org/reports/africa-energy-outlook

KPMG. (2023). *Indian Healthcare Sector: Navigating the Future.* Retrieved from https://kpmg.com/in/en/insights/2023/11/indian-healthcare-sector-investment-trends-to-look-for.html

National Health Profile, India. (2023). *Annual Reports.* Ministry of Health and Family Welfare, Government of India. Retrieved from https://cbhidghs.mohfw.gov.in

Noor Ouarzazate Solar Complex. (2023). *Renewable Energy in North Africa.* Retrieved from https://www.masen.ma/en/our-projects/noor-ouarzazate

Porter, M. E. (1980). *Competitive Strategy: Techniques for Analyzing Industries and Competitors.* Free Press.

Slack, N., Chambers, S., & Johnston, R. (2023). *Operations Management.* Pearson Education.

The World Bank. (2023). *Energy Access in Sub-Saharan Africa.* Retrieved from https://www.worldbank.org/en/programs/africa-energy

United Nations Framework Convention on Climate Change (UNFCCC). (2022). *Paris Agreement Commitments and African Climate Action.* Retrieved from https://unfccc.int/process-and-meetings/the-paris-agreement

World Economic Forum. (2023). *Africa's Energy Future and Sustainable Development.* Retrieved from https://www.weforum.org/agenda/2023/10/renewable-energy-africa

Chapter 5
On the Causal Dynamics Between REC, NREC, and CO2 Emission in Türkiye by Considering Structural Break

Mikail Pehlivan
https://orcid.org/0000-0002-7273-333X
Trakya University, Turkey

Umut Akduğan
https://orcid.org/0000-0001-8659-541X
Trakya University, Turkey

ABSTRACT

Türkiye's limited use of renewable energy increases reliance on fossil fuels, widening the trade deficit and heightening energy security risks. This dependency also restricts economic growth, job creation, and worsens environmental damage due to higher carbon dioxide (CO2) emissions. This study examines the impact of renewable and non-renewable energy consumption (REC and NREC) on CO2 emissions in Türkiye from 1965 to 2023, using the asymmetric bootstrap nonlinear cointegration test. Results show that, in the long term, NREC significantly raises CO2 emissions more than REC, while in the short term, only NREC affects emissions. Based on these findings and Türkiye's renewable energy potential, key policy implications are identified. In conclusion, increasing domestic renewable energy production is a vital solution for reducing the trade deficit and energy security risks, aligning with the broader goal of a low-carbon future.

DOI: 10.4018/979-8-3373-1549-2.ch005

INTRODUCTION

There has been a strong opinion in the literature and global political sphere about decreasing the use of fossil energy sources, i.e. oil, natural gas, and coal, in the past few decades. Main reasons of this fact are (i) concerns about depletion of fossil sources since they are non-renewable energy sources, (ii) environmental effects and the urgent need to take action against global climate change, and (iii) the volatility of their prices (Lau et al., 2012; Nejat et al., 2015; Bilgili et al., 2017a, 2017b; Bulut, 2017; Bulut and Muratoglu, 2018).

In Türkiye's case, there are two additional factors that highlight the drawbacks of continued reliance on fossil fuels: (i) the substantial energy imports, which contribute to the growing trade imbalances, and (ii) the country's limited energy independence due to the current availability of domestic resources. Türkiye's energy output is inadequate to satisfy the nation's continuously rising energy needs, resulting in its status as a net energy importer. This reliance on foreign energy significantly impacts the trade balance. In other words, Türkiye 's dependence on imported energy is a key contributor to its large trade and current account deficits. These challenges have emphasized the urgent need for Türkiye to rapidly transition to renewable energy alternatives (Bulut and Muratoglu, 2018). Türkiye's oil and lignite coal reserves are insufficient to meet the nation's fuel demands. Over half of Türkiye's energy needs are met through imports. In other words, the gap between supply and demand, coupled with rising consumption, drives the country's reliance on energy imports, which results in significant import dependency and large trade deficits. Moreover, relying exclusively on fossil fuels and a limited number of major fossil fuel-exporting nations weakens Türkiye's energy independence (Ozcan, 2018).

Türkiye has significant potential to generate renewable energy, which is sustainable, unlike finite fossil fuels. Numerous studies in the literature argue that renewable energy should play a larger role in Türkiye's energy mix and that the country's energy policy needs to be redesigned with a strong emphasis on renewables, given the concerns mentioned earlier (Ediger and Kentel, 1999; Kaygusuz and Sari, 2003; Ocak et al., 2004; Ozgur, 2008; Bilgen et al., 2008; Yuksel and Kaygusuz, 2011; Capik et al., 2012; Baris and Kucukali, 2012; Simsek and Simsek, 2013; Toklu, 2013; Ozcan, 2018).

As Fang (2011) points out, both policymakers and the public typically have two key expectations regarding renewable energy. The first is to satisfy energy demand, ensuring the continued functioning of economic activities, while the second is to address the environmental issues arising from fossil fuel use. In light of Türkiye's reliance on energy imports, Bulut and Muratoglu (2018) highlight a third expectation for the country. In this regard, policymakers may also hope that renewable energy will reduce Türkiye's dependence on energy imports and help diminish its trade deficit. While it is essential to develop an empirical model to assess whether renewable energy can fulfill the first two expectations, no intricate econometric analysis is needed to determine whether an increase in renewable energy production can lower Türkiye's import dependence and trade imbalance. This is because greater renewable energy generation can directly reduce energy imports and improve the nation's trade balance.

Building on the previous discussion, this study focuses on the first two expectations for Türkiye. In other words, it examines the impact of both renewable and non-renewable energy consumption (REC and NREC, respectively) on carbon dioxide (CO_2) emissions in Türkiye using annual data from 1965 to 2023. The paper contributes to the energy economics literature in three key ways. First, although the body of empirical research on the relationship between renewable energy and economic growth (EG) is growing, few studies have specifically explored this connection for Türkiye, as seen in the Literature Review section. Thus, it can be argued that there is a notable gap in research in this area. Second, the study assesses the sign and magnitude of the coefficients concerning the growth hypothesis in the causal relationship between REC, NREC, and CO_2. Third, it analyzes the combined effects of REC and NREC on CO_2 emissions in Türkiye, providing insights to inform future energy policies.

In addition to these contributions, the originality of the paper stems from two key aspects: (i) methodology and (ii) the discussion section on policy implications. (ia) Most studies have used panel data sets covering short periods to investigate the relationship between REC and EG. A common critique of these studies is the use of panel data and panel selection, particularly the potential heterogeneity among countries in the panel and cross-sectional dependence. This issue can be addressed with estimation methods designed for heterogeneous panels with cross-sectional dependence. However, this does not make heterogeneous panel data analyses with different country characteristics more robust than time-series analyses with larger numbers of observations. (ib) This study estimates long-term output elasticities for Türkiye. Conducting these analyses separately on a country-specific basis offers an advantage over panel data analyses, as time-series techniques yield more accurate results. These elasticity estimates reflect long-term demand for REC and NREC in the CO_2 process, providing valuable tools for policy purposes. (ic) The empirical analysis employs a novel technique based on bootstrap simulations that account for structural breaks in the series. The economy is highly susceptible to disruptions in energy supply, spikes in energy costs, shifts in institutional frameworks, and abrupt fluctuations in macroeconomic factors. Based on the reviewed time-series data, it seems likely that Türkiye has experienced at least one such change in the past. Studies by Acaravci (2010) and Saatci and Dumrul (2012), which explore the connection between EG and energy consumption (EC) in Türkiye, confirm that structural changes exist within the data. Therefore, overlooking the presence of structural breaks in time-series analysis can lead to inaccurate conclusions and forecasting mistakes. (iia) Another uniqueness of this paper comes from having assessed CO_2 reduction policies on a sectoral basis, specifically for Türkiye, based on the results obtained from the empirical analysis in the Discussion section.

The rest of the paper is organized as follows: Section 2 presents the developments in renewable energy in Türkiye. Literature review is given in Section 3. Section 4 introduces the model and data. Methodology is presented in Section 5. Section 6 reports the findings. Section 7 implements a dicussion and Section 8 concludes the paper with a summary of main findings and some policy proposals.

RENEWABLE ENERGY UTILIZATION IN TÜRKİYE

First of all, Türkiye holds significant potential in renewable energy resources. Its geographic positioning, combined with a humid and warm climate, offers favorable conditions for the extensive use of renewable energy (Ediger and Kentel, 1999). The country's feasible renewable energy capacity represents 13% of the total renewable energy potential within the European Union (EU)-27. However, this renewable potential has not been fully exploited. A study reveals that Türkiye has tapped into around 70% of its viable hydropower capacity, 8.9% of its wind energy potential, 0.45% of its solar energy capacity, 30.7% of its geothermal resources, and 17.3% of its biomass potential (Ozcan, 2018). Hydroelectric plants, biomass for heating, and geothermal energy have been the primary renewable energy sources integrated into Türkiye's energy mix. In recent years, technologies such as solar and wind energy-based electricity generation have also started to gain traction in the sector (Bulut and Muratoglu, 2018).

Türkiye is an developing/emerging economy and EC of Türkiye has rapidly grown especially in recent years (Bölük and Mert, 2014). By referring to Figure 1 and 2, Türkiye's REC increased from 0.58 million kg of oil equivalent (mkgoe) in 1965 to 31.2 mkgoe in 2023; on the other hand, NREC increased from 7.3 kgoe in 1965 to 136 mkgoe in 2023. In addition, the real gross domestic product (GDP) per capita increased from $3176 in 1970 to $14713 in 2023. In Türkiye, the rapid growth of urban areas, favorable demographic shifts, economic expansion, and rising per capita income are driving up energy demand. It is projected by Turkish Electricity Transmission Company (TEIAS) that annual growth rate of electricity demand is 3.2% in low scenario or 4.1% in high scenario (TEIAS, 2024). As emphasized by Cleveland et al. (1997), energy serves as a crucial driver of development in the context of privatization and globalization, particularly in developing countries.

Figure 1. Türkiye's renewable and non-renewable energy consumption

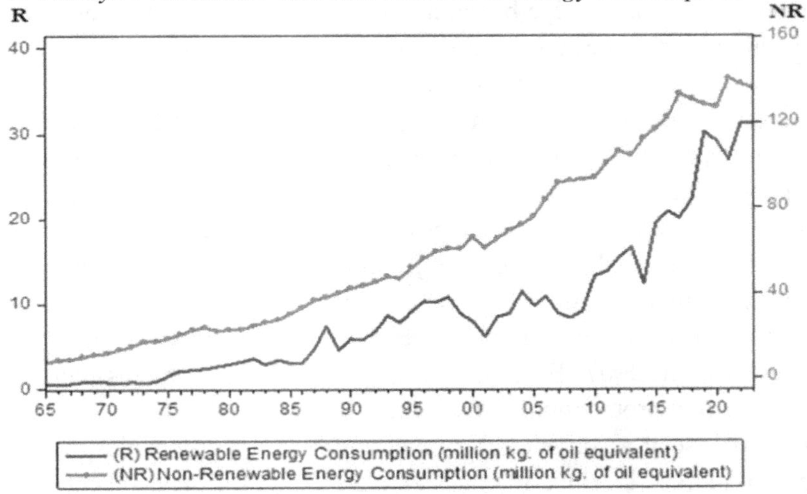

Source: *Energy Institute database*

Figure 2. Türkiye's CO_2 emissions and GDP per capita

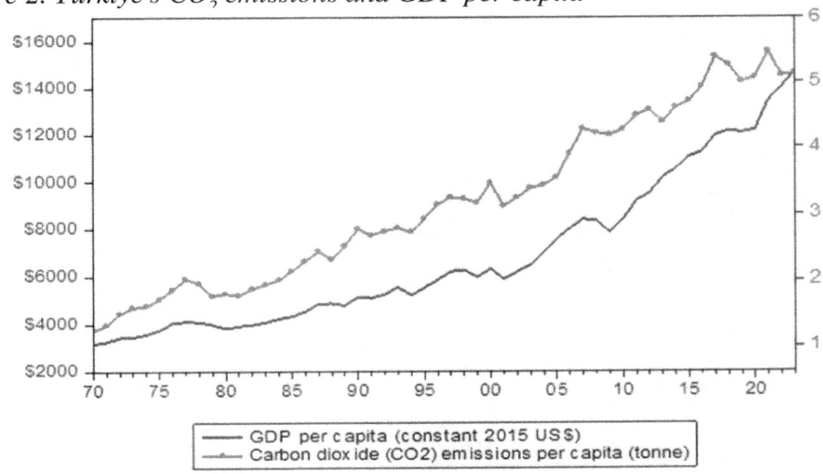

Source: *World Bank database*

Türkiye became a signatory of the Kyoto Protocol on August 26, 2009. In order to meet the requirements of the protocol, Türkiye must reduce its emissions. This is part of the country's efforts to align with EU standards as part of its EU membership aspirations. The EU's Renewable Energy Directive sets a comprehensive framework to encourage the use of renewable energy. As part of this directive, 40%

of the total energy consumption must be derived from renewable sources by 2030 (EU Directive, 2023). Türkiye began aligning its policies with renewable energy goals and energy efficiency concerns (Gaupp, 2008). The country places significant emphasis on energy and climate finance to support sustainable development. To achieve these objectives, Türkiye has implemented several reforms in the energy sector and has started promoting renewable energy initiatives. The first, enacted in 2001[1], encourages electricity generation from renewable sources, while the 2005[2] law supports biomass, geothermal, hydro, solar, and wind energy. The government aims to increase renewable energy's share in electricity generation and develop a local renewable industry. In 2008[3], the electricity market law was amended, and in 2007[4], a law was passed to regulate geothermal energy production and protect water resources. The 2007 "Energy Efficiency Law[5]" also promotes energy savings and environmental protection. In 2011, the "Renewable Energy Law[6]" introduced new incentives, such as subsidies for solar and biomass companies, research and development support, and investment incentives. Key government incentives include licensing exemptions, grid connections, feed-in tariffs (until 2011), purchase guarantees, and land acquisitions (Özkivrak, 2005; Baris and Kucukali, 2012; Capik et al., 2012; Simsek and Simsek, 2013).

Table 1. Türkiye's total electricity generation by the proportion of energy sources (%)

	Oil	Gas	Coal	Hydro	Biofuels+Geothermal+Solar+Wind
1985	20.70	0.17	43.92	35.22	0.00
1990	6.85	17.71	35.07	40.37	0.00
1995	6.69	19.22	32.52	41.57	0.00
2000	7.45	37.00	30.57	24.94	0.04
2005	3.39	45.35	26.67	24.54	0.05
2010	1.03	46.47	26.06	26.38	0.06
2015	0.85	37.90	29.09	31.96	0.20
2020	0.11	23.13	34.50	41.85	0.42
2021	0.08	33.22	30.89	35.42	0.39
2022	0.12	22.86	34.63	41.98	0.42
2023	0.21	21.24	36.06	42.05	0.44

Source: Energy Institute database

The data on Türkiye's electricity generation by the proportion of energy sources (Table 1) highlights key insights into the country's energy mix:

i) Fossil energy sources have consistently accounted for a larger share of electricity generation than renewable energy sources.

(ii) Since 1995, the proportion of renewables in electricity generation experienced a significant decline.

(iii) In recent decades, the share of hydroelectric power in Türkiye's energy mix has decreased, while the contributions from biofuels, geothermal, solar, and wind energy have seen modest increases.

(iv) Imported fossil fuels make up the majority of Türkiye's primary energy supply, which creates a major obstacle to achieving energy independence (Figure 3).

In summary, Türkiye's limited use of renewable energy leads to greater reliance on fossil fuels, driving up the trade deficit and energy security risks. This also hinders economic growth, reduces job opportunities, and worsens environmental impact through higher CO_2 emissions.

Figure 3. Energy import rates and foreign trade deficit between 1996 and 2024

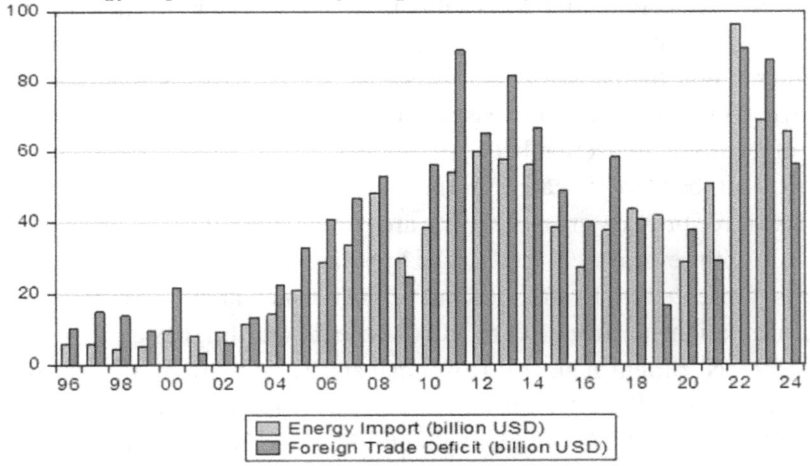

Source: *Turkish Central Bank database*

LITERATURE REVIEW

EC is essential for production processes, as energy is required to power various industries, factories, and other sectors of the economy. As EC rises, the output of goods and services typically increases, which in turn drives EG. However, all production processes, especially those relying on fossil fuels, generate CO_2 emissions as a byproduct. Therefore, an increase in EC, which leads to higher production output and economic growth, often correlates with a rise in CO_2 emissions. In this context, an increase in CO_2 emissions can be seen as an indicator of economic growth, as it reflects the higher levels of production and energy usage that fuel economic activity.

Studies on the link between EC and EG have primarily focused on examining four central hypotheses (Squalli, 2007; Payne, 2009; Ozturk, 2010): (i) The growth hypothesis proposes a unidirectional causal relationship from EC to EG. It suggests that an increase in EC can stimulate EG, while limitations on energy use may obstruct economic progress. This theory emphasizes that EC, along with labor and capital, is crucial for EG, both directly and indirectly, especially through industrial activities. Since energy is a critical factor for EG, any disruption in its supply is likely to have a detrimental effect on economic growth. (ii) The conservation hypothesis argues for a one-way causality from EG to EC. This view posits that in certain economies, especially those with lower energy dependence, policies aimed at reducing EC can be implemented without significantly harming EG. If higher real GDP leads to greater EC, this would provide support for the conservation hypothesis. (iii) The feedback hypothesis asserts a bidirectional causality between EC and EG, where both factors mutually influence one another. (iv) The neutrality hypothesis claims there is no substantial relationship between EC and EG. According to this hypothesis, changes in EC—whether increases or decreases—do not notably affect EG, suggesting the absence of a causal connection between the two.

The existing body of research on the connection between EC and EG can be divided into three primary areas. The first area investigates the relationship between EC and EG, with several well-established hypotheses. The first is the conservation hypothesis, presented by Kraft and Kraft (1978), which argues that there is a one-way causal relationship from EG to EC (Abosedra and Baghestani, 1989; Cheng and Lai, 1997; Nazlioglu et al., 2011; Akhmat and Zaman, 2013; Bloch et al., 2015; Jaforullah and King, 2015; Destek and Aslan, 2017; Koçak and Şarkgüneşi, 2017; Yazdi and Shakouri, 2017; Gokmenoglu and Kaakeh, 2018; Haseeb et al., 2018; Ozcan and Ozturk, 2019; Pegkas, 2020; Gyimah et al., 2022). The second is the growth hypothesis, which suggests that EG is driven by increased EC, implying causality runs from EC to EG (Stern, 1993; Soytas and Sari, 2003; Wolde-Rufael and Menyah, 2010; Rafiq and Salim, 2011; Tugcu et al., 2012; Chang et al., 2013; Naser, 2015; Omri et al., 2015; Bekun et al., 2019; Banday and Aneja, 2020). The third is the feedback hypothesis, which proposes a two-way causality between EC and EG (Pao and Fu, 2013; Wang et al., 2016; Alabi et al., 2017; Mbarek et al., 2017; Liu and Hao, 2018; Can and Korkmaz, 2019; Piłatowska et al., 2020). Finally, the neutrality hypothesis suggests that there is no causal relationship between EC and EG (Eden and Jin, 1992; Rafiq and Salim, 2009; Ozturk and Acaravci, 2010; Yildirim et al., 2012; Alper and Oguz, 2016; Saidi and Mbarek, 2016; Bulut and Muratoglu, 2018; Zhe et al., 2021). The second area looks at the relationship between REC and EG. Numerous studies support this connection, such as the growth hypothesis found by Tiwari (2011), Mezghani and Ben Haddad (2017), Bilgili et al. (2017a), and Chien and Hu (2007). Similarly, Sadorsky (2009) suggests the conservation hypothesis, while the feedback hypothesis has been confirmed by researchers like Gozgor et al. (2018), Inglesi-Lotz (2016), and Bhattacharya et al. (2016). The third area, which this paper focuses on, explores the broader relationship between EC and EG, incorporating both REC and NREC.

In the pursuit of a sustainable energy future, it is crucial to examine the interaction between REC, NREC and EG. Overall, while findings differ across various contexts, the existing body of research consistently emphasizes the complex and context-specific nature of the energy-growth relationship. This highlights the necessity for further research to gain a deeper understanding of the subtleties of these connections in diverse economic and geographical environments. Since the paper focuses on the growth hypothesis, a time-series analysis literature review related to this topic is provided in Table 2 for comparison purposes.

As observed in the empirical research on the relationship between REC and EG in Türkiye, several findings emerge. First, Soytas and Sari (2003) supports the growth hypothesis. Second, Salim and Rafiq (2012) investigate the applicability of the feedback hypothesis. Third, Ocal and Aslan (2013) and Dogan (2016) identify that the conservation hypothesis prevails in their studies. Finally, Dogan (2015), Koçak and Şarkgünesi (2017) and Bulut and Muratoglu (2018) present evidence supporting the neutrality hypothesis. Additionally, Bölük and Mert (2014) showed that REC decreases CO_2 in Türkiye, as Sulaiman et al. (2013) demonstrated for Malaysia and Bento and Moutinho (2016) for Italy. On the other hand, Bulut (2017) stated that CO_2 emissions are positively related to renewable energy in Türkiye. Moreover, Menyah and Wolde-Rufael (2010) found that REC is insignificant in decreasing CO_2 in the US, as Al-Mulali et al. (2015) observed in Vietnam.

Table 2. Summary of studies on renewable energy-economic growth nexus

Study	Country	Period	Method
Payne (2011)	The US	1949–2007	TY
Chang et al. (2013)	Canada, Germany, Italy, Japan, the US	1990-2011	BPGC
Bloch et al. (2015)	China	1977-2013	ARDL, VECM
Jaforullah and King (2015)	The US	1965-2012	JCO, GC

continued on following page

Table 2. Continued

Study	Country	Period	Method
Omri et al. (2015)	Hungary, India, Japan, Netherlands, Sweden	1990-2011	GMM
Alper and Oguz (2016)	Bulgaria	1990-2009	ARDL, HGC
Bhattacharya et al. (2016)	Austria, Bulgaria, Canada, Chile, China, Czech Republic, Denmark, Finland, France, Germany, Greece, Italy, Kenya, South Korea, Morocco, Netherlands, Norway, Peru, Poland, Portugal, Romania, Spain and the UK (positive effect). India, Ukraine, the US, Israel (negative effect).	1991-2012	PCA
Yazdi and Shakouri (2017)	Iran	1979-2014	ARDL, GC
Destek and Aslan (2017)	Peru	1980-2012	BPGC
Koçak and Şarkgüneşi (2017)	Bulgaria, Greece, Macedonia, Russia, Ukraine	1990–2012	HPCA
Cai et al. (2018)	Canada Germany, the US	1965-2015	BARDL
Haseeb et al. (2018)	Malaysia	1980-2016	ARDL
Lee and Jung (2018)	South Korea (negative effect)	1990-2012	ARDL
Luqman et al. (2019)	Pakistan	1990-2016	NARDL
Ozcan and Ozturk (2019)	Poland	1990–2016	BARDL
Banday and Aneja (2020)	Russia	1990-2017	BPC
Pegkas (2020)	Greece	1990-2016	ARDL, VAR
Shahbaz et al. (2020)	Ukraine, Sweden, Slovenia, Poland, Peru, Netherlands, Mexico, South Korea, Kenya, Japan, Ireland, Greece, Germany, Finland, Denmark, Czech Republic, China, Chile, Bulgaria, Brazil, Austria, Australia	1990-2018	DOLS, FMOLS, HNA
Gyimah et al. (2022)	Ghana	1990-2015	GC

Notes: ARDL: autoregressive distributed lag; BARDL: bootstrap autoregressive distributed lag; BPC: bootstrap panel causality test; BPGC: bootstrap panel Granger causality test; DOLS: the dynamic ordinary least squares; FMOLS: fully modified ordinary least squares; GC: Granger causality; GMM: generalised method of moments; HGC: Hatemi-J Granger causality test; HNA: heterogeneous non-causality approaches; HPCA: heterogeneous panel causality analysis; JCO: Johansen cointegration test; NARDL: nonlinear autoregressive distributed lag; PCA: panel cointegration analysis; VAR: vector autoregression; VECM: vector error correction model.

MODEL AND DATA

This study focuses on the potential shift between renewable and non-renewable energy consumption in the context of carbon dioxide emissions in Türkiye. The general form of the model used to investigate the relationship between these variables is as follows:

$$CO_2 = f(REC, NREC) \qquad (1)$$

Where, REC is the renewable energy consumption (tonnes of oil equivalent - toe), NREC is the non-renewable energy (oil, gas and coal) consumption (tonnes of oil equivalent - toe) and CO_2 is the carbon dioxide emissions (million tonnes of carbon dioxide). The series of the REC, NREC and CO_2 are obtained from "2024 Energy Institute Statistical Review of World Energy". This study focuses on the long-term relationships among the variables in Eq. (1). In this context, the model specifications for examining the long-term relationship among the variables are as follows:

$$ln\,CO_{2t} = \gamma_0 + @trend + \gamma_1 ln\,REC_t + \gamma_2 ln\,NREC_t + u_t \qquad (2)$$

where $t = 1, ..., T$ represents the time period, u_t is white noise stochastic disturbance term, assumed to be normally distributed. In order to have consistent empirical results, and considering the possibility of heteroscedasticity in data, all variables are transformed into natural logarithmic forms. Eq. (2) includes the logarithm of CO_2 ($lnCO_2$) as the dependent variable, and the logarithms of *REC (lnREC)* and *NREC (lnNREC)* as independent variables. The long-term relationships between these variables will be investigated for each country based on Eq. (2). In the case of identifying a cointegration relationship, the next step will involve estimating the long-term coefficients. Here, γ_1 and γ_2 indicates the long-term elasticity of CO_2 with respect to independent variables. The signs and absolute values of these coefficients, which represent the elasticities, will allow drawing inferences regarding the impact of the substitution between REC and NREC on carbon dioxide emissions. In testing the hypotheses and examining the relationships between the variables, annual data over the period 1965-2023 for Türkiye has been used.

METHODOLOGY

The theoretical foundation of the cointegration and unit root tests utilized in this research is covered in this title. The methodological background of the conventional linear unit root tests and the nonlinear unit root test method utilizing the bootstrap simulations, which is employed in this research, will first be covered in detail. Subsequently, theoretical information regarding the bootstrap nonlinear cointegration test used to analyze the long-term relationship between the variables will be provided. The econometric analysis that applied in the paper follows a precise flow which is shown by Figure 4.

Figure 4. The flow of the empirical analysis

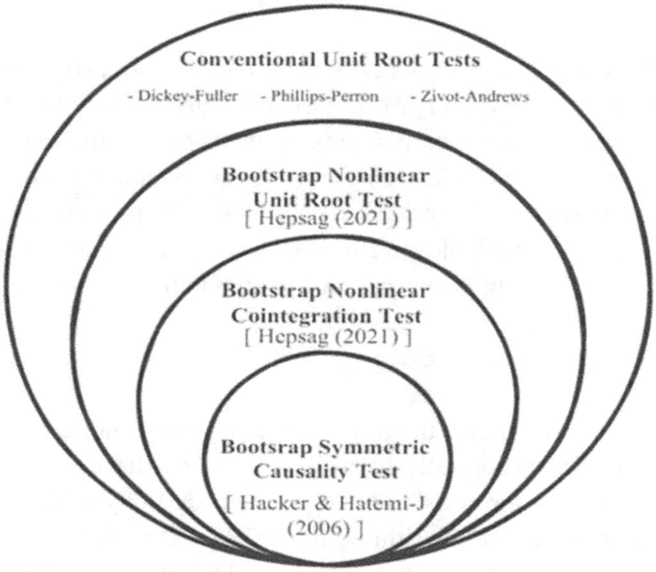

Bootstrap Hepsag (2021) Nonlinear Unit Root Test

There is a substantial number of empirical studies that examine the impact of structural changes (breaks) in time series on the unit root process from various perspectives. It is well known that conventional unit root tests, such as Dickey-Fuller type DF-GLS or ADF-GLS, are considered weak tests due to their failure to account for structural breaks. The weakness of these tests and the criticisms directed at them stem from their inability to accommodate existing structural breaks, leading to a tendency to fail to reject the null hypothesis of a unit root and causing the tests to be biased. This is because a time series with a structural break may be stationary, yet due to the use of conventional unit root tests, it may be incorrectly concluded that the series has a unit root (Cheung and Chinn, 1997). On the other hand, empirical evidence indicates that as the number of breaks in the series rises, these tests are more likely to fail in rejecting the null hypothesis of a unit root (Silvapulle, 1996).

To address this issue, Perron (1989) proposed a unit root test that allows for a structural break, which is either externally specified or accounted for in the Augmented Dickey-Fuller test. Subsequently, due to the potential personal biases involved in identifying breakpoints, Zivot and Andrews (1992) and Perron (1997) recommended an approach where the breakpoint is determined "endogenously" from the data (Glynn et al. 2007). In Perron (1989), structural breaks are incorporated into the process through a dummy variable, assuming that structural changes occur abruptly. On the other hand, when structural breaks are addressed through a shadow variable, it is assumed that the structural changes occur suddenly. However, Leybourne et al. (1998) argued that it is irrational to assume that changes in economic variables occur instantaneously in time series data. In this regard, Leybourne et al. (1998), along with Harvey and Mills (2002), developed a unit root test with gradual structural breaks, based on the assumption that structural changes may occur more slowly and smoothly. In the unit root test developed by Leybourne et al. (1998) (hereafter LNV (1998)), structural breaks are modeled using a nonlinear logistic smooth transition function instead of a shadow variable. Thus, the LNV (1998) test serves as both a unit root test with structural breaks and a nonlinear unit root test.

In the nonlinear unit root test developed by Hepsag (2021), both soft structural breaks and nonlinearity are considered simultaneously. In this test, soft structural breaks are modeled using the nonlinear logistic smooth transition function, as proposed by LNV (1998), while nonlinearity is modeled through the ESTAR (Exponential Smooth Transition Autoregressive) process, as suggested by Kruse (2011). The following is the formula for the ESTAR process (Kruse, 2011):

$$\Delta y_t = \alpha y_{t-1} + \varphi y_{t-1}\left(1 - exp\left\{-\gamma\left(y_{t-1} - c\right)^2\right\}\right) + \varepsilon_t \tag{3}$$

Where $\varepsilon_t \sim iid(0, \sigma^2)$ and γ is smoothness parameter. The ESTAR model transforms into a linear AR(1) model when the smoothness parameter γ gets closer to zero. Kapetanios et al. (2003) show the case where the ESTAR model transforms into a random walk process under the constraint α=0 when γ=0 as follows:

$$\Delta y_t = \varphi y_{t-1}\left(1 - exp\left\{-\gamma\left(y_{t-1} - c\right)^2\right\}\right) + \varepsilon_t \tag{4}$$

In Eq. (4), the random walk model is obtained under the constraint φ=0. This indicates that there is no way to directly test the unit root hypothesis. Because in order to test H_0: γ=0 and *vice versa*, φ must be defined. A common method for avoiding unnecessary parameter presence under the null hypothesis is to approximate the smooth transition function with a Taylor approximation. On the other hand, for the application of the first order Taylor approximation, an auxiliary regression is needed as follows:

$$\Delta y_t = \beta_1 y_{t-1}^3 + \beta_2 y_{t-1}^2 + \sum_{i=1}^{p} \rho_i \Delta y_{t-i} + u_t \tag{5}$$

In Eq. (5), u_t is a noise term depends on ε_t and φ. At the same time u_t is the residual of the Taylor expansion. The null hypothesis (H_0: $\beta_1 = \beta_2 = 0$), which indicates that the series has a unit root, is tested against the alternative hypothesis (H_1: $\beta_1 < 0$, $\beta_2 \neq 0$), which suggests that the series is stationary and follows an ESTAR process. The test statistic proposed by Kruse (2011) is as follows:

$$\tau = t^2_{\beta_2=0} + 1(\hat{\beta}_1 < 0) t^2_{\beta_1=0} \tag{6}$$

The models considered in the Hepsag (2021) test, where the nonlinear logistic smooth transition function is combined with the ESTAR process, are as follows:

Model A:

$$y_t = \alpha_1 + \alpha_2 S_t(\lambda, \tau) + v_t \tag{7}$$

Model B:

$$y_t = \alpha_1 + \beta_1 t + \alpha_2 S_t(\lambda, \tau) + v_t \tag{8}$$

Model C:

$$y_t = \alpha_1 + \beta_1 t + \alpha_2 S_t(\lambda, \tau) + \beta_2 t S_t(\lambda, \tau) + v_t \tag{9}$$

Model A represents the case where the soft break occurs in the constant term, Model B refers to the case where the soft break occurs in the constant term in the presence of a deterministic trend, and Model C denotes the case where the soft break occurs in both the constant term and the deterministic trend. In these models, $S_t(\lambda, \tau)$ is the logistic smooth transition function, which is defined as follows (Hepsag, 2021):

$$S_t(\lambda, \tau) = [1 + exp\{-\lambda(t - \tau T)\}]^{-1}, (\lambda > 0) \tag{10}$$

In the logistic smooth transition function presented above, the parameter λ represents the transition speed between the two regimes, τ denotes the point at which the soft structural break occurs, t is the deterministic trend variable, and T is the total number of observations. In this test, a two-stage estimation method is followed. In the first stage, the models expressed by Eq. (7), Eq. (8), or Eq. (9) are estimated using the nonlinear ordinary least squares (nonlinear OLS) method, and the residuals of these models are obtained as follows (Hepsag, 2021):

Model A:

$$\hat{v}_t = y_t - \hat{\alpha}_1 - \hat{\alpha}_2 S_t(\hat{\lambda}, \hat{\tau}) \tag{11}$$

Model B:

$$\hat{v}_t = y_t - \hat{\alpha}_1 - \hat{\beta}_1 t - \hat{\alpha}_2 S_t(\hat{\lambda}, \hat{\tau}) \tag{12}$$

Model C:

$$\hat{v}_t = y_t - \hat{\alpha}_1 - \hat{\beta}_1 t - \hat{\alpha}_2 S_t(\hat{\lambda}, \hat{\tau}) - \hat{\beta}_2 t S_t(\hat{\lambda}, \hat{\tau}) \tag{13}$$

In the second stage of the test, it is assumed that the residuals, expressed by Eq. (11), Eq. (12), or Eq. (13), follow the ESTAR process proposed by Kruse (2011):

$$\Delta\hat{v}_t = \gamma \hat{v}_{t-1}\left(1 - exp\left\{-\theta\left(\hat{v}_{t-1} - c\right)^2\right\}\right) + \varepsilon_t \tag{14}$$

Using Eq. (14), the null hypothesis (H_0: θ=0) representing the presence of a unit root can be tested against the alternative hypothesis of soft threshold ESTAR stationarity (H_0: θ>0). However, since the other parameter of the ESTAR process, γ, it is not possible to directly test for the presence of a unit root through Eq. (14). Therefore, by applying a first-order Taylor expansion to Eq. (14), the following test regression is derived, which allows for testing of the unit root hypothesis. This test regression is then estimated using the method of OLS (Hepsag, 2021).

$$\Delta\hat{v}_t = \delta_1 \hat{v}^3_{t-1} + \delta_2 \hat{v}^2_{t-1} + \sum_{i=1}^{k} \alpha_i \Delta\hat{v}_{t-i} + \varepsilon_t \tag{15}$$

In Hepsag (2021), the null hypothesis (H_0: δ_1=δ_2=0) represents the presence of a unit root, while the alternative hypothesis (H_1: δ_1=δ_2≠0) corresponds to soft threshold ESTAR stationarity. In the alternative hypothesis, it can be observed that one-tailed hypotheses are formulated for the δ_1 parameter and two-tailed hypotheses for the δ_2 parameter. Therefore, an F-test based on constraints cannot be applied. To address this issue, the test statistics derived by orthogonalizing δ_1 with respect to δ_1 are computed as follows (Hepsag, 2021):

$$\tau_{SNL\alpha} = \tau_{SNL\alpha(\beta)} = \tau_{SNL\alpha\beta} = \left(\widehat{\Psi}_{22} - \frac{\widehat{\Psi}_{21}^2}{\widehat{\Psi}_{11}}\right)\left(\hat{\delta}_2 - \hat{\delta}_1 \frac{\widehat{\Psi}_{21}}{\widehat{\Psi}_{11}}\right)^2 + 1\left(\hat{\delta}_1 < 0\right)\frac{\hat{\delta}_1^2}{\widehat{\Psi}_{11}} \quad (16)$$

The test statistics $\tau_{SNL\alpha}$, $\tau_{SNL\alpha(\beta)}$ and $\tau_{SNL\alpha\beta}$ which appear in Eq. (16), are valid for Model A, Model B, and Model C, respectively. These test statistics are then compared with the critical values obtained by Hepsag (2021) to test the null hypothesis.

In this study, however, the test statistics are compared not with the critical values from Hepsag (2021), but with those derived from the bootstrap procedure for each test. This approach allows the investigation of the stationarity processes of the series through the Bootstrap Hepsag (2021) nonlinear unit root test. Given the bootstrap's strong performance with finite samples in stationary processes, its application to non-stationary series has recently become more popular. The bootstrap procedure accounts for factors like sample size, initial conditions, and error distribution, offering potentially better finite sample properties than methods that rely on asymptotic approaches, which do not consider these factors. The steps of the bootstrap procedure are outlined as follows (Palm et al., 2006):

(i) First, the residuals of the unit root test regression are obtained. Fit an AR(p) model to $\hat{u}_t$, where $\hat{u}_t = \Delta y_t$ if there is only a constant term as the deterministic component, and $\hat{u}_t = \Delta y_t - n^{-1}\sum_{t=1}^{n}\Delta y_t$ if there is both a constant term and a linear time trend as the deterministic component.

$$\hat{\varepsilon}_t = \hat{u}_t - \sum_{j=1}^{p}\widehat{\Phi}_j \hat{u}_{t-j}, \; t = 1 + p, \ldots, n \quad (17)$$

(ii) With the help of Eq. (18), an i.i.d. sample ε_t^* is created by randomly drawing from the residuals.

$$\hat{\varepsilon}_t - (n-p)^{-1}\sum_{t=1+p}^{n}\hat{\varepsilon}_t \quad (18)$$

(iii) Bootstrap errors are obtained by successive iterations of Eq. (19).

$$u_t^* = \sum_{j=1}^{p}\widehat{\Phi}_j u_{t-j}^* + \varepsilon_t^* \quad (19)$$

(iv) The bootstrap sample y_t^* of the time series used in the unit root test is obtained by Eq. (20) and Eq. (21). If there are no deterministic components or only a constant term, Eq. (20) should be used. However, if both a constant term and a linear trend are present, Eq. (21) should be applied.

$$y_t^* = y_{t-1}^* + u_t^* \tag{20}$$

$$y_t^* = n^{-1}\sum_{t=1}^{n}\Delta y_t + y_{t-1}^* + u_t^* \tag{21}$$

(v) The coefficient and test statistics of the unit root test are calculated using the bootstrap sample for the previously specified deterministic specification.

(vi) Steps (ii) to (v) are repeated *N* times to obtain the bootstrap distributions, test statistics and critical values, where *N* denotes the number of bootstrap iterations.

In this context, bootstrap Hepsağ (2021) nonlinear unit root test was employed to assess the stationarity properties of nonlinear series, based on the results of the linearity test. Unlike many studies that rely on traditional nonlinear unit root tests using Monte Carlo simulations, this study is the first to integrate the nonlinear unit root test with the bootstrap procedure, marking a significant innovation in the literature.

Bootstrap Hepsag (2021) Nonlinear Cointegration Test

In the nonlinear cointegration test developed by Kapetanios et al. (2006), it is assumed that the error correction mechanism is consistent with the ESTAR process. This assumption implies that, after determining the cointegration relationship, positive and negative shocks occurring in the short term have the same effect in the long-term equilibrium adjustment process, meaning that a symmetric effect exists in the long-term adjustment process. In the nonlinear cointegration test developed by Hepsag (2021), this assumption is extended, positing that positive and negative shocks in the short term will have an asymmetric effect in the long-term equilibrium adjustment process (Hepsag, 2021).

Hepsag (2021) nonlinear cointegration test is a two-stage procedure. In the first stage, the regression model formed with the level values of the variables is estimated using the OLS method, and the residuals of the model are obtained. It is assumed that the residuals follow an asymmetric nonlinear exponential smooth transition autoregressive (AESTAR) process. Therefore, in the second stage of the test, the following asymmetric nonlinear exponential smooth transition autoregressive error correction model (AESTAR-ECM) is considered:

$$\Delta y_t$$

$$= G_t(\theta_1, u_{t-1})\{S_t(\theta_2, u_{t-1})\gamma_1 + (1 - S_t(\theta_2, u_{t-1}))\gamma_2\}u_{t-1} + \psi'\Delta \mathbf{x}_t + \sum_{t=1}^{p}\omega_1'\Delta \mathbf{z}_{t-1} + \varepsilon_t \tag{22}$$

The model expressed by Eq. (22) is defined as:

$$\Delta \mathrm{x}_{\mathrm{t}} = \sum_{t=1}^{p} \Gamma_1' \Delta \mathrm{z}_{t-1} + \eta_{\mathrm{t}}$$

$$G_t(\theta_1, u_{t-1}) = 1 - \exp(-\theta_1(u_{t-1}^2)), \theta_1 \geq 0$$

$$S_t(\theta_2, u_{t-1}) = [1 + \exp(-\theta_2 u_{t-1})]^{-1}, \theta_2 \geq 0$$

In addition, $\mathbf{z}_{\mathrm{t}} = (y_t, \mathrm{x}_{\mathrm{t}}')'$ is a first-order stationary $n\times1$ vector of variables, where $x_t = (x_{1t},\ldots,x_{kt})$ is a $k\times1$ vector of independent variables, and u_t is the error term of the long-term model, expressed as $u_t = y_t - \beta_{\mathrm{x}}' \mathrm{x}_t$.

In Hepsag (2021), in testing the cointegration relationship with the model expressed by Eq. (22), the null hypothesis (θ_1=0), which indicates the absence of cointegration, is tested against the alternative hypothesis (θ_1>0), which suggests the existence of nonlinear symmetric or asymmetric ESTAR cointegration. However, the parameters θ_2, γ_1 and γ_2 in this model are not defined under the null hypothesis. Therefore, it is not possible to directly test for the existence of cointegration. To address this issue, a first-order Taylor expansion is applied to the AESTAR-ECM in Eq. (22). As a result of the first-order Taylor expansion, the test regression that allows for the testing of cointegration existence is as follows (Hepsag, 2021):

$$\Delta y_t = \phi_1 \hat{u}_{t-1}^3 + \phi_2 \hat{u}_{t-1}^4 + \psi' \Delta \mathbf{x}_{\mathrm{t}} + \sum_{i=1}^{p} \omega_{\mathrm{i}}' \Delta \mathbf{z}_{\mathrm{t-i}} + v_t \qquad (23)$$

In the test regression expressed by Eq. (23), $\hat{u}_t$ represents the residuals obtained from the long-term model in the first stage of the test, and is calculated as $\hat{u}_t = \hat{y}_t - \hat{\beta}_{\mathrm{x}}' \mathrm{x}_t$. Using the auxiliary regression model in Eq. (23), the null hypothesis (ϕ_1=ϕ_2=0) indicating the absence of a cointegration relationship is formulated and tested against the alternative hypothesis (ϕ_1=ϕ_2≠0). At this stage, if the $\mathrm{F}_{\mathrm{ANEC}}$ test statistic, computed based on the constraints imposed on the coefficients, is greater than the critical values obtained by Hepsağ (2021), the null hypothesis, which indicates the absence of cointegration, is rejected in favor of the existence of nonlinear symmetric or asymmetric ESTAR cointegration. On the other hand, the second test regression developed by Hepsağ (2021) takes into account the residuals from the long-term model predicted in the first stage and is expressed as follows:

$$\Delta \hat{u}_t = \phi_1 \hat{u}_{t-1}^3 + \phi_2 \hat{u}_{t-1}^4 + \sum_{i=1}^{p} \lambda_i \Delta \hat{u}_{t-i} + \xi_t \qquad (24)$$

Using the test regression in Eq. (24), the null hypothesis ($\phi_1=\phi_2=0$) is formulated and tested against the alternative hypothesis ($\phi_1=\phi_2\neq0$). Based on these constraints, the F_{ANEG} test statistic is computed and compared with the critical values obtained by Hepsağ (2021). However, in this study, the test statistics calculated will not be compared with the critical values obtained from Hepsağ (2021), but with the critical values derived using the bootstrap procedure. This approach allows for the analysis of the cointegration process through the Bootstrap Hepsağ (2021) nonlinear cointegration test. The aforementioned bootstrap procedure has been explained in detail in the previous section.

On the other hand, in the asymmetric nonlinear cointegration test developed by Hepsağ (2021), after determining the cointegration relationship between the variables, it is also possible to test symmetric ESTAR cointegration against asymmetric ESTAR cointegration. To test symmetric ESTAR cointegration against asymmetric ESTAR cointegration, the test regressions in Eq. (17) and Eq. (18) are considered, and the null hypothesis $\phi_2=0$ is tested against the alternative hypothesis $\phi_2\neq0$. For this test, an F-test statistic is computed and compared with the critical values from the standard F distribution. In the case of determining the existence of symmetric ESTAR cointegration, positive and negative shocks occurring in the short term will have the same effect in the long-term equilibrium adjustment process, whereas in the case of determining the existence of asymmetric ESTAR cointegration, they will not have the same effect.

Hacker and Hatemi-J (2006) Bootstrap Causality Test

The causality test developed by Granger (1969) is based on asymptotic distribution theory and a vector autoregressive model of order p (VAR(p)). However, Granger and Newbold (1974) pointed out that in the case of non-stationary variables, Monte Carlo simulations based on asymptotic distribution theory in regression analysis do not yield accurate results. Additionally, Sims et al. (1990) argued that asymptotic distribution theory cannot be applied when testing causality between non-stationary variables using vector autoregressive (VAR) models. Toda and Yamamoto (1995) developed a modified Wald (MWALD) test that, regardless of the integration degree or cointegration properties of the variables in the model, has an asymptotic chi-square distribution. However, Hacker and Hatemi-J (2006) noted that the MWALD test does not perform well for data that do not conform to a normal distribution with autoregressive conditional heteroscedasticity (ARCH) effects. Therefore, they developed a bootstrap causality test that provides consistent results under both non-normal distributions and heteroscedasticity.

Hacker and Hatemi-J (2006) improved the MWALD test's size properties using bootstrap resampling. Their results showed that the bootstrapped empirical size is more accurate than the asymptotic distribution, especially when the extra lags match or exceed the integration order of both variables. They also found that ARCH errors cause greater size distortions in the asymptotic distribution, but with leverage adjustments in the bootstrap method, these distortions were minimal (Ozcan and Ari, 2015).

The Toda-Yamamoto procedure is based on a lag(s) augmentation of the VAR model. Toda and Yamamoto (1995) suggest the following augmented VAR(p+d) model that can be used for tests of causality between integrated variables (Hacker and Hatemi-J, 2006):

$$y_t = \hat{v} + \widehat{A}_1 y_{t-1} + \ldots + \widehat{A}_p y_{t-p} + \ldots + \widehat{A}_{p+d} y_{t-p-d} + \hat{\varepsilon}_t \tag{25}$$

It is assumed that the order *p* of the process is known and *d* is equal to the maximum order of integration of the variables. The VAR(p+d) in Eq. (25) could be written as:

$$Y = \widehat{D} Z + \hat{\delta} \tag{26}$$

The terms in Eq. (26) for a sample size of *T* are defined as follows (Hacker and Hatemi-J, 2006):

$Y = (y_1, \ldots, y_T)$, an (n x T) matrix;

$\widehat{D} = (\hat{v}, \widehat{A}_1, \ldots, \widehat{A}_p, \ldots, \widehat{A}_{p+d})$, an (n x (1+n(p+d))) matrix;

$Z_t = \begin{bmatrix} 1 \\ y_t \\ y_{t-1} \\ \vdots \\ y_{t-p-d+1} \end{bmatrix}$, a (1+n(p+d))x1) matrix, for t=1,…,T ;

$Z = (Z_0, \ldots, Z_{T-1})$, a (1+n(p+d))xT) matrix;

$\hat{\delta} = (\hat{\varepsilon}_1, \ldots, \hat{\varepsilon}_T)$, an (n x T) matrix.

The MWALD test statistic, proposed by Toda and Yamamoto (1995) to test for non-Granger causality between one variable in y_t and another in y_t, can be expressed as follows:

$$MWALD = (C\hat{\beta})' \left[C\left((Z'Z)^{-1} \oplus S_U \right) C' \right]^{-1} (C\hat{\beta}) \tag{27}$$

where $\oplus$ is the Kronecker product, $\hat{\beta} = vec(\hat{D})$ (*vec* denotes the column-stacking operator), C is a $p \times n(1+n(p+d))$ matrix. Each of the p rows of C corresponds to a restriction that sets one parameter in β to zero. S_U is the variance-covariance matrix of the residuals in Eq. (26). Using Eq. (27), the null hypothesis for non-Granger causality is expressed as follows:

$$H_0: C\beta = 0$$

The MWALD test statistic is asymptotically χ^2 distributed with the number of degrees of freedom equal to p, the number of restrictions to be tested. The bootstrap critical values were calculated in this study by performing 10,000 bootstrap simulations, with the MWALD test statistic generated each time. This process created the empirical distribution for the Wald test statistic. The final step is to compute the MWALD statistic using the original data. As a result, in the Hacker-Hatemi-J (2006) bootstrap causality test, the Toda-Yamamoto (1995) causality test is applied to determine the causality between variables. However, due to the potential non-normality of the residuals, the critical values are validated using bootstrap simulations.

FINDINGS

The integration order of the variables is determined first, before examining the long-term relationship between them. Since the variables need to be integrated at first order in order to perform the cointegration tests. In determining the stationarity characteristics of the series, the Augmented Dickey-Fuller (ADF) and Phillips-Perron (PP) unit root tests, which are commonly used in the literature, were applied, along with the Zivot-Andrews unit root test, which allows for a structural break, for comparison purposes. The results of these tests are presented in Table 3.

Table 3. The results of ADF, PP and ZA unit root tests

			$\ln CO_2$	lnREC	lnNREC
			t-statistic	**t-statistic**	**t-statistic**
Panel A					
ADF Test	Level	Intercept	-2.37	-1.22	-3.73*
		Intercept+Trend	-2.31	-2.65	-2.25
	First Differences	Intercept	-7.49*	-8.59*	-
		Intercept+Trend	-8.36*	-8.57*	-4.55*

continued on following page

Table 3. Continued

			$lnCO_2$		lnREC		lnNREC	
			t-statistic		t-statistic		t-statistic	
PP Test	Level	Intercept	-2.11		-1.27		-4.05*	
		Intercept+Trend	-2.27		-2.65		-2.23	
	First Differences	Intercept	-7.49*		-9.24*		-	
		Intercept+Trend	-8.40*		-9.32*		-7.98*	
Panel B								
ZA	Level	Intercept	-3.01	1982	-4.96**	1975	-3.08	1985
		Intercept+Trend	-3.11	1990	-3.73	1981	-3.21	2008
	First Differences	Intercept	-8.73*	1978	-	1989	-8.17*	1980
		Intercept+Trend	-8.45*	1980	-8.85*	1977	-8.65*	1982

* and ** denote the rejection of the null hypothesis at the 1% and 5% levels, respectively.

The results of the ADF, PP, and ZA unit root tests presented in Table 3 can be summarized as follows:

(i) The $lnCO_2$ variable exhibits a unit root at the level in both the intercept and intercept+trend models in all three tests. When the first difference of the $lnCO_2$ variable is computed, it becomes stationary, indicating that it is first-order integrated [$I(1)$] at the 1% significance level.

(ii) The lnREC variable exhibits a unit root at the level in both the intercept and intercept+trend models in the ADF and PP tests. According to these tests, the lnREC variable becomes stationary after taking the first difference. Additionally, in the ZA test, the lnREC variable shows a unit root at the level in the intercept+trend model but is stationary at the level in the intercept model.

(iii) The lnNREC variable is stationary at the level in the intercept model in both the ADF and PP tests. However, it exhibits a unit root at the level in the intercept+trend models. In the ZA test, the lnNREC variable shows a unit root at the level in both the intercept and intercept+trend models, but becomes stationary at the first difference.

As mentioned previously, there is a growing consensus that conventional unit root tests give biased results due to the tendency to fail to reject the unit root hypothesis because they do not allow for structural breaks. On the other hand, the ZA test only considers the possibility of a single endogenous structural break in the series. In addition, all of these tests assume that the series are linear. Therefore, in the case of series that exhibit nonlinear behavior at the level, the results of traditional unit root tests will be misleading. For this reason, this paper examines the stochastic properties of the series using the nonlinear unit root test, which accounts for smooth or sharp breaks and other nonlinearities. The results of the bootstrap Hepsag (2021) nonlinear unit root test, applied to the level values of the series, are presented in Table 4.

Table 4. The results of bootstrap Hepsag (2021) nonlinear unit root tests

Variables	Bootstrap Test Statistics	Citical Values			Bootstrap p-value	Optimal Lag
		99%	95%	90%		
$lnCO_2$	20.18	34.47	24.69	20.92	0.1138	0
lnREC	18.85	49.04	38.70	33.32	0.5122	0
lnNREC	13.85	38.11	28.72	24.59	0.4424	0

When the test statistics for the $lnCO_2$, lnREC, and lnNREC variables presented in Table 4 are compared with the critical values obtained from bootstrap simulations, the null hypothesis suggesting the presence of a unit root cannot be rejected. Therefore, according to the bootstrap Hepsag (2021) nonlinear unit root test, all the series are found to have a unit root at their level values. Therefore, the results of conventional unit root tests will be taken into consideration. This is because the application of the nonlinear unit root tests to differenced series is generally not recommended. The rationale behind this is that structural breaks are sought in the deterministic components (such as intercept and trend), and when the series is differenced, these deterministic components are removed. In this context, based on the results of the ADF, PP, and ZA unit root tests presented in Table 3, it is concluded that the series are stationary at the first difference [$lnREC_t \sim I(1)$; $lnNREC_t \sim I(1)$; $lnCO_{2t} \sim I(1)$]. The second step is to determine whether there is a long-term equilibrium relationship between the variables. After it has been determined that all of the series are integrated at first order. In this regard, the results of the applied bootstrap Hepsag (2021) nonlinear cointegration test are shown in Table 5.

Table 5. The results of bootstrap Hepsag (2021) nonlinear cointegration test

Test	Bootstrap Test Statistics	Citical Values			Bootstrap p-value	Symmetric ESTAR vs. Asymmetric ESTAR	
		99%	95%	90%		F-stat	p-value
F_{ANEC}	163.91*	20.12	14.66	12.28	0.0000	4.0072	0.1503
F_{ANEG}	192.75*	33.53	24.98	21.22	0.0000	2.5621	0.1150

* indicates the null hypothesis is rejected at the 1% level.

The results of the bootstrap Hepsag (2021) nonlinear cointegration test show that both F_{ANEC} and F_{ANEG} test statistics are greater than the 1% significance level critical values, leading to the rejection of the null hypothesis of no cointegration. This indicates the presence of either symmetric or asymmetric ESTAR cointegration among the analyzed variables. To determine whether the cointegration is symmetric or asymmetric ESTAR, an F-test was applied, with results presented in Table 5. Based on the F statistics' p-values, both F_{ANEC} and F_{ANEG} tests fail to reject the null hypothesis of symmetric ESTAR cointegration, confirming its existence. This suggests that positive and negative shocks in the short term have the same (symmetric) effect on the long-term equilibrium adjustment process.

When there is a cointegration relationship between the variables, long-term parameters can also be estimated using various estimation methods. In this regard, the long-term parameters were estimated using the OLS method, and the results are presented in Table 6. According to these findings, the long-term elasticity of CO_2 emissions with respect to REC is 0.0247, and this parameter is statistically significant at the 5% significance level. Beside this, the long-term elasticity of CO_2 emissions with respect to NREC is 0.9333, and this parameter is statistically significant at the 1% significance level. This result shows that a 1% increase in REC leads to a 0.02% increase in CO_2 emissions, while a 1% increase in NREC results in a 0.93% increase in CO_2 emissions.

Table 6. The estimated long-term coefficients

Variable	Coefficient	SE	t-statistic
constant	-11.8586	0.3469	-34.1771*
@trend	-0.0008	0.0010	0.4423
lnREC	0.0247	0.0104	2.3550**
lnNREC	0.9333	0.0246	37.8085*

* and ** denote the statistically significance of the coefficients at the 1% and 5% significance level.

Finally, the results of the symmetric bootstrap causality test developed by Hacker and Hatemi-J (2006) are presented in Table 7. According to these results, no causality relationship is found between CO_2 emissions and REC. On the other hand, the null hypothesis stating "CO_2 emissions do not cause NREC" cannot be rejected, while the null hypothesis stating "NREC does not cause CO_2 emissions" is rejected at the 5% significance level. Therefore, a unidirectional causality relationship is detected from NREC to CO_2 emissions.

Table 7. Hacker and Hatemi-J (2006) bootstrap causality test results

Null Hypothesis	Wald test stat.	Critical Values		
		1%	5%	10%
$lnCO_2 \nRightarrow lnREC$	0.007	7.173	4.114	2.791
$lnREC \nRightarrow lnCO_2$	0.003	7.135	4.123	2.907
$lnCO_2 \nRightarrow lnNREC$	1.660	7.060	3.947	2.830
$lnNREC \nRightarrow lnCO_2$	6.540**	7.489	4.022	2.893

** indicates the null hypothesis is rejected at the 5% level.

DISCUSSION

There is a significant gap in empirical research that explores the causal links between REC, NREC and CO_2 emissions, particularly using advanced time series econometrics techniques for both short- and long-term analysis. The importance of studying this relationship lies in the central role renewable energy plays in current debates surrounding environmental conservation and sustainable development. By investigating this dynamic in Türkiye, insights may be drawn that are applicable to other nations with similar socio-economic conditions. This is especially relevant given the global push to identify effective strategies for reducing emissions and the contribution of various energy sources toward reaching this goal.

The energy security challenges faced by countries that rely on energy imports are similarly significant (Hedenus et al., 2010). The concentration of energy resources in the unstable Middle East presents risks for numerous nations regarding the consistency and dependability of their energy supply (Gnansounou, 2008). Both energy security concerns and environmental issues are pushing many countries to seek alternatives to fossil fuels. Renewables are seen as potential solutions to address both energy security and environmental deterioration. As a result, numerous nations have invested in renewables technologies to decrease their reliance on imported oil, enhance energy security, reduce the price volatility linked to fossil fuel imports, and lower CO_2 emissions (Toth and Rogner, 2006; Vaillancourt et al., 2008; Adamantiades and Kessides, 2009).

The results indicate that, in the long term, NREC increases CO_2 emissions significantly more than REC, while in the short term, only NREC has an effect on CO_2 emissions. The findings suggest that the growth hypothesis holds true for both REC and NREC, and therefore, these results align with other studies in the literature that support the growth hypothesis. Moreover, when examining studies specific to Türkiye, the findings are consistent with those of Soytas and Sari (2003), Salim and Rafiq (2012), and Bulut (2017) in the context of EC and CO_2/EG. However, the findings of this study diverge from the studies by Bölük and Mert (2014), Dogan (2015), Koçak and Şarkgüneşi (2017), and Bulut and Muratoglu (2018).

As noted by Apergis et al. (2010), Menyah and Wolde-Rufael (2010), and Jaforullah and King (2015), the literature provides three explanations for why increased use of renewable energy sources may not reduce CO_2 emissions. First, the modest reduction in CO_2 emissions is attributed to the small increase in the share of renewable energy in total energy consumption during the study period. Second, the share of renewables in the energy supply of the countries examined has not yet reached a level high enough to significantly reduce CO_2 emissions. Lastly, it is noted that most renewable energy sources have intermittent output, which makes storage challenging. As a result, renewable energy may not be reliable enough to meet peak demand, necessitating the use of backup energy from fossil fuels. This implies that renewable energy may not effectively reduce emissions. However, while the limitations of renewable energy may limit the reduction in fossil fuel dependence, this does not rule out the possibility of emissions reductions from renewables during off-peak times. In other words, even if fossil fuels cannot be entirely replaced by renewables, each additional unit of renewable energy still reduces the need for energy generated from fossil fuels.

Overall, due to the differences in the levels, proportions, and growth rates of industrial, agricultural, and residential renewable and non-renewable energy consumption across various sectors, it can be concluded that the short- and long-term effects of energy consumption on CO_2 emissions and economic growth – and thus the related policy implications – cannot be the same for all sectors or energy sources.

As it stated by Table 8, to reduce CO_2 emissions, Türkiye needs a sector-specific approach to renewable energy integration. The industrial, transport, and power sectors require the most immediate action due to their high emissions. The agriculture, building, and waste sectors also need targeted policies to integrate renewable energy, though their impact on total emissions may be smaller. By focusing on sector-specific changes, Türkiye can significantly reduce its CO_2 emissions while fostering economic growth through the transition to cleaner energy sources.

Table 8. CO_2 emissions in different energy-consuming sectors in Türkiye (million tonne)

Sector	1970	1980	1990	2000	2010	2020	2021	2022
1. Agriculture	0.72	0.97	1.42	1.60	2.11	2.31	2.27	2.27
2. Building	12.15	16.46	29.16	30.36	67.99	71.71	71.05	68.32
3. Fugitive Emissions	2.62	4.04	7.00	8.15	12.46	17.89	17.43	17.07

continued on following page

Table 8. Continued

Sector	1970	1980	1990	2000	2010	2020	2021	2022
4. Ind. Combust. and Process	14.11	29.27	52.78	63.43	92.89	135.58	124.18	124.54
5. Power Industry	9.06	14.36	35.57	69.83	113.26	143.70	128.70	131.08
6. Transport	11.10	16.87	26.56	33.31	44.00	89.46	90.11	95.04
7. Waste	0.01	0.02	0.03	0.03	0.01	0.00	0.00	0.00
CO_2 emissions (total)	49.76	81.98	152.52	206.72	332.73	460.66	433.75	438.32

Notes: (1) Farming, livestock, rice paddies, and fertilizer use, mostly methane (CH_4) and nitrous oxide (N_2O). (2) homes and buildings (heating, cooling, appliances), mainly CO_2. (3) Leaks of gases like methane from natural gas, oil, and coal infrastructure. (4) Burning fossil fuels and chemical reactions in industrial processes like cement and steel production, releasing CO_2. (5) Emissions from burning fossil fuels (coal, oil, gas) for electricity generation. (6) Emissions from fuel combustion in cars, trucks, planes, ships, and trains. (7) Emissions from landfill decomposition (methane) and waste incineration.

Source: World Bank database

CONCLUSION

Given the growing concerns over climate change, as well as the economic and political challenges stemming from fossil fuel reliance, a strategy centered on renewable energy has emerged as a key solution to address both the impacts of climate change and the rising demand for energy. Global policymakers anticipate that renewable energy will (i) fulfill energy demands for sustainable development, (ii) reduce environmental issues like global warming by fossil fuel use, and (iii) enhance energy security. In Türkiye, policymakers have additional expectations for renewable energy, believing that replacing fossil fuels with renewable sources can lower energy import dependence and strengthen the country's current account balance. This paper evaluates whether renewable energy can effectively meet the first two expectations outlined above.

The study, which examines the potential changes between REC and NREC in the context of CO_2 emissions in Türkiye, analyzes the impact of the REC and NREC variables on CO_2 emissions. In this regard, the stationarity properties of the relevant series were investigated using traditional unit root tests, along with the Hepsag (2021) bootstrap nonlinear unit root test, which accounts for soft structural breaks. Subsequently, the long-term relationship between the variables, as well as the asymmetric effects of positive and negative shocks in the short-term on the long-term equilibrium process, were examined using the Hepsag (2021) asymmetric bootstrap nonlinear cointegration test. Additionally, causality relationships between the variables were determined through the Hacker and Hatemi-J (2006) bootstrap causality test. The results indicate that, in the long term, NREC increases CO_2 emissions significantly more than REC, while in the short-term, only NREC has an effect on CO_2 emissions. The findings suggest that the growth hypothesis holds true for both REC and NREC. Additionally, it can be inferred that the neutrality hypothesis holds for REC in the short term, as the lack of significance may be attributed to the fact that renewable energy sources require advanced human capital and substantial technological investments. The positive effects of these investments are likely to materialize in the future.

As mentioned, to tap into the vast potential of renewable energy, the Turkish government has (i) introduced key legislative reforms for renewable energy, (ii) implemented various policies to promote it, and (iii) increased competition within the energy sector. Consequently, the share of renewable energy in the total energy mix has started to rise in recent years. However, it remains relatively low in comparison to fossil fuels. Furthermore, it can be deduced that investing in renewable energy may improve the effectiveness of resource allocation in Türkiye. By boosting the use of renewable energy and enhancing the current account balance, the country could allocate more resources, along with technology and capital, to the renewable energy sector, which would, in turn, promote economic growth. As a result, this paper supports the transition from fossil fuels to renewable energy and highlights that renewable energy can achieve these objectives if the Turkish government provides stronger incentives and takes a more active role in advancing the sector. To this end, as compromised in the studies by Simsek and Simsek (2013), Kaygusuz (2002), Kaygusuz and Sari (2003), Baris and Kucukali (2012), Toklu (2013), Bulut and Muratoglu (2018), and Ozcan (2018), the following policy recommendations can be suggested:

(i) Energy policies should aim to attract both private and foreign investment to fund the renewable energy sector. Government funds should prioritize infrastructure development, as strong infrastructure reduces production costs, ensures availability, and attracts more private investment in renewables.
(ii) Since renewable energy production is still evolving, investing in efficient, cost-effective technologies is crucial. Turkish government should increase research and development funding, especially for advancements in energy storage.
(iii) While hydropower is Türkiye's primary renewable source, the government should diversify and fully exploit other renewable energy options.
(iv) Carbon taxes on fossil fuels should be introduced to create a fund that supports renewable energy growth, in line with the double dividend hypothesis in Türkiye (Fullerton and Metcalf, 1997; Akkaya and Bakkal, 2005).
(v) The use of fossil fuels in renewable energy production should be minimized as much as possible.

(vi) Implementing subsidies and incentives for the industrial, transport, and power sectors to adopt renewable energy and energy-efficient practices is crucial for reducing CO_2 emissions. By raising awareness through these actions, society will become more engaged in the sustainability movement. Ultimately, this policy approach will not only reduce emissions but also empower resilient communities, setting the stage for a sustainable, low-carbon future.

Taking into account the economic, political, and environmental factors, producing renewable energy locally to fulfill a significant portion of the nation's energy needs appears to be the most logical and ultimately necessary solution for Türkiye. As a matter of fact, there is growing recognition at the government level regarding Türkiye's renewable energy potential, the importance of generating renewable energy, and the necessity to boost the proportion of renewables in the nation's overall energy production and consumption. Efforts have been made to support the renewable energy sector and align legislative frameworks with these goals. However, a more urgent, all-encompassing, and robust strategy is required, along with a significant shift, to reach the country's renewable energy targets and address the economic and environmental challenges linked to fossil fuel usage.

REFERENCES

Abosedra, S., & Baghestani, H. (1989). New evidence on the causal relationship between United States energy consumption and gross national product. *The Journal of Energy and Development*, •••, 285–292.

Acaravci, A. (2010). Structural breaks, electricity consumption and economic growth: Evidence from Turkey. *Journal for Economic Forecasting*, *2*, 140–154.

Adamantiades, A., & Kessides, I. (2009). Nuclear power for sustainable development: Current status and future prospects. *Energy Policy*, *37*(12), 5149–5166. DOI: 10.1016/j.enpol.2009.07.052

Akhmat, G., & Zaman, K. (2013). Nuclear energy consumption, commercial energy consumption and economic growth in South Asia: Bootstrap panel causality test. *Renewable & Sustainable Energy Reviews*, *25*, 552–559. DOI: 10.1016/j.rser.2013.05.019

Akkaya, Ş., & Bakkal, U. (2005). Çevre Vergileri ve Çifte Yarar. *İstanbul Üniversitesi İktisat Fakültesi Mecmuası, 55*(2), 1–22.

Al-Mulali, U., Saboori, B., & Ozturk, I. (2015). Investigating the environmental Kuznets curve hypothesis in Vietnam. *Energy Policy*, *76*, 123–131. DOI: 10.1016/j.enpol.2014.11.019

Alabi, O., Ackah, I., & Lartey, A. (2017). Re-visiting the renewable energy–economic growth nexus: Empirical evidence from African OPEC countries. *International Journal of Energy Sector Management*, *11*(3), 387–403. DOI: 10.1108/IJESM-07-2016-0002

Alper, A., & Oguz, O. (2016). The role of renewable energy consumption in economic growth: Evidence from asymmetric causality. *Renewable & Sustainable Energy Reviews*, *60*, 953–959. DOI: 10.1016/j.rser.2016.01.123

Apergis, N., Payne, J., Menyah, K., & Wolde-Rufael, Y. (2010). On the causal dynamics between emissions, nuclear energy, renewable energy, and economic growth. *Ecological Economics*, *69*(11), 2255–2260. DOI: 10.1016/j.ecolecon.2010.06.014

Banday, U., & Aneja, R. (2020). Renewable and non-renewable energy consumption, economic growth and carbon emission in BRICS: Evidence from bootstrap panel causality. *International Journal of Energy Sector Management*, *14*(1), 248–260. DOI: 10.1108/IJESM-02-2019-0007

Baris, K., & Kucukali, S. (2012). Availibility of renewable energy sources in Turkey: Current situation, potential, government policies and the EU perspective. *Energy Policy*, *42*, 377–391. DOI: 10.1016/j.enpol.2011.12.002

Bekun, F., Emir, F., & Sarkodie, S. (2019). Another look at the relationship between energy consumption, carbon dioxide emissions, and economic growth in South Africa. *The Science of the Total Environment*, *655*, 759–765. DOI: 10.1016/j.scitotenv.2018.11.271 PMID: 30476856

Bento, J., & Moutinho, V. (2016). CO2 emissions, non-renewable and renewable electricity production, economic growth, and international trade in Italy. *Renewable & Sustainable Energy Reviews*, *55*, 142–155. DOI: 10.1016/j.rser.2015.10.151

Bhattacharya, M., Paramati, S., Ozturk, I., & Bhattacharya, S. (2016). The effect of renewable energy consumption on economic growth: Evidence from top 38 countries. *Applied Energy*, *162*, 733–741. DOI: 10.1016/j.apenergy.2015.10.104

Bilgen, S., Keles, S., Kaygusuz, A., Sari, A., & Kaygusuz, K. (2008). Global warming and renewable energy sources for sustainable development: A case study in Turkey. *Renewable & Sustainable Energy Reviews*, *12*(2), 372–396. DOI: 10.1016/j.rser.2006.07.016

Bilgili, F., Kocak, E., Bulut, U., & Kuloglu, A. (2017b). The impact of urbanization on energy intensity: Panel data evidence considering cross-sectional dependence and heterogeneity. *Energy*, *133*, 242–256. DOI: 10.1016/j.energy.2017.05.121

Bilgili, F., Koçak, E., Bulut, Ü., & Kuşkaya, S. (2017a). Can biomass energy be an efficient policy tool for sustainable development? *Renewable & Sustainable Energy Reviews*, *71*, 830–845. DOI: 10.1016/j.rser.2016.12.109

Bloch, H., Rafiq, S., & Salim, R. (2015). Economic growth with coal, oil and renewable energy consumption in China: Prospects for fuel substitution. *Economic Modelling*, *44*, 104–115. DOI: 10.1016/j.econmod.2014.09.017

Bölük, G., & Mert, M. (2014). Fossil & renewable energy consumption, GHGs (greenhouse gases) and economic growth: Evidence from a panel of EU (European Union) countries. *Energy*, *74*, 439–446. DOI: 10.1016/j.energy.2014.07.008

Bulut, U. (2017). The impacts of non-renewable and renewable energy on CO2 emissions in Turkey. *Environmental Science and Pollution Research International*, *24*(18), 15416–15426. DOI: 10.1007/s11356-017-9175-2 PMID: 28508336

Bulut, U., & Muratoglu, G. (2018). Renewable energy in Turkey: Great potential, low but increasing utilization, and an empirical analysis on renewable energy-growth nexus. *Energy Policy*, *123*, 240–250. DOI: 10.1016/j.enpol.2018.08.057

Cai, Y., Sam, C., & Chang, T. (2018). Nexus between clean energy consumption, economic growth and CO2 emissions. *Journal of Cleaner Production, 182*, 1001–1011. DOI: 10.1016/j.jclepro.2018.02.035

Can, H., & Korkmaz, O. (2019). The relationship between renewable energy consumption and economic growth: the case of Bulgaria. *International Journal of Energy Sector Management*, 1-18.

Capik, M., Yılmaz, A., & Cavusoglu, I. (2012). Present situation and potential role of renewable energy in Turkey. *Renewable Energy, 46*, 1–13. DOI: 10.1016/j.renene.2012.02.031

Chang, T., Gupta, R., Simokengne, B., Smithers, D., & Trembling, A. (2013). The causal relationship between renewable energy consumption and economic growth: Evidence from the G7 countries. *Applied Economics Letters, 23*(1), 1–9.

Cheng, B., & Lai, T. (1997). An investigation of co-integration and causality between energy consumption and economic activity in Taiwan. *Energy Economics, 19*(4), 435–444. DOI: 10.1016/S0140-9883(97)01023-2

Cheung, Y.-W., & Chinn, M. (1997). Further Investigation of the Uncertain Unit Root in GNP. *Journal of Business & Economic Statistics, 15*(1), 68–73. DOI: 10.1080/07350015.1997.10524688

Chien, T., & Hu, J. (2007). Renewable energy and macroeconomic efficiency of OECD and non-OECD economies. *Energy Policy, 35*(7), 3606–3615. DOI: 10.1016/j.enpol.2006.12.033

Cleveland, C., Costanza, R., Hall, C., & Kaufmann, R. (1997). Energy and the US economy: A biophysical perspective. *International Library of Critical Writings in Economics, 75*, 295–302.

Destek, M., & Aslan, A. (2017). Renewable and non-renewable energy consumption and economic growth in emerging economies: Evidence from bootstrap panel causality. *Renewable Energy, 111*, 757–763. DOI: 10.1016/j.renene.2017.05.008

Directive, E. U. (2023). Revised Directive EU/2023/2413. *Official Journal of the European Union*, 1-77.

Dogan, E. (2015). The relationship between economic growth and electricity consumption from renewable and non-renewable sources: A study of Turkey. *Renewable & Sustainable Energy Reviews, 52*, 534–546. DOI: 10.1016/j.rser.2015.07.130

Dogan, E. (2016). Analyzing the linkage between renewable and non-renewable energy consumption and economic growth by considering structural break in time-series data. *Renewable Energy, 99*, 1126–1136. DOI: 10.1016/j.renene.2016.07.078

Eden, S., & Jin, J. (1992). Cointegration tests of energy consumption, income, and employment. *Resources and Energy, 14*(3), 259–266. DOI: 10.1016/0165-0572(92)90010-E

Ediger, V., & Kentel, E. (1999). Renewable energy potential as an alternative to fossil fuels in Turkey. *Energy Conversion and Management, 40*(7), 743–755. DOI: 10.1016/S0196-8904(98)00122-8

Energy Institute database. (2024). 2024 Statistical Review of World Energy. <https://www.energyinst.org/statistical-review> (last visited on 19.03.2025).

Fang, Y. (2011). Economic welfare impacts from renewable energy consumption: The China experience. *Renewable & Sustainable Energy Reviews, 15*(9), 5120–5128. DOI: 10.1016/j.rser.2011.07.044

Fullerton, D., & Metcalf, G. (1997). *Environmental taxes and the double-dividend hypothesis: did you really expect something for nothing?* Cambridge: NBER Working Paper No 6199.

Gaupp, D. (2008). Turkey's new law on renewable energy sources within the context of the accession negotiations with the EU. *German Law Journal, 08*(04), 413–416. DOI: 10.1017/S2071832200005666

Glynn, J., Perera, N., & Verma, R. (2007). Unit root tests and structural breaks: A survey with applications. *Journal of Quantitative Methods for Economics and Business Administration, 3*(1), 63–79.

Gnansounou, E. (2008). Assessing the energy vulnerability: Case of industrial countries. *Energy Policy, 36*(10), 3734–3744. DOI: 10.1016/j.enpol.2008.07.004

Gokmenoglu, K., & Kaakeh, M. (2018). Causal relationship between nuclear energy consumption and economic growth: Case of Spain. *Strategic Planning for Energy and the Environment, 37*(3), 58–76. DOI: 10.1080/10485236.2018.11958660

Gozgor, G., Lau, C., & Lu, Z. (2018). Energy consumption and economic growth: New evidence from the OECD countries. *Energy, 153*, 27–34. DOI: 10.1016/j.energy.2018.03.158

Granger, C. (1969). Investigating causal relations by econometric models and cross-spectral methods. *Econometrica, 37*(3), 424–438. DOI: 10.2307/1912791

Granger, C., & Newbold, P. (1974). Spurious regressions in econometrics. *Journal of Econometrics*, *2*(2), 111–120. DOI: 10.1016/0304-4076(74)90034-7

Gyimah, J., Yao, X., Tachega, M., Hayford, I., & Opoku-Mensah, E. (2022). Renewable energy consumption and economic growth: New evidence from Ghana. *Energy*, *248*, 1–11. DOI: 10.1016/j.energy.2022.123559

Hacker, R., & Hatemi-J, A. (2006). Tests for causality between integrated variables using asymptotic and bootstrap distributions: Theory and application. *Applied Economics*, *38*(13), 1489–1500. DOI: 10.1080/00036840500405763

Harvey, D., & Mills, T. (2002). Unit roots and double smooth transitions. *Journal of Applied Statistics*, *29*(5), 675–683. DOI: 10.1080/02664760120098739

Haseeb, M., Abidin, I., Hye, Q., & Hartani, N. (2018). The impact of renewable energy on economic well-being of Malaysia: Fresh evidence from auto regressive distributed lag bound testing approach. *International Journal of Energy Economics and Policy*, *9*(1), 269–275.

Hedenus, F., Azar, C., & Johansson, D. (2010). Energy security policies in EU-25: The expected cost of oil supply disruptions. *Energy Policy*, *38*(3), 1241–1250. DOI: 10.1016/j.enpol.2009.01.030

Hepsag, A. (2021). A unit root test based on smooth transitions and nonlinear adjustment. *Communications in Statistics. Simulation and Computation*, *50*(3), 625–632. DOI: 10.1080/03610918.2018.1563154

Inglesi-Lotz, R. (2016). The impact of renewable energy consumption to economic growth: A panel data application. *Energy Economics*, *53*, 58–63. DOI: 10.1016/j.eneco.2015.01.003

Jaforullah, M., & King, A. (2015). Does the use of renewable energy sources mitigate CO2 emissions? A reassessment of the US evidence. *Energy Economics*, *49*, 711–717. DOI: 10.1016/j.eneco.2015.04.006

Kapetanios, G., Shin, Y., & Snell, A. (2003). Testing for a unit root in the nonlinear STAR framework. *Journal of Econometrics*, *112*(2), 359–379. DOI: 10.1016/S0304-4076(02)00202-6

Kapetanios, G., Shin, Y., & Snell, A. (2006). Testing for cointegration in nonlinear smooth transition error correction models. *Econometric Theory*, *22*(2), 279–303. DOI: 10.1017/S0266466606060129

Kaygusuz, K. (2002). Environmental impacts of energy utilisation and renewable energy policies in Turkey. *Energy Policy*, *30*(8), 689–698. DOI: 10.1016/S0301-4215(02)00032-0

Kaygusuz, K., & Sari, A. (2003). Renewable energy potential and utilization in Turkey. *Energy Conversion and Management*, *44*(3), 459–478. DOI: 10.1016/S0196-8904(02)00061-4

Koçak, E., & Şarkgüneşi, A. (2017). The renewable energy and economic growth nexus in Black Sea and Balkan countries. *Energy Policy*, *100*, 51–57. DOI: 10.1016/j.enpol.2016.10.007

Kraft, J., & Kraft, A. (1978). On the Relationship Between Energy and GNP. *The Journal of Energy and Development*, *3*(2), 401–403.

Kruse, R. (2011). A new unit root test against ESTAR based on a class of modified statistics. *Statistische Hefte*, *52*(1), 71–85.

Lau, L., Lee, K., & Mohamed, A. (2012). Global warming mitigation and renewable energy policy development from the Kyoto Protocol to the Copenhagen Accord-A comment. *Renewable & Sustainable Energy Reviews*, *16*(7), 5280–5284. DOI: 10.1016/j.rser.2012.04.006

Lee, S., & Jung, Y. (2018). Causal dynamics between renewable energy consumption and economic growth in South Korea: Empirical analysis and policy implications. *Energy & Environment*, *29*(7), 1298–1315. DOI: 10.1177/0958305X18776546

Leybourne, S., Newbold, P., & Vougas, D. (1998). Unit roots and smooth transitions. *Journal of Time Series Analysis*, *19*(1), 83–97. DOI: 10.1111/1467-9892.00078

Liu, Y., & Hao, Y. (2018). The dynamic links between CO2 emissions, energy consumption and economic development in the countries along "the Belt and Road". *The Science of the Total Environment*, *645*, 674–683. DOI: 10.1016/j.scitotenv.2018.07.062 PMID: 30031325

Luqman, M., Ahmad, N., & Bakhsh, K. (2019). Nuclear energy, renewable energy and economic growth in Pakistan: Evidence from non-linear autoregressive distributed lag model. *Renewable Energy*, *139*, 1299–1309. DOI: 10.1016/j.renene.2019.03.008

Mbarek, M., Saidi, K., & Amamri, M. (2017). The relationship between pollutant emissions, renewable energy, nuclear energy and GDP: Empirical evidence from 18 developed and developing countries. *International Journal of Sustainable Energy*, *37*(6), 597–615. DOI: 10.1080/14786451.2017.1332060

Menyah, K., & Wolde-Rufael, Y. (2010). CO2 emissions, nuclear energy, renewable energy and economic growth in the US. *Energy Policy*, *38*(6), 2911–2915. DOI: 10.1016/j.enpol.2010.01.024

Mezghani, I., & Haddad, H. (2017). Energy consumption and economic growth: An empirical study of the electricity consumption in Saudi Arabia. *Renewable & Sustainable Energy Reviews*, *75*, 145–156. DOI: 10.1016/j.rser.2016.10.058

Naser, H. (2015). Can Nuclear Energy Stimulates Economic Growth? Evidence from Highly Industrialised Countries. *International Journal of Energy Economics and Policy*, *5*(1), 164–173.

Nazlioglu, S., Lebe, F., & Kayhan, S. (2011). Nuclear energy consumption and economic growth in OECD countries: Cross-sectionally dependent heterogeneous panel causality analysis. *Energy Policy*, *39*(10), 6615–6621. DOI: 10.1016/j.enpol.2011.08.007

Nejat, P., Jomehzadeh, F., Taheri, M., Gohari, M., & Majid, M. (2015). A global review of energy consumption, CO2 emissions and policy in the residential sector (with an overview of the top ten CO2 emitting countries). *Renewable & Sustainable Energy Reviews*, *43*, 843–862. DOI: 10.1016/j.rser.2014.11.066

Ocak, M., Ocak, Z., Bilgen, S., Keles, S., & Kaygusuz, K. (2004). Energy utilization, environmental pollution and renewable energy sources in Turkey. *Energy Conversion and Management*, *45*(6), 845–864. DOI: 10.1016/S0196-8904(03)00192-4

Ocal, O., & Aslan, A. (2013). Renewable energy consumption–economic growth nexus in Turkey. *Renewable & Sustainable Energy Reviews*, *28*, 494–499. DOI: 10.1016/j.rser.2013.08.036

Omri, A., Mabrouk, N., & Sassi-Tmar, A. (2015). Modeling the causal linkages between nuclear energy, renewable energy and economic growth in developed and developing countries. *Renewable & Sustainable Energy Reviews*, *42*, 1012–1022. DOI: 10.1016/j.rser.2014.10.046

Ozcan, B., & Ari, A. (2015). Nuclear Energy Consumption-Economic Growth Nexus in OECD: A Bootstrap Causality Test. *Procedia Economics and Finance*, *30*, 586–597. DOI: 10.1016/S2212-5671(15)01271-X

Ozcan, B., & Ozturk, I. (2019). Renewable energy consumption-economic growth nexus in emerging countries: A bootstrap panel causality test. *Renewable & Sustainable Energy Reviews*, *104*, 30–37. DOI: 10.1016/j.rser.2019.01.020

Ozcan, M. (2018). The role of renewables in increasing Turkey's self-sufficiency in electrical energy. *Renewable & Sustainable Energy Reviews, 82*, 2629–2639. DOI: 10.1016/j.rser.2017.09.111

Ozgur, M. (2008). Review of Turkey's renewable energy potential. *Renewable Energy*, *33*(11), 2345–2356. DOI: 10.1016/j.renene.2008.02.003

Özkivrak, Ö. (2005). Electricity restructuring in Turkey. *Energy Policy*, *33*(10), 1339–1350. DOI: 10.1016/j.enpol.2003.12.010

Ozturk, I. (2010). A literature survey on energy–growth nexus. *Energy Policy*, *38*(1), 340–349. DOI: 10.1016/j.enpol.2009.09.024

Ozturk, I., & Acaravci, A. (2010). CO2 emissions, energy consumption and economic growth in Turkey. *Renewable & Sustainable Energy Reviews*, *14*(9), 3220–3225. DOI: 10.1016/j.rser.2010.07.005

Palm, F., Smeekes, S., & Urbain, J. (2006). *Bootstrap unit root tests: comparison and extensions*. Maastricht: METEOR, Maastricht University School of Business and Economics. METEOR Research Memorandum No. 015.

Pao, H., & Fu, H. (2013). Renewable energy, non-renewable energy and economic growth in Brazil. *Renewable & Sustainable Energy Reviews*, *25*, 381–392. DOI: 10.1016/j.rser.2013.05.004

Payne, J. (2009). On the dynamics of energy consumption and output in the US. *Applied Energy*, *86*(4), 575–577. DOI: 10.1016/j.apenergy.2008.07.003

Payne, J. (2011). On biomass energy consumption and real Output in the U.S. *Energy Sources. Part B, Economics, Planning, and Policy*, *6*(1), 47–52. DOI: 10.1080/15567240903160906

Pegkas, P. (2020). The impact of renewable and non-renewable energy consumption on economic growth: The case of Greece. *International Journal of Sustainable Energy*, *39*(4), 380–395. DOI: 10.1080/14786451.2019.1700261

Perron, P. (1989). The great crash, the oil price shock, and the unit root hypothesis. *Econometrica*, *57*(6), 1361–1401. DOI: 10.2307/1913712

Perron, P. (1997). Further Evidence on Breaking Trend Functions in Macroeconomic Variables. *Journal of Econometrics*, *80*(2), 355–385. DOI: 10.1016/S0304-4076(97)00049-3

Piłatowska, M., Geise, A., & Włodarczyk, A. (2020). The effect of renewable and nuclear energy consumption on decoupling economic growth from CO2 emissions in Spain. *Energies*, *13*(9), 1–18. DOI: 10.3390/en13092124

Rafiq, S., & Salim, R. (2009). Temporal causality between energy consumption and income in six Asian emerging countries. *Applied Economics Quarterly*, *55*(4), 335–350. DOI: 10.3790/aeq.55.4.335

Rafiq, S., & Salim, R. (2011). The linkage between energy consumption and income in six emerging economies of Asia: An empirical analysis. *International Journal of Emerging Markets*, *6*(1), 50–73. DOI: 10.1108/17468801111104377

Saatci, M., & Dumrul, Y. (2012). The relationship between energy consumption and economic growth: Evidence from a structural break analysis for Turkey. *International Journal of Energy Economics and Policy*, *3*(1), 20–29.

Sadorsky, P. (2009). Renewable energy consumption, CO2 emissions and oil prices in the G7 countries. *Energy Economics*, *31*(3), 456–462. DOI: 10.1016/j.eneco.2008.12.010

Saidi, K., & Mbarek, M. (2016). Nuclear energy, renewable energy, CO2 emissions, and economic growth for nine developed countries: Evidence from panel Granger causality tests. *Progress in Nuclear Energy*, *88*, 364–374. DOI: 10.1016/j.pnucene.2016.01.018

Salim, R., & Rafiq, S. (2012). Why do some emerging economies proactively accelerate the adoption of renewable energy? *Energy Economics*, *34*(4), 1051–1057. DOI: 10.1016/j.eneco.2011.08.015

Shahbaz, M., Raghutla, C., Chittedi, K., Jiao, Z., & Vo, X. (2020). The effect of renewable energy consumption on economic growth: Evidence from the renewable energy country attractive index. *Energy*, *207*, 1–14. DOI: 10.1016/j.energy.2020.118162

Silvapulle, P. (1996). Testing for a unit root in a time series with mean shifts. *Applied Economics Letters*, *3*(10), 629–635. DOI: 10.1080/135048596355826

Sims, C., Stock, J., & Watson, M. (1990). Inference in linear time series models with some unit roots. *Econometrica*, *58*(1), 113–144. DOI: 10.2307/2938337

Simsek, H., & Simsek, N. (2013). Recent incentives for renewable energy in Turkey. *Energy Policy*, *63*, 521–530. DOI: 10.1016/j.enpol.2013.08.036

Soytas, U., & Sari, R. (2003). Energy consumption and GDP: Causality relationship in G-7 countries and emerging markets. *Energy Economics*, *25*(1), 33–37. DOI: 10.1016/S0140-9883(02)00009-9

Squalli, J. (2007). Electricity consumption and economic growth: Bounds and causality analyses of OPEC members. *Energy Economics*, *29*(6), 1192–1205. DOI: 10.1016/j.eneco.2006.10.001

Stern, D. (1993). Energy and economic growth in the USA: A multivariate approach. *Energy Economics*, *15*(2), 137–150. DOI: 10.1016/0140-9883(93)90033-N

Sulaiman, J., Azman, A., & Saboori, B. (2013). The potential of renewable energy: Using the environmental Kuznets curve model. *American Journal of Environmental Sciences*, *9*(2), 103–112. DOI: 10.3844/ajessp.2013.103.112

TEIAS. (2024). Turkish electrical energy 10-year generation capacity projection. 1-57. <https://www.teias.gov.tr/ilgili-raporlar> (last visited on 15.03.2025).

Tiwari, A. (2011). A structural VAR analysis of renewable energy consumption, real GDP and CO2 emissions: Evidence from India. *Economic Bulletin*, *31*(2), 1793–1806.

Toda, H., & Yamamoto, T. (1995). Statistical inference in vector autoregressions with possibly integrated processes. *Journal of Econometrics*, *66*(1-2), 225–250. DOI: 10.1016/0304-4076(94)01616-8

Toklu, E. (2013). Overview of potential and utilization of renewable energy sources in Turkey. *Renewable Energy*, *50*, 456–463. DOI: 10.1016/j.renene.2012.06.035

Toth, F., & Rogner, H. (2006). Oil and nuclear power: Past, present, and future. *Energy Economics*, *28*(1), 1–25. DOI: 10.1016/j.eneco.2005.03.004

Tugcu, C., Ozturk, I., & Aslan, A. (2012). Renewable and non-renewable energy consumption and economic growth relationship revisited: Evidence from G7 countries. *Energy Economics*, *34*(6), 1942–1950. DOI: 10.1016/j.eneco.2012.08.021

Turkish Central Bank database. (2025). Electronic Data Delivery System. <https://evds2.tcmb.gov.tr/index.php?> (last visited on 19.03.2025).

Vaillancourt, K., Labriet, M., Loulou, R., & Waaub, J.-P. (2008). The role of nuclear energy in long-term climate scenarios: An analysis with the World-TIMES model. *Energy Policy*, *36*(7), 2296–2307. DOI: 10.1016/j.enpol.2008.01.015

Wang, S., Li, Q., Fang, C., & Zhou, C. (2016). The relationship between economic growth, energy consumption, and CO2 emissions: Empirical evidence from China. *The Science of the Total Environment*, *542*, 360–371. DOI: 10.1016/j.scitotenv.2015.10.027 PMID: 26520261

Wolde-Rufael, Y., & Menyah, K. (2010). Nuclear energy consumption and economic growth in nine developed countries. *Energy Economics*, *32*(3), 550–556. DOI: 10.1016/j.eneco.2010.01.004

World Bank database. (2025). World Development Indicators. <https://databank.worldbank.org/source/world-development-indicators> (last visited on 19.03.2025).

Yazdi, S., & Shakouri, B. (2017). Renewable energy nonrenewable energy consumption, and economic growth. *Energy Sources. Part B, Economics, Planning, and Policy*, *12*(12), 1038–1045. DOI: 10.1080/15567249.2017.1316795

Yildirim, E., Saraç, Ş., & Aslan, A. (2012). Energy consumption and economic growth in the USA: Evidence from renewable energy. *Renewable & Sustainable Energy Reviews*, *16*(9), 6770–6774. DOI: 10.1016/j.rser.2012.09.004

Yuksel, I., & Kaygusuz, K. (2011). Renewable energy sources for clean and sustainable energy policies in Turkey. *Renewable & Sustainable Energy Reviews*, *15*(8), 4132–4144. DOI: 10.1016/j.rser.2011.07.007

Zhe, L., Yüksel, S., Dinçer, H., Mukhtarov, S., & Azizov, M. (2021). The Positive Influences of Renewable Energy Consumption on Financial Development and Economic Growth. *SAGE Open*, *11*(3), 1–10. DOI: 10.1177/21582440211040133

Zivot, E., & Andrews, D. (1992). Further evidence on the great crash, the oil-price shock, and the unit-root hypothesis. *Journal of Business & Economic Statistics*, *10*(3), 25–44. DOI: 10.1080/07350015.1992.10509904

KEY TERMS AND DEFINITIONS

Asymmetric Nonlinear Cointegration: It refers to a situation where two or more time series are linked in the long term, but their relationship is nonlinear and behaves differently during periods of positive or negative changes.

Carbon Dioxide Emission: It is the release of carbon dioxide into the atmosphere from burning fossil fuels, transportation, and industrial processes, and it is a key contributor to climate change by trapping heat in the atmosphere.

Nonlinearity: It refers to situations where the relationship between past and current values is not proportional or constant. This means the effect of past data on future values can change depending on factors like regime shifts, thresholds, or time-varying conditions, making the model more complex than linear ones.

Non-Renewable Energy: It refers to energy derived from sources that cannot be replenished on a human timescale, such as coal, oil, and natural gas.

Renewable Energy: It refers to energy derived from naturally replenishing resources like solar, wind, hydro, and geothermal power, which produce little to no environmental impact and are essential for reducing carbon emissions and combating climate change.

Smooth Structural Break: It refers to a gradual shift in the underlying data generation process over time, where the relationship between variables evolves smoothly, often due to factors like changing economic conditions, policy shifts, or other long-term influences, rather than an abrupt change.

Sustainable Development: It refers to meeting the needs of the present without compromising future generations' ability to meet their own needs, by balancing economic growth, environmental protection, and social equity to ensure long-term well-being for both people and the planet.

ENDNOTES

1 Electricity Market Law (No: 4628).

2 Law on Utilization of Renewable Energy Sources for the Purpose of Generating Electrical Energy (No: 5346).

3 Amendments to the Electricity Market Law (No: 5784).

4 Geothermal Resources and Natural Mineral Waters Law (No: 5686).

5 Energy Efficiency Law (No: 5627).

6 Amended law on Renewable Energy Law (No: 6094).

Chapter 6
Global Regulatory Framework and Guidelines for Biomedical Waste Disposal and Management

Bhupinder Singh
https://orcid.org/0009-0006-4779-2553
Sharda University, India

ABSTRACT

Biomedical Waste is coming from medical institutions and its proper disposal and management is important for public health, environmental protection and occupational safety. International bodies such as the World Health Organization (WHO) and the United Nations Environment Programme (UNEP) and National Regulators have widely provided comprehensive frameworks and guidelines for the standardization of biomedical waste management. The WHO's "Safe Management of Wastes from Healthcare Activities" holds important protocols for segregating, treating and disposing of hazardous medical waste. The Basel Convention sets controls on the trans-boundary movement of hazardous waste to facilitate environmentally sound disposal and disposal methods worldwide. Regulation is further ensured through public awareness, the training of healthcare workers, and stringent monitoring by environmental agencies. In the age of Industry 6.0, international cooperation and strict regulations will ensure that the disposal of biomedical waste complies with environmental and public health requirements.

DOI: 10.4018/979-8-3373-1549-2.ch006

1. INTRODUCTION

Biomedical waste management includes waste minimization, source segregation, safe transport and environmentally friendly disposal methods like autoclaving, incineration and chemical treatment. Regulatory frameworks are incorporating smart innovations like IoT devices for waste tracking, artificial intelligence (AI) for waste sorting and blockchain for transparency (Li et al., 2021). The harmonized strategy across all countries that can mitigate the potential risks associated with such biomedical waste, in order to avoid the transmission of diseases and provide sustainability, is just vital. Improperly managed biomedical waste (BMW) poses a grave threat to human health and the environment (Namdech et al., 2025). Biomedical waste management includes, among other things, infectious, pathological, chemical, pharmaceutical and radioactive waste, and is generated principally by healthcare facilities e.g. hospitals, laboratories, research centres and veterinary clinics. Biomedical waste, if not disposed of properly, poses a serious threat to public health, causing infectious diseases, pollution, and occupational health hazards to the handlers of biomedical waste. The explosion in the amount of healthcare waste produced during global pandemics such as COVID-19 has highlighted the importance of having a robust regulatory system in place that governs the collection, treatment, transportation and disposal of BMW in a safe manner. Additionally, sensible waste management practices therein 9олит solely reduce these riskspring 10ocolinally computing side virtually (Dash et al., 2019).

A well-functioning global regulatory schema of Biomedical waste management plays an important role to ensure global waste disposal practices, mitigate environmental problems, and minimize public health impact. Guidelines for the management of hazardous medical waste have been created by a variety of international organizations, such as the World Health Organization (WHO), United Nations Environment Programme (UNEP), Basel Convention, Stockholm Convention, and Minamata Convention. These are the aspects of waste disposal technologies discussed in these frameworks: waste segregation, eco-friendly treatment technologies and compliance monitoring for reduction of pollution. However, while these are global in nature, a problem arises because national regulatory mechanisms are vastly different when comparing various countries, which results in a lack of uniformity when it comes to the management of waste and waste standards. Industrialized countries, including the United States, Germany, and Japan, have strict regulations and the advanced technologies necessary to dispose of hazardous waste, while developing countries face challenges related to limited infrastructure, funding for waste disposal, and effective implementation of environmentally appropriate policies.

Mismanagement of biomedical solid waste has serious health and environmental impact, marking for air and water pollution and proliferation of antimicrobial-resistant pathogens, along with contamination to natural ecosystems. E.g. Healthcare waste is open dumped or incinerated uncontrolled releasing toxic pollutants like dioxins and furans which are in highly carcinogenic (Zhou et al., 2021). So, improperly disposed medical waste, such as syringes and pharmaceutical products, can be found on the illegal market, creating further public health issues. There needs to be international treaties and cooperation with national governments in order to create a consistent and enforceable way to manage biomedicine waste. Advances, for example AI during the Waste Collection and IoT during waste segregation enable well-organized Ways and Processing with enhanced efficiency and sustainable processes.

This chapter explores the international regulatory landscape surrounding biomedical waste disposal and management and how they can enforce compliance. The chapter delves into the global waste crisis and provides an in-depth analysis of the current challenges in managing waste, the effectiveness of existing regulations, technological innovations related to waste treatment and policy recommendations on how to improve global waste governance. Through cross-regional comparisons of regulatory frameworks, this paper seeks to elucidate implementation lakes, suggest best practices for biomedical waste management, and underscore the need for harmonized policies to effectively address the dangers posed by biomedical waste. The facts clearly call for strengthening global cooperation, investment in sustainable waste management technologies and better enforcement of the regulations as critical measures to ensure a safer and more environmentally sound healthcare waste management system everywhere (Muthu et al., 2020).

1.1 Definition of Biomedical Waste and its Sources: Urgency of Proper Biomedical Waste Disposal Due to Health and Environmental Risks

Biomedical waste is defined as any waste created during the diagnosis, treatment, or immunization of humans and animals, in research activities, or the production or testing of biological materials (Bhattacharyya et al., 2023). This includes infectious materials, human tissues, pathological waste, used syringes and needles, unused medicines, chemical waste, and radioactive substances. Healthcare one of the biggest hazardous waste generators, Biomedical waste mostly produced by hospitals, clinics, laboratories, veterinary facilities, pharmaceutical companies, and research institutions. The healthcare systems across the globe have experienced rapid growth which has led to an exponential increase in the amount of biomedical waste, thus posing serious challenges to ensure its safe and effective handling and disposal.

Inadequate management of biopharmaceutical waste can create serious health and environmental hazards for healthcare personnel, patients, waste treatment employees and the general population. Exposure to infectious waste can cause the spread of deadly diseases like HIV, hepatitis B and C, and tuberculosis. If the sharps are not handled properly, waste handlers and scavengers can be accidentally injured and infected. Additionally, pharmaceutical and chemical waste from the hospital can also leach into water bodies, polluting drinking water sources and harming aquatic organisms. Incineration of biomedical waste without regulation causes the release of toxic air pollutants like dioxins and furans into the atmosphere, resulting in air and environmental pollution that in turn lead to further respiratory ailments. In many low- and middle-income nations, proper waste management infrastructure is not developed, allowing for open dumping and burning which sustains pollutant and public health threats. Given these threats, the importance of strict policies of biomedical waste management cannot be overemphasized. Waste generated during health care processes poses a vital health threat to both the individual and the community (Malik & Tyagi, 2022).

1.2 World Health Organization (WHO) Mandates

The international guidelines and global regulations of the WHO (World Health Organization) were designed to prevent risk at points of generation and to ensure the safe, effective and environmentally safe management of biomedical waste. Biomedical waste comprising infectious waste, sharps, pathological waste, residues of pharmaceuticals, etc. constitutes a potential hazard to human health and the environment, if it is not adequate treatment. WHO's Safe Management of Wastes from Health-care Activities includes best practice and key principle guidance regarding reduction of waste, safety of segregation at the place of generation and the correct treatment technologies. Various advises highlight the need for dividing hazardous from non-hazardous waste by color-coded containers, storing, handling, and transport measures, and the use of treatment procedures such as autoclaving, chemical disinfection, microwaving, incineration where required.

The rules also put an emphasis on cradle-to-grave, meaning tracking and taking care of waste at every stage of its life cycle. Furthermore, occupational safety precautions like the use of personal protective equipment (PPE), education of staff and vaccination e.g., against Hepatitis B have to be observed to keep the personnel on the hospitals and waste transporters safe as well. Environmental protection is actively pursued, with an emphasis on non-burn technologies and waste minimization policies to reduce dioxin PMercury emissions in accordance with international agreements such as the Basel, Stockholm and Minamata Conventions. The WHO also endorses the establishment of national policies and regulations to include management of biomedical waste in the context of health systems strengthening and public health strategies. It is recommended that on-going monitoring, auditing, and consultation with affected parties are promoted to facilitate compliance and improvement over time. Through evidence-based, user-friendly recommendations, WHO assists countries to develop strong biomedical waste management systems that safeguard public health, maintain environmental sustainability, and support achievement of Sustainable Development Goals (SDGs), specifically target SDG 3 (Good Health and Well-being) and SDG 12 (Responsible Consumption and Production).

2. CLASSIFICATION OF BIOMEDICAL WASTE

Broadly biomedical waste can be classified according to its source directly or indirectly, composition and the potential health hazards foreshadowing for the human beings and environment at large. This is an important step to ensure that it is handled, disposed of, and treated correctly in order to reduce the hazard of contamination and health risks. Biomedical waste is classified into 4 or 5 categories according to the World Health Organization (WHO) or national regulatory authorities, and each category demands different handling and disposal methods. Types of Biomedical Waste: Infectious Waste, Pathological Waste, Sharps Waste, Pharmaceutical Waste, Chemical Waste, Radioactive Waste, and General Healthcare Waste Each of these categories has different risks and requires special treatment to avoid environmental and human health impacts (Ahmed et al., 2019).

Biomedical waste classification is a critical step in its safe handling, treatment, and disposal. The identification and appropriate segregation of infectious, pathological, sharps, pharmaceutical, chemical, radioactive and general healthcare wastes reduce health risks, protect the environment and comply with legal and regulatory frameworks. Collaboration between governments and healthcare facilities is essential to addressing biomed waste disposal and segregation, but cooperation from regulatory agencies and waste disposal companies is also necessary. Enhanced segregation of biomedical waste through global best-practice methods will significantly reduce biomedical waste-related risks and protect public health and environmental sustainability.

2.1 Infectious Waste

Infectious waste material that is contaminated with biological agents that are capable of causing disease in humans and animals, including blood, boddily fluids and human or animal tissue. This includes bloodied bandages, gauze, cotton swabs, culture dishes, used gloves and protective equipment such as personal protective equipment (PPE). A lot of infectious waste is produced on daily basis from hospitals, clinics and diagnostic laboratories. Without proper disposal, it can become a significant vehicle for disease transmission, transmitting infectious diseases including hepatitis, HIV, and tuberculosis. Infectious waste is disposed of after treating it through autoclaving, incineration, or chemical disinfection.

2.2 Pathological Waste

Pathological waste can be classified as human tissues or organs, body parts or animal carcasses generated as result of research for clinical procedures. It's a biohazard and has the potential for infectious disease. And they generate a significant amount of pathological waste in veterinary clinics, surgical centers and even forensic laboratories. Such waste must be treated with the utmost caution to avoid exposure and contamination. Common disposal modes for pathological waste are deep burial, incineration, and alkaline hydrolysis.

2.3 Sharps Waste

It comprises needles, syringes, scalpels, glass slides, lancets, broken glass and other sharp objects used in medical procedures. This type is extremely dangerous due to the potential for injury and diseases that can be passed through blood. Healthcare workers, sanitation personnel and waste handlers, are at risk of accidental needlestick injuries leading to HIV, hepatitis B and other infections. Containerized in puncture proof containers, followed by incineration or sterilization (autoclave) as some areas offer such safe needle disposal programs through specialized recycling (Panesar, 2019).

2.4 Pharmaceutical Waste

Pharmaceutical waste refers to medicines, vaccines, antibiotics, and drugs that have expired, been unused, or become contaminated, and are thrown away by hospitals, pharmacies, and research institutions. Such waste can be dangerous to the environment, as they can contaminate water sources, endanger wildlife, and contribute to antibiotic resistance. Incineration, chemical neutralization and encapsulation to keep toxic chemicals from leaking into the environment are among the most common disposal techniques. Some pharmaceutical waste, if not disposed of properly flushing medicines down the drain, for example can seep into water supplies and become a public health threat.

2.5 Chemical Waste

Solid medical waste includes general waste, reusable waste, and some types of waste from laboratories and medical facilities (reagents, disinfectants, solvents, heavy metals (mercury from broken measuring devices such as thermometers or sphygmomanometers), cleaning agents. Many of these chemicals are toxic, corrosive, or flammable, and are harmful to human health and the environment. Mercury-containing waste is of special concern due to its effects as a neurotoxin, which has led to international treaties, such as the Minamata Convention on Mercury, which restricts the use and disposal of mercury-laden medical devices. Some waste, such as chemical waste, needs special treatment, which includes neutralization, chemical precipitation, and secure landfill disposal. For example, some cytotoxic drugs used in chemotherapy are decimated only through high-temperature incineration.

2.6 Radioactive Waste

The source of radioactive waste is nuclear medicine, radiotherapy and medical imaging units that operate using radioactive isotopes for diagnosis and treatment. This includes dirty materials including gloves, syringes, vials, and unused radioactive substances (Sengan et al., 2020). There are serious health effects from exposure to radioactive waste including radiation poisoning, genetic mutations and cancer. Radioactive waste disposal guidelines which are determined by international authorities such as IAEA recommend, storing the waste on site in shielded containers, storage waiting until the radioactivity has decayed, and controlled burial in radioactive waste disposal sites.

2.7 General Healthcare Waste

General healthcare waste is made up of non-hazardous waste from medical facilities, including paper, food waste, packaging and other non-contaminated items. Although this category is not a direct health hazard, it still requires appropriate segregation and disposal to avoid mixing with hazardous waste. General health care waste may be recycled, composted, or incinerated, or sent to municipal landfill sites depending on the waste management protocols in the facility. But if biomedical waste is not segregated properly, it gets mixed up with common waste leading to potential contamination of the environment.

3. CORRECTLY CLASSIFYING AND DISPOSING OF MATERIALS

Biomedical Waste Classification Biomedical waste classification is important for the effective management of waste as well as adherence to regulations to maintain human and environmental health. Even though there is potential hazard from infectious organisms, toxic agents, and environmental degradation through improper waste segregation. For instance, if hazardous waste was mixed with general waste, it could be disposed of it the wrong way, causing water, soil and air pollution. Moreover, wrong type of waste treatment causes additional pressure to the health system and waste companies due to wrong classification (Rodrigues et al., 2021).

The poor awareness, lack of infrastructure and feeble regulatory compliance are further contributors to poor biomedical waste management especially in developing countries. Open dumping, uncontrolled burning and improper incineration of waste can release toxic pollutants including dioxins and furans, both known carcinogens. To tackle the problem, stringent regulations for the segregation, treatment and disposal of biomedical waste must be implemented at the level of international organizations and national governments. Healthcare institutions must establish all-encompassing waste management plans, which should include training of healthcare workers, acquisition of environmentally friendly disposal technologies, and engagement of artificial intelligence (AI) and Internet of Things (IoT) solutions to provide real-time biomedical waste monitoring.

4. BIOMEDICAL WASTE GENERATION AND ASSOCIATED RISKS

The generation of biomedical waste is a biomedical waste consequence of medical and health related activities, the volume and complexity of which are increasing at a significant rate as a result of advances in modern medicine, the growing population and emerging public health emergencies. Medical waste is generated from hospitals, clinics, laboratories, research facilities, veterinary services, pharmaceutical industries and home healthcare, including but not limited to infectious waste, sharps, pharmaceutical, and radioactive waste. Biomedical waste is the continuously increasing category of waste due to an escalation in healthcare services subsequent to medical innovations and population dynamics, and it poses serious health and environmental threats if not handled properly. The growing number of chronic illnesses and the escalating reliance on medical treatments such as dialysis, chemotherapeutic agents, and organ transplantation result in an unbroken rise of dangerous biomedical waste. The excessive mobility and disposal of biomedical waste has dangerous consequences, where improper segregation and disposal of waste results in the exposure of healthcare workers, waste handlers, and the general population to infectious diseases, chemical poisoning, and causes environmental degradation (Pramanik et al., 2022).

Biomedical waste includes an extensive range of wastes, such as sharps, infectious, chemical, pharmaceutical, and genotoxic waste, among others, and each presents a unique risk to health or the environment. Infectious waste, which includes used bandages, gloves, and culture dishes, contains pathogens that can cause illnesses like tuberculosis, hepatitis B and C, and HIV. Human tissues, organs and body fluids are collectively termed as pathological waste and require special handling to avoid any infections. Sharps waste includes needles, scalpels, or broken glass that might cause a significant risk of injury and transmission of bloodborne diseases if not disposed of in puncture-resistant containers.

Improper disposal of pharmaceutical waste expired medications, antibiotics, or vaccines can also cause antimicrobial resistance by polluting water sources and speeding up the evolution of drug-resistant bacteria. The chemical and radioactive waste resulting from diagnostic laboratories, the destruction of biological material through radiation therapy, and the use of disinfectants; contribute to long-term toxicological effects that affect human health and ecosystems, while exposure to heavy metals such as mercury and lead to entering the food chain which affects neurological pathways and causes organ failure. Improper management of such heterogeneous waste streams exacerbates public health risks, positioning appropriate biomedical waste management as an urgent global imperative (Sengan et al., 2020).

The health hazards posed by biomedical waste are grave, it is not only medical professionals but also waste handlers and the larger societies that are being affected. In cases of poor sanitation and medical waste disposal infrastructure, infectious waste gives rise to cross-contamination, the increases the spread of contagious diseases. Despite the fact that they are a common hazard in occupational health, we will focus on needle-stick injuries in the healthcare setting. Thousands of cases of viral transmission, such as hepatitis and HIV, occur each year because of misuse of sharps disposal. Waste handlers not equipped with adequate protective tools or training face greater risk of direct exposure to biohazards and suffer from higher rates of infections, respiratory diseases and chemical poisoning. The unregulated dumping of antibiotics and pharmaceutical residues in water systems lead to antimicrobial resistance, which is an emerging global health threat. The prevalence of pathogenic organisms resistant to drugs exacerbates infection treatment, rendering current medical treatments less efficacious and thereby escalating healthcare costs.

In addition to human health issues, the improper management of biomedical waste leads to devastating environmental impacts. In several parts of the globe, biomedical waste is burnt in an uncontrolled manner or poorly incinerated, releasing toxic pollutants such as dioxins, furans, and heavy metals to the atmosphere. These contaminants cause respiratory illnesses, cancer risks and climate change, compromising air quality and the growing environmental devastation. Burning plants—near or even if there are nobody around particularly in poor countries lack emission control methods and thus releasing toxins to the environment. Biomedical waste that gets dumped in open landfills or water bodies pollutes the ground waters, rivers and oceans leading to the death of marine animals, jeopardising biodiversity (Elbadawi et al., 2021).

Dumping chemicals and radioactive waste creates ruptures in ecosystems; soil loses productivity, water sources are poisoned and hazardous compounds accumulate in the food chain. The increase in disposable plastics in healthcare, from IV bags to syringes to face masks, also contributes to the growing global plastic pollution crisis, with non-degradable medical waste finding its way into landfills and bodies of water. In the absence of robust regulations and sustainable biomedical waste management practices, the environmental impact of biomedical waste will compound over time, aggravating ecological and public health concerns.

Strategies to mitigate biomedical waste risks must include stringent regulations, technological innovations and widespread awareness. The management of biomedical waste is governed by guidelines set forth by the World Health Organization (WHO) and various international treaties like the Basel Convention and the Stockholm Convention, which detail best practices for the segregation, disposal and environmental safety of biomedical waste. The proper classification of waste and the division of specific waste in separate repositories are achieved through colour-coded systems, which lsit waste as non-infectious or infectious, thereby ensuring non-contamination, and allowing for prevention in infectious waste by ensuring its safe disposal. Energetic waste treatment technologies, including but not limited to autoclaving, microwave treatment, and plasma pyrolysis, can be more environmentally friendly, eliminating toxic emissions and lessening the environmental footprint of biomedical waste disposal compared to incineration (Cabello et al., 2020). So, integrating AI, blockchain, and IoT can facilitate real-time waste tracking, compliance monitoring, and data-informed decision-making in biomedical waste management.

Public education and training programs are also fundamental to reducing biomedical waste hazards. The healthcare workers and waste handlers should be trained properly on safe disposal of wastes such as, sharps, infectious and chemical waste. The importance of awareness campaigns for the public on the hazards of improper biomedical waste disposal is to be emphasized, especially in light of well-established home care where auto-medicaments generate increasing volumes of hazardous waste. These efforts involve communication, policy-making, and cross-sector collaboration with healthcare providers, regulatory agencies, and environmental organizations to create a culture of responsible biomedical waste management.

The generation of biomedical waste is an inevitable outcome of modern healthcare systems, with its volume on the rise due to medical advancements, global health crises and booming populations. Disease transmission, occupational hazards, environmental contamination, antimicrobial resistance, etc. The improper management of biomedical waste has serious implications on human health, the environment, and sustainable development. Biomedical waste can be effectively managed by waste segregation, safe disposal technologies, strict regulatory compliance, and heightened public awareness campaigns. With sustainable waste management practices in place and technological innovations plugged into the system, global healthcare systems can achieve a fine balance between managing the risks posed by biomedical waste and preserving public health and the ecosystem for the coming generations.

Figure 1. Specifies the points on biomedical waste generation and associated risks (Source- Self Prepared- Original)

5. INTERNATIONAL REGULATIONS AND GUIDELINES FOR BIOMEDICAL WASTE MANAGEMENT

International Regulations and Guidelines on Biomedical Waste Management Biomedical waste is a major concern in the medical field, as it is produced from healthcare activities, laboratories, research facilities, and pharmaceutical industries, and includes infectious substances, toxic chemicals, radioactive elements and sharps (Tiwari et al., 2022). However due to this waste's mismanaged disposal, it poses a major threat to human health and ecosystems, requiring strict international regulations on its safe handling, treatment and disposal. Throughout the years, organizations like the World Health Organization (WHO), the United Nations Environment Programme (UNEP) and many domestic regulatory authorities have created frameworks to help promote consistency in the management of biomedical waste from country to country. It governs all aspects of waste management including classification, segregation, treatment, and disposal methods to meet health, environmental and occupational safety requirements.

5.1 World Health Organization (WHO) Guidelines

WHO has extensive protocols on healthcare waste management in both disposal for safety purposes and for environment sustainability. Medical waste can be divided by the World Health Organization (WHO) into types: infectious, pathological, pharmaceutical, chemical, radioactive, and sharps waste and then provide recommendations for their segregation. According to the WHO, it should be made sure that the waste is segregated into different bins which are color coded, which ensures a combination of both types of waste so that special attention and handling of hazardous materials (Devika & Karegowda, 2021). The organization is promoting environmentally friendly treatment technologies, such as autoclaving, microwave disinfection and chemical disinfection, to cut down on the use of incineration, which contributes to air pollution and the release of toxic dioxins and furans. The WHO's Safe Management of Wastes from Health-Care Activities is the global reference for best practice, offering guidance for health care facilities, policymakers, and regulatory authorities on the correct management of waste within the health sector.

5.2 Basel Convention on Hazardous Waste

The international treaty related to the management and disposal of hazardous biomedical waste is the Basel Convention on the Control of Transboundary Movements of Hazardous Wastes and Their Disposal. The Basel Convention, established in 1989, seeks to avoid the dumping of toxic waste including biomedical waste in the poorer countries where it serves the economic interest of companies and individuals taking advantage of less stringent disposal regulations. The treaty requires the country of origin to obtain prior informed consent before exporting hazardous biomedical waste to another country to ensure that waste disposal does not threaten human health or the environment. To ensure safe management of biomedical waste, the Basel Convention also promotes waste minimization, recycling and environmentally sound disposal of wastes. A number of countries have incorporated the principles of the Basel Convention into their national biomedical waste regulations.

5.3 Stockholm Convention on Persistent Organic Pollutants (POPs)

While much of the framework for regulating biomedical waste disposal comes from the Basel Convention on the Control of Transboundary Movements of Hazardous Wastes and Their Disposal (Basel Convention), the Stockholm Convention on Persistent Organic Pollutants (POPs) is also an important factor, particularly relating to byproducts of incineration of biomedical wastes. Adopted in 2001, the convention is designed to eliminate or limit the production and use of persistent organic pollutants, including dioxins and furans, which are emitted when biomedical waste is incinerated (Sukeshini et al., 2021). By sparking concern that all medical waste is likely to be potentially profitable, the Stockholm Convention is pushing healthcare facilities and waste management authorities around the world to make alternative treatment methods such as steam sterilization, pyrolysis and plasma gasification part of reducing toxic pollutant emissions. The arbitrary exception of biomedical waste incineration facilities from regulation under the Stockholm Convention may create detrimental impacts on public health and the environment, both at a local and global level.

5.4 European Union (EU) Directives on Bio Medical Waste

The European Union (EU) Waste Framework Directive provides the framework for managing biomedical waste as part of the EU's legislative amendments adopted in recent years. The directive sets specific rules for the classification, treatment, and disposal of waste, with a focus on the waste hierarchy, which prioritizes prevention, reuse, recycling, recovery, and disposal, respectively. This essentially makes it mandatory for healthcare institutions to segregate hazardous biomedical waste at the point of generation, appropriately treated and disposed of as per regulations. The EU also imposes monitoring and reporting requirements on healthcare waste incinerators, which must comply with emission limits to reduce environmental impact. Furthermore, the EU Directive on Hazardous Waste has special rules for handling infectious and toxic biomedical waste, which demand special precaution in treatment procedures and disposal methods to ensure protection of public health and the environment.

6. TECHNOLOGIES AND INNOVATIONS IN BIOMEDICAL WASTE TREATMENT

Biomedical waste management is regulated by various agencies, including the Environmental Protection Agency (EPA), the Occupational Safety and Health Administration (OSHA), and the Centers for Disease Control and Prevention (CDC) in the United States. The Resource Conservation and Recovery Act (RCRA), implemented through the EPA, govern the disposal of hazardous biomedical waste and chemical and pharmaceutical waste. The comprehensive legislation controls hazardous types of waste, creating strict rules for the management of wastes that are harmful to the environment and population, such as biomedical waste in healthcare institutions (Reddy, 2023). OSHA, in contrast, promulgates occupational safety and health regulations to protect healthcare workers and waste handlers against potential exposure to dangerous biomedical waste. OSHA's Bloodborne Pathogen Standard requires engineering controls, personal protective equipment (PPE), and proper disposing of sharps and infectious waste to prevent disease transmission in healthcare workers. Furthermore, and biomedical waste disposal, and notes best practices for infection control, outlining effective waste management practices for healthcare facilities preparing for the spread of WNV.

This situation is aggravated by the fact that developing countries don't enforce biomedical waste management regulations well, also lack adequate waste treatment plants and citizens don't seem to be aware of the issue. Developing nations have looked to the United Nations Environment Programme (UNEP) to help devise regulatory guidelines and implement waste management policies to tackle this issue (Keikhosrokiani, 2022). The UN Environment Program (UNEP)'s Global Partnership on Waste Management offers practical technical assistance, capacity building initiatives, and funding support to improve biomedical waste disposal practices in low-resource contexts. Similar initiatives have also been assisted outside of Brazil, with agencies like the World Bank and the Global Environment Facility (GEF) sponsoring waste management schemes in developing nations, aiding in the adoption of environmentally friendly waste treatment technologies and bolstering regulatory compliance.

The disposal of radioactive biomedicine waste from medical imaging, radiation therapy, and nuclear medicine is along the lines of legislation under International Atomic Energy Agency (IAEA) guidelines. The IAEA's Code of Practice on the

Safety and Security of Radioactive Sources provides guidelines for proper handling, storage, transport and disposal of radioactive biomedical waste to minimize radiation exposure and environmental contamination. Healthcare institutions that use radioactive materials must adhere to strict protocols for the disposal of radioactive waste, so that the waste is properly contained and treated before its disposal. The IAEA provides training programs and technical guidance on radiation safety and waste management practices, in partnership with national regulatory agencies.

Even though these international regulations and guidelines are in place there are still difficulties in guaranteeing their effective adoption and local application. Compliance is a significant barrier for most healthcare facilities, especially in developing countries, due to limitations in funds, lack of waste disposal plants, and poor regulation. Improving biomedical waste management practices across the globe requires strengthening national policies, investing in technologies for waste treatment, and developing greater collaboration among countries. We are already seeing new regulatory requirements related to e-waste management practices, and the convergence of digital technologies like blockchain for waste track and trace, as well as AI for waste segregation, will further enforce regulatory compliance, while also improving waste management economy and effectiveness.

Comprehensive alternative classification and treatment/disposal frameworks can be found in organizations such as WHO, UNEP, the Basel and Stockholm Conventions, the EU and individual national regulatory bodies. But practicing these rules requires focus, money, and technological solutions. Through increased adoption of such technologies and strengthening of global cooperation, the world can explore solutions to the growing threat of biomedical waste generation leaving the path to a safer and sustainable future for all (Ariffin et al., 2021).

Increasing healthcare waste generation, strict environmental regulations, and demand for sustainable waste management solutions have led to significant advances in technologies and innovations in biomedical waste treatment. Biomedical waste comprises of infectious waste, pathological waste, pharmaceutical waste, chemical waste, radioactive waste and sharps waste and if not properly treated and disposed then it becomes a threat to human health and environment. Conventional waste treatment methods, including incineration and landfilling, have been extensively deployed, but issues, including air pollution, groundwater contamination, and greenhouse gas production, have led to the necessity for the creation of innovative and sustainable waste treatment technologies. This has led to the development of new technologies such as autoclaving, microwave treatment, plasma gasification and chemical disinfection, along with AI-driven and robotic biomedical waste management systems.

Autoclaving is a commonly used biomedical waste treatment technology that uses high-pressure steam at temperatures of 121°C to 134°C to sterilize infectious waste. This method works particularly well for microbiological waste, used personal protective equipment (PPE) and sharps, which are rendered non-infectious before their disposal (Bao et al., 2019). This method is an environmentally friendly alternative to incineration; it destroys pathogens while not releasing harmful emissions. Due to its effectiveness and environmental compliance, many health care facilities have adopted autoclaves. Autoclaves are receiving updates too: some with advanced automation and real-time monitoring systems that provide greater efficiency in treatment and confirm decontamination.

Microwave treatment is another type of advanced technology, which applies high-frequency electromagnetic waves to heat and kill microorganisms in biomedical waste. Medicinal treatment, a new method of disposal of healthcare waste, takes lower energy than immersive. Microwave treatment is ideal for infectious waste since the volume is reduced (by at least 80%) while microbial decontamination is also ensured (Zhou et al., 2021). To avoid the need for hauling waste to central disposal, and on-site treatment, a number of microwave based portable systems have been developed for small scale healthcare facilities. Why Go for a Microwave Treatment for Waste? Further, some advanced microwave treatment units are combined with shredders to minimize the volume of waste, making it more convenient for disposal.

Plasma gasification is a new biomedical waste treatment technology for environmentally friendly treatment of waste through its conversion into syngas and vitrified slag. Utilizing plasma torches, this molecular-level breakdown occurs through incredibly high temperatures (over 3000°C), converting toxic biomedical waste into synthetic gas and other useful byproducts that can be used for energy. In contrast to incineration, plasma gasification does not generate toxic dioxins or furans as by-products, making it an environmentally safe alternative to the conventional treatment of biomedical waste. If plasma gasificatio does require hights capital invested, it is considered a advantage for generating energy from wastes and therefore aligned to world sustainability goals, therefore so many studies are doning to improve and make this tecnology viable for everyones.

Chemical disinfection: A common biomedical waste treatment method for liquid waste, pharmaceutical waste and laboratory-generated biohazardous materials (Padmaja et al., 2022). These strong disinfectants for example, chlorine compounds, peracetic acid, and hydrogen peroxide are used to neutralize infective agents in biomedical waste. Chemical treatment is frequently used in addition to shredding to improve volume reduction of waste and improve safe disposal. Moreover, advancements in chemical disinfection technologies paved way for biodegradable disinfectants that reduce environmental toxicity factors and align with green waste management strategies. For industrial use, some contemporary chemical disinfection systems are installed with real-time surveillance and automatic dosing systems that deliver the right amount of treatment.

Thermal decomposition: It is a defined method known as pyrolysis, it is being practiced as biomedical waste treatment technology, as it is performed in the absence of oxygen. This process then derives biochar, syngas, and bio-oil from organic matter, which can be applied for energy recovery. Pyrolysis is ideal because it occurs at lower temperatures than incineration, minimizing toxic emissions while contributing to material recovery. Researchers are also finding novel methods to maximize the efficiency of pyrolysis, such as incorporating catalysts and utilizing AI for improved waste to energy conversion rates (Thapa & Camtepe, 2021).

With the advent of digitalization and automation, AI and robotics have changed the scenario of Biomedical Waste Management. AI-powered solutions can automate the segregation of waste by locating and classifying varying varieties of biomedical waste to be treated accordingly. With its machine learning algorithms, waste composition data is analyzed to optimize waste treatment process while ensuring compliance with relevant regulatory standards (Behura et al., 2021). Conversely, robotics are being used in waste handling, cutting human exposure to dangerous materials. AI-driven autonomous waste-sorting robots are able to sort biomedical waste in hospitals and laboratories, reducing cross-contamination opportunities.

7. COMPLIANCE, MONITORING AND ENFORCEMENT MECHANISMS

The compliance, monitoring and enforcement tools are essential to guarantee that hospitals, other health care establishments, laboratories, pharmaceutical industries and waste treatment centers comply with the regulatory guidelines governing biomedical waste. Because biomedical waste is dangerous and poses a threat to the public and the environment, it requires strict supervision mechanisms to prohibit dumping, reduce risks and promote practices of sustainable waste disposal. The compliance stands for adherence to national and international regulations ranging from waste segregation, treatment, transportation and final disposal. Monitoring entails keeping tabs on waste creation, transfer and disposal methods, while enforcement mechanisms hold offenders legally accountable and impose fines/penalties, upholding public safety and environmental protections (Rudin, 2019).

Regulatory compliance to the guidelines provided by the different organization like the World Health Organization (WHO), United Nations Environment Programme (UNEP) and country-specific regulatory bodies like the Environmental Protection Agency (EPA) in the US and the Central Pollution Control Board (CPCB) in the India is the main aspects of compliance. These rules specify procedures for waste categorization, management, storage, treatment, and disposal in order to decrease health hazards and environmental contamination. Healthcare facilities and waste management companies should have the required licenses and permits to ensure waste treatment practices follow legal policies. Audits, self-reporting, and submissions to regulators help show the regulatory body that the facility is following proper disposal of waste protocols.

They ensure that waste is tracked during its generation, transport and disposal, and have a monitoring mechanism to ensure responsible disposal of biomedical waste. Governments and institutions have implemented electronic waste tracking systems using barcoding, GPS tracking and blockchain technology at all touchpoints of waste from point of creation to final disposal. These systems enhance transparency, deter illegal dumping and enable real-time compliance reporting. For example, waste management companies employ digital manifest systems that enable regulatory bodies to authenticate or verify how waste is segregated at source and how it is treated downstream as well as identifying violators (Saadat & Shuaib, 2020). AI tools help monitor compliance with safety protocols and guidelines in healthcare settings too, including AI-powered surveillance cameras, IoT-enabled waste bins.

Statewide enforcement mechanisms hold facilities accountable and institute penalties, followed by license suspension, or even litigation against repeat offenders of the biomedical waste regulations (Shahid, 2021). Before the penalties makes it the responsibility of citizens, in USA for instance punishment of citizens is strict to deter people from improper waste disposal and citizens and Environmental organization can also leverage public reporting systems to report violations. Enforcement range from surprise facility inspections to laboratory sampling of waste streams and or compliance and from these checks, and treatment facilities are routinely required to assess their operating status and conformity with mandated environmental standards. In the event of non-compliance, facilities may be subject to fines, legal penalties, or ordered to take corrective actions, such as upgrading treatment infrastructure or adopting additional safety measures. In extreme instances, legal action can lead to the shuttering of facilities or criminal charges, especially if violations create public health risks or environmental contamination.

International cooperation and capacity-building programs are also strengthening compliance through global efforts. For example, the WHO, UNEP, and Basel Convention Secretariat offer technical assistance, training programs, and financial support to aid developing countries in strengthening their biomedical waste management frameworks. Governments, private actors in the waste management sector, and non-governmental organizations (NGOs) working together encourage information exchange and help to establish uniform enforcement measures across borders (Monteriro et al., 2021). So, compliance, monitoring and enforcement mechanisms can play a crucial role in ensuring the proper management of biomedical waste to minimize risks and safeguard public health and the environment. Governments and institutions must promote responsible biomedical waste disposal and encourage sustainable waste management practices through regulatory frameworks, national digital monitoring tools and strict enforcement policies.

8. CHALLENGES AND BARRIERS IN GLOBAL BIOMEDICAL WASTE MANAGEMENT

Proper biomedical waste management is a major challenge worldwide due to multiple regulatory, infrastructural, technological, financial, and socio-cultural barriers. Such events pose a serious health and environmental threat as e.g., dissemination of infectious diseases and chemical pollution and ecosystem degradation. Many countries, especially low- and middle-income ones, face challenges with the safe collection, treatment and disposal of biomedical waste despite international guidelines and national regulations. Models exist for this, but their misinformation can lead to detrimental environmental consequences which are exacerbated by health services expansion, hazardous medical waste from SARS-like pandemic outbreaks and novel forms of waste from complex new medical devices. An act of uniformity and consistent enforcement of regulations is one of the biggest challenges faced in the task of biomedical waste management (Qureshi et al., 2023). There are already a number of guidelines set in place at international organizations, such as the World Health Organization (WHO) and within the United Nations Environmental Programme (UNEP), but there is no universal accountability mechanism. Laws vary from country to country, which results in inconsistent waste classification, management and disposal processes. Developing countries often do not have laws governing biomedical waste or do not enforce existing laws due to lack of regulation and corrupt practices. This results in bad practices regarding waste disposal such as open dumping, incineration of hazardous wastes and mixing biomedical waste with municipal waste, thus amplifying contamination risks.

The other major hurdle is lack of facilities and sufficient treatment infrastructure. Proper waste segregation systems, storage units, and disposal technologies for treatment such as autoclaving, incineration, and chemical disinfection, for manage medical waste are rarely available in health care facilities, especially in rural and underserved areas. In certain countries, this biomedical waste is thrown away in open lands, polluting the soil, water and the air as well. Without new waste treatment plants, hospitals and clinics are often reduced to old, ineffective or environment-unfriendly disposal, like uncontrolled incineration, which spews toxic contaminants such as dioxins and furans into the air (Leung et al., 2021).

Economic constraints are another impediment in implementing the management of biomedical waste, more so in developing economies. The cost involved in establishing and maintaining waste treatment infrastructure, involved hiring trained personnel and compliance with regulators can be high influencing developing counties to explore better methods of waste disposal that are cheaper and suitable. Most healthcare institutions have tight budgets and waste management isn't the immediate priority. The private waste disposal companies might impose prohibitive costs for their waste collection and treatment services, rendering smaller clinics and rural hospitals unable to afford an adequate disposal system. Investment in sustainable waste disposal methods is not encouraged by a lack of financial incentives and government subsidies.

Biomedical waste hazard and the role of healthcare workers and waste handlers do not aware about biomedical waste management. A third of the participants said the existing hospitals and clinics did not adequately train them on the need to segregate waste, or how to handle hazardous waste safely (Teikari et al., 2019). This leads to a higher rate of diseases transmission as the medical workers1 inadvertently combine a sharps disposal (infectious waste) with a common garbage. They often collect recyclable materials on dumpsites, which puts them at serious risk of disease through their direct contact with contaminated medical waste.

Technological constraints are also limiting efficient management of biomedical waste. There are advanced treatment methods, such as plasma gasification, pyrolysis, and AI-based waste monitoring systems, but their use is complicated by high implementation costs and technical difficulties, hindering widespread adoption in low-income areas. The lack of proper maintenance of the existing waste treatment equipment results in low operational efficiency and environmental pollution.

Biomedical waste management, globally, is a matter of concern with multiple inter-related challenges of regulatory challenges, infrastructural deficits, financial limitations, incompetent awareness, and technological roadblocks. Overcoming these barriers will require a multi-stakeholder approach involving governments, health care institutions, the private sector and international organizations (Chakraborty & Bhattacharya, 2020). Well-designed regulations/ financial support schemes/ technological innovations and capacity building initiatives will very important to safeguard the safe, sustainable and efficient management of biomedical waste.

Figure 2. Express the challenges and barriers in global biomedical waste management (Source- Self Prepared- Original)

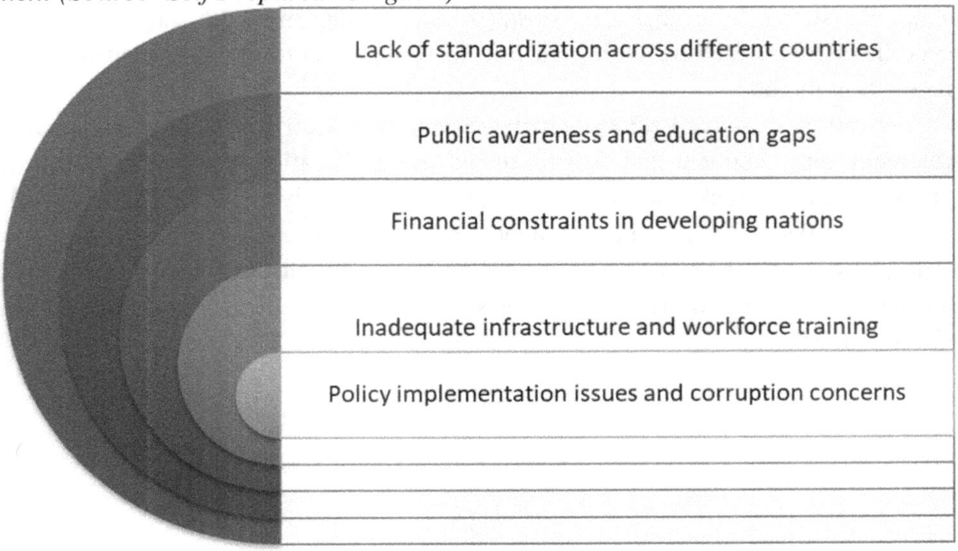

9. CONCLUSION AND FUTURE SCOPE

Blockchain technology is being applied to increase the traceability of biomedical waste and enhance regulatory compliance. This not only enhances the efficiency of the waste management process but also encourages compliance with regulations and can facilitate audits Blockchain based platforms for smart contracts enable automated checks of compliance with regulations to ensure strict compliance and avoid violations for waste disposal at each stage. Large healthcare facilities and pharmaceuticals that produce a large volume of biomedical waste especially find this technology useful. To improve sustainability in the lab, 3D printing is being used to manufacture biodegradable containers for biomedical waste and devices for treatment. Scientist are working on 3D printed waste bins that contain bacteria inhibiting materials that reduce the growth of harmful bacteria and make the storage of waste safer. 3D printing is being used to recycle biomedical waste into valuable materials like biodegradable prosthetics and medical instruments, reducing the overall waste produced in hospital systems.

Although innovative technologies showing promise for biomedical waste treatment, their large-scale applications still need to overcome challenges especially on the part of developing countries. Although in many areas throughout the world, waste disposal methods such as open burning and uncontrolled landfilling remain widespread and cause environmental and health pollution. Multifaceted challenges is necessitated by global teamwork, greater funding in health care waste task force infrastructure along with plan reforms that promote the use of sustainable medication methods.

Innovations in biomedical waste management from steam sterilization, microwave treatment, and chemical disinfection to plasma gasification, pyrolysis, AI-sation automation, and Blockchain traceability are progressively reshaping the approach towards biomedical waste management across the globe. Although challenges to widespread adoption persist, ongoing research, investment, and international cooperation are essential in delivering safe and sustainable biomedical waste management practices for the future.

REFERENCES

Ahmed, Y. A. E., Yue, B., Gu, Z., & Yang, J. (2023). An overview: Big data analysis by deep learning and image processing. *International Journal of Quantum Information, 21*(7), 2340009. DOI: 10.1142/S0219749923400099

Ariffin, N. A., Yunus, A. M., & Kadir, I. K. (2021). The role of big data in the healthcare industry. *Journal of Islamic, 6*(36), 235–245.

Bao, Y., Tang, Z., Li, H., & Zhang, Y. (2019). Computer vision and deep learning–based data anomaly detection method for structural health monitoring. *Structural Health Monitoring, 18*(2), 401–421. DOI: 10.1177/1475921718757405

Behura, A., Sahu, S., & Kabat, M. R. (2021). Advancement of Machine Learning and Cloud Computing in the Field of Smart Health Care. *Machine Learning Approach for Cloud Data Analytics in IoT*, 273-306.

Bhattacharyya, D., Stephen Neal Joshua, E., & Thirupathi Rao, N. (2023). Medical Image Analysis of Lung Cancer CT Scans Using Deep Learning with Swarm Optimization Techniques. *Machine Intelligence, Big Data Analytics, and IoT in Image Processing: Practical Applications*, 23-50.

Cabello, J. C., Karimipour, H., Jahromi, A. N., Dehghantanha, A., & Parizi, R. M. (2020). Big-data and cyber-physical systems in healthcare: Challenges and opportunities. *Handbook of Big Data Privacy*, 255-283.

Chakraborty, C., & Bhattacharya, S. (2020). Healthcare data monitoring under Internet of Things. In *Green Computing and Predictive Analytics for Healthcare* (pp. 1–18). Chapman and Hall/CRC. DOI: 10.1201/9780429317224-1

Dash, S., Shakyawar, S. K., Sharma, M., & Kaushik, S. (2019). Big data in healthcare: Management, analysis and future prospects. *Journal of Big Data, 6*(1), 1–25. DOI: 10.1186/s40537-019-0217-0

Devika, G., & Karegowda, A. G. (2021). Deep Learning in IoT: Introduction, Applications, and Perspective in the Big Data Era. In *Deep Learning Applications and Intelligent Decision Making in Engineering* (pp. 1-54). IGI Global.

Elbadawi, M., Gaisford, S., & Basit, A. W. (2021). Advanced machine-learning techniques in drug discovery. *Drug Discovery Today, 26*(3), 769–777. DOI: 10.1016/j.drudis.2020.12.003 PMID: 33290820

Keikhosrokiani, P. (Ed.). (2022). *Big data analytics for healthcare: datasets, techniques, life cycles, management, and applications*. Academic Press.

Leung, P. H., Chui, K. T., Lo, K., & de Pablos, P. O. (2021). A support vector machine–based voice disorders detection using human voice signal. In *Artificial Intelligence and Big Data Analytics for Smart Healthcare* (pp. 197–208). Academic Press. DOI: 10.1016/B978-0-12-822060-3.00014-0

Li, W., Chai, Y., Khan, F., Jan, S. R. U., Verma, S., Menon, V. G., & Li, X. (2021). A comprehensive survey on machine learning-based big data analytics for IoT-enabled smart healthcare system. *Mobile Networks and Applications*, *26*(1), 234–252. DOI: 10.1007/s11036-020-01700-6

Malik, S., & Tyagi, A. K. (Eds.). (2022). *Intelligent Interactive Multimedia Systems for E-Healthcare Applications*. CRC Press. DOI: 10.1201/9781003282112

Monteiro, A. C. B., França, R. P., Arthur, R., & Iano, Y. (2021). An overview of medical Internet of Things, artificial intelligence, and cloud computing employed in health care from a modern panorama. *The Fusion of Internet of Things, Artificial Intelligence, and Cloud Computing in Health Care*, 3-23.

Muthu, B., Sivaparthipan, C. B., Manogaran, G., Sundarasekar, R., Kadry, S., Shanthini, A., & Dasel, A. (2020). IOT based wearable sensor for diseases prediction and symptom analysis in healthcare sector. *Peer-to-Peer Networking and Applications*, *13*(6), 2123–2134. DOI: 10.1007/s12083-019-00823-2

Namdech, P., Rotjanawasuthorn, S., Singh, B., & Kaunert, C. (2025). Harnessing artificial intelligence for lifelong education: opportunities and challenges. *Asian Education and Learning Review*, *3*(2), 4–4.

Padmaja, M., Shitharth, S., Prasuna, K., Chaturvedi, A., Kshirsagar, P. R., & Vani, A. (2022). Grow of artificial intelligence to challenge security in IoT application. *Wireless Personal Communications*, *127*(3), 1829–1845. DOI: 10.1007/s11277-021-08725-4

Panesar, A. (2019). *Machine learning and AI for healthcare*. Apress. DOI: 10.1007/978-1-4842-3799-1

Pramanik, P. K. D., Pal, S., & Mukhopadhyay, M. (2022). Healthcare big data: A comprehensive overview. *Research anthology on big data analytics, architectures, and applications*, 119-147.

Qureshi, R., Irfan, M., Ali, H., Khan, A., Nittala, A. S., Ali, S., & Alam, T. (2023). Artificial Intelligence and Biosensors in Healthcare and its Clinical Relevance: A Review. *IEEE Access : Practical Innovations, Open Solutions*, *11*, 61600–61620. DOI: 10.1109/ACCESS.2023.3285596

Reddy, B. M. (2023). Amalgamation of Internet of Things and Machine Learning for Smart Healthcare Applications–A Review. *Int. J Comp. Eng. Sci. Res, 5*, 08-36.

Rodrigues, J. F.Jr, Florea, L., de Oliveira, M. C., Diamond, D., & Oliveira, O. N.Jr. (2021). Big data and machine learning for materials science. *Discover Materials*, *1*(1), 1–27. DOI: 10.1007/s43939-021-00012-0 PMID: 33899049

Rudin, C. (2019). Stop explaining black box machine learning models for high stakes decisions and use interpretable models instead. *Nature Machine Intelligence*, *1*(5), 206–215. DOI: 10.1038/s42256-019-0048-x PMID: 35603010

Saadat, M. N., & Shuaib, M. (2020). Advancements in deep learning theory and applications: Perspective in 2020 and beyond. *Advances and Applications in Deep Learning, 3*.

Sengan, S., Kamalam, G. K., Vellingiri, J., Gopal, J., Velayutham, P., & Subramaniyaswamy, V. (2020). Medical information retrieval systems for e-Health care records using fuzzy based machine learning model. *Microprocessors and Microsystems*, •••, 103344.

Shahid, Z. (2021). Distributed Machine Learning for Anomalous Human Activity Recognition using IoT Systems.

Sukeshini, S., P., Ved, M., Chintalapti, J., & Pal, S. N. (2021). Big data analytics and machine learning technologies for HPC applications. In Evolving Technologies for Computing, Communication and Smart World: Proceedings of ETCCS 2020 (pp. 411-424). Springer Singapore.

Teikari, P., Najjar, R. P., Schmetterer, L., & Milea, D. (2019). Embedded deep learning in ophthalmology: Making ophthalmic imaging smarter. *Therapeutic Advances in Ophthalmology, 11*, 2515841419827172. DOI: 10.1177/2515841419827172 PMID: 30911733

Thapa, C., & Camtepe, S. (2021). Precision health data: Requirements, challenges and existing techniques for data security and privacy. *Computers in Biology and Medicine*, *129*, 104130. DOI: 10.1016/j.compbiomed.2020.104130 PMID: 33271399

Tiwari, S. K., Kaur, J., Singla, P., & Hrisheekesha, P. N. (2022, September). A Comprehensive Review of Big Data Analysis Techniques in Health-Care. In *International Conference on Emergent Converging Technologies and Biomedical Systems* (pp. 401-420). Singapore: Springer Nature Singapore.

Zhou, S., Zhang, R., Chen, D., & Zhu, X. (2021). A novel framework for bringing smart big data to proactive decision making in healthcare. *Health Informatics Journal*, *27*(2), 14604582211024698. DOI: 10.1177/14604582211024698 PMID: 34159834

Zhou, S. K., Greenspan, H., Davatzikos, C., Duncan, J. S., Van Ginneken, B., Madabhushi, A., Prince, J. L., Rueckert, D., & Summers, R. M. (2021). A review of deep learning in medical imaging: Imaging traits, technology trends, case studies with progress highlights, and future promises. *Proceedings of the IEEE*, *109*(5), 820–838. DOI: 10.1109/JPROC.2021.3054390 PMID: 37786449

Chapter 7
Urban Wetlands and Climate Resilience: A Case Study of Surajpur Wetland, Greater Noida, Uttar Pradesh, India

Laxmi Kant Bhardwaj
https://orcid.org/0000-0001-7518-4199
Amity Institute of Environmental Toxicology, Safety, and Management, Amity University, Noida, India

K. Dhanaraj
Amity Institute of Geoinformatics and Remote Sensing, Amity University, Noida, India

Prangya Rath
https://orcid.org/0000-0002-6764-1625
Amity Institute of Environmental Sciences, Amity University, Noida, India

Vartika Singh
https://orcid.org/0000-0003-4400-8664
Amity Institute of Global Warming and Ecological Study, Amity University, Noida, India

Moharana Choudhury
Environmental Research and Management Division, Voice of Environment (VoE), Guwahati, India

ABSTRACT

Urban wetlands serve an important role in cooling cities, sequestering carbon,

DOI: 10.4018/979-8-3373-1549-2.ch007

and offering recreational and educational opportunities for urban residents. They enhance water quality by filtering contaminants and removing surplus nutrients, conserving aquatic habitats. They reduce the emission of GHGs into an urban environment. The current study investigates the diverse significance of the Surajpur wetland. It extends much-needed habitat to a wide range of species in the region. It is significant due to its proximity to metro expansion in Delhi and adjacent districts. However, these benefits are faced with challenges like development pressure and pollution in urban wetlands, which, as a result, require good management and restoration. This concern chapter mainly deals with water quality indices for the year 2024 in the Surajpur wetland. As a constant case, this case study attempts to implement wetlands in city life for a resilient, sustainable city. This chapter discloses that the Surajpur wetland is a fundamental structure that needs to be protected for ecological and human beings.

1. INTRODUCTION

Wetlands are one of the most beneficial regions of our ecological system (Kumar et al., 2020; Kumar & Choudhury, 2021; Bhardwaj et al., 2025a). They appear in all locations, ranging from tropical to temperate areas. They occur between the open water organizations and terrestrial environments and form a main category of aquatic environments (Mitsch et al., 2009). They share properties of both dry land and open water systems. They may remain either waterlogged or under shallow water permanently (Chakraborty et al., 2023a). They are a breeding habitat for various migratory and domiciled birds (Yamian et al., 2012). They are the fields of marshes, swamps, mangroves, fen, water, shallow lakes, and floodplains (Butt et al., 2021; Choudhury et al., 2021).

Calhoun et al. (2017) stated that these lands may be artificial, natural, temporary, or permanent. As per the United States Geological Survey study, these lands are permanently or seasonally waterlogged (Subcommittee, 2013). They are the most threatened territories in the world (Xu et al., 2019). They remain a resource of sustenance for local populations, especially in developing territories, and are extremely valued by many established cultures (Gopal, 2013). Such landscapes are identified as low carbon sequestration, characterized by near homogeneity and high diversity (Gogoi et al., 2024). They indirectly and directly assist millions of people, providing goods and services to them. The direct benefits of wetlands are products such as fuel wood, livestock grazing, fish, energy resources, timber, medicinal plants, agriculture, tourism, wildlife resources, water storage, and transport.

Groundwater recharge, storm protection, and flood control are the indirect benefits arising from the tasks within the system (Boutwell & Westra, 2016). They perform other valuable functions like water storage, water purification, floral and faunal habitats, protection of shorelines, gene pools, and recreation, besides requiring yields of commercial value and economic sustenance to the inhabitants (Verma et al., 2001). They also act as a natural reservoir, absorbing excess rainfall to reduce urban flooding (Acreman & Holden, 2013). Its biological roles in controlling the microclimate, boosting biodiversity, and serving as a carbon sink are essential to the region's ability to withstand climate change (Choudhury et al., 2021). They also help keep the local weather moderate (Ramchandra, 2001). They keep water during dry-out periods, thus keeping the water table superior and relatively constant (Phukan & Saikia, 2014).

In metropolitan cities and urbanized areas, more heat is generated due to construction, human activity, impermeable surfaces, and less vegetation cover (Bhardwaj et al., 2024). For example, it has presented problems for the Surajpur Wetland in recent years. It impacts the wetland's biodiversity, water levels, and ecological processes. Wetland vegetation provides a defense against soil erosion that is important in managing shorelines and accounts for storm bearing (Ford et al., 2016). They assist in advancing all species that have their habitat in wetlands. Some of these processes include transporting water over the wetland into the ocean or streams, decomposition of organic substances, the issue of nitrogen (N), sulphur (S), and carbon (C) into the air, and out of water that is in transit through the wetland. They control the water, nutrient inputs, outputs, and nature's physical and biological rhythms (Zedler & Kercher, 2005).

However, these roles depend mainly upon their hydrological attributes. While the importance of wetlands has been understood for centuries, the importance of wetlands for sustaining life processes is not well understood. These systems cover approximately 6 percent of the total terrestrial area (Fluet-Chouinard et al., 2023). India's enormous geographical stretch and varied topography and environmental conditions have enabled a wide spectrum of inland and coastal wetland ecosystems (Bassi et al., 2014). The classification is depicted in Table 1, which shows the inland and coastal wetlands (Costanza et al., 2021).

The wetlands in India are allotted in numerous conservation territories (Bassi et al., 2014). A total of 67,429 wetlands in India cover ~ 4.1 million hectares of land (Chakraborty et al., 2023a). Out of 4.1 million hectares of land, 1.5 million hectares of wetland are made naturally, and 2.6 million hectares of wetland are artificial. In contrast, the coastal wetlands are present at 6,750 sq km and are dominated by mangroves (Ramchandra, 2001). The distribution of wetland areas in percentage in India is shown in Figure 2 based on their hydrology, ecological functions, physical characteristics, and vegetation. These lands can be divided into the following types (Sarkar & Das, 2016).

(i) Freshwater Marshes

Shallow, standing, or slow-moving water with mostly herbaceous plants, including reeds, sedges, and cattails, defines freshwater marshes (Pisani et al., 2018). Low-lying areas, floodplains, and river and lake margins all contain them, providing vital habitat for various species, including amphibians, birds, and fish. Environmental health depends seriously on these marshes since they help prevent floods, filter water, and store carbon dioxide. Despite their ecological consequence, freshwater marshes are threatened by drainage, pollution, and climate change, so conservation initiatives are even more valuable to conserve these priceless territories.

(ii) Saltwater or Coastal Marshes

Found along coastlines where saltwater and freshwater unite, saltwater or coastal marshes are wetland ecosystems creating a distinctive home for varied plant and animal groups (Hopkinson et al., 2019). Salt-tolerant vegetation, such as cordgrass, mangroves, and saltbush, dominates in these marshes; these serve to steady shorelines and decrease coastal destruction. Along with providing surroundings for migrating birds, they are vital breeding and nursery grounds for fish, crabs, and shellfish. Filtering pollutants, boosting water quality, and functioning as natural buffers versus storm surges and flooding differ in coastal marshes in substantial segments (Choudhury et al., 2019). However, rising sea levels, climate change, and human activity, including land reclamation and pollution, increasingly endanger them.

(iii) Freshwater Swamps

Water-tolerant trees, shrubs, and vegetation define freshwater swamps, wetlands dominated by slow-moving or standing waters (Chakraborty et al., 2023b). Usually observed in low-lying spots, floodplains, or next to rivers or lakes, these ecological units are noteworthy for flood control, water filtering, and territory for a variety of groups. Common vegetations are cypress, mangroves, and willows; wildlife like amphibians, reptiles, fish, and birds flourish in these nutrient-dense surroundings. Additionally, as carbon sinks, freshwater marshes help control temperature by storing carbon and lowering greenhouse gas (GHG) emissions.

(iv) Saltwater or Coastal Swamps

Near coastlines, saltwater, or coastal swamps are wetland ecosystems where freshwater from rivers and streams meets saline ocean water (White & Kaplan, 2017). These ecosystems are home to salt-tolerant plants like mangroves, which provide a good environment for many marine and land varieties. They also serve as nurseries for fish and additional aquatic life and naturally guard against coastal erosion, storm surges, and flooding. Coastal marshes, which accumulate vast capacities of carbon in their soils and plants, are also significant for carbon sequestration, helping slow global warming.

(v) Ombrotrophic Bogs

Nutrient-poor, acidic wetlands, known as ombrotrophic bogs, get their water and nutrients from precipitation rather than groundwater or surface runoff (Gopal & Ghosh, 2008). Usually dominated by sphagnum mosses, these bogs help explain their high acidity and slow decomposition rates, which, over time, help to accumulate peat. Ombrotrophic bogs sustain a distinct ecosystem with specialized plants, such as carnivorous species, that offset nutrient shortages because they are isolated from mineral-rich sources. By accumulating significant volumes of atmospheric carbon in their peat layers, they significantly contribute to carbon storage and control of the temperature.

(vi) Minerotrophic Bogs

They are more nutrient-rich than ombrotrophic bogs and are also known as fens and get water and nutrients from either surface or groundwater (Money et al., 2009). Higher mineral richness made possible by this outside water source supports a wider variety of plant life, including some shrubs, reeds, and sedges. Because of the effect of calcium and other dissolved minerals, minerotrophic bogs often have a more neutral to somewhat alkaline pH than acidic bogs. These wetlands are vital for water filtering, carbon sequestration, and habitat provision for rare plant and animal species.

(vii) Prairie Potholes

Found largely in the north-central United States and south-central Canada, Prairie Potholes is a distinctive wetland habitat spanning portions of North Dakota, South Dakota, Minnesota, Iowa, and Manitoba (Forcey, 2006). Retreating glaciers during the last Ice Age created these shallow, water-filled depressions vital for water retention, groundwater recharge, and flood management. For many waterfowl species, they provide essential breeding and feeding sites. The main emphasis of conservation activities is restoring and safeguarding these important habitats to support animals and preserve ecological balance.

(viii) Playa Lakes

Mostly in the southwestern United States, Mexico, and portions of Central Asia, playa lakes are shallow, seasonal water bodies found in dry and semi-arid areas (Scuderi et al., 2010). Usually, in wet seasons, these lakes develop in natural depressions where rainfall or runoff gathers momentarily before evaporating or seeping into the ground. Usually lined with clay, they lack permanent inflow or outflow and serve to hold water for some duration. Important homes for migratory birds and other animals, such as playa lakes, promote biodiversity and significantly help recharge groundwater. Human activity and climate change, which are highly susceptible, can thus degrade these lakes.

(ix) Mangrove Forests

Found in tropical and subtropical climates, mangrove forests are coastal ecosystems in which salt-tolerant plants and shrubs flourish in intertidal zones (Kathiresan & Bingham, 2001). While offering essential homes for many marine and terrestrial species, these forests are significant barriers against coastal erosion, storm surges, and tsunamis. With their roots and soil capturing and storing substantial carbon, mangroves are vital for carbon sequestration (Alongi, 2012). By providing tools such as timber, petrol, and fishing, they also help nearby areas. However, mangrove forests are in danger from pollution, deforestation, and climate change, so their existence and ecological balance depend on conservation initiatives.

(x) Fens

These waterlogged wetlands are rich in peat and are nourished by groundwater instead of direct rain (Middleton et al., 2006). They have a higher pH than bogs, which are acidic and sustain a greater variety of plant and animal life, including rare orchids, sedges, and reeds. Important habitats for biodiversity also function as carbon sinks, controlling temperature change.

(xi) Constructed Wetlands

They are planned systems replicating natural wetlands for habitat restoration, stormwater management, and wastewater treatment (Malaviya & Singh, 2012; Stefanakis, 2019). They filter and eliminate nutrients, heavy metals, and pathogens from polluted water via natural mechanisms, including wetland vegetation, soil, and microbial activity. These systems fall into surface or subsurface flow wetlands based on water movement. The environmental advantages constructed wetlands offer are better water quality, more biodiversity, lower flood risks, and carbon sequestration. For environmentally friendly water management, they are affordable, sustainable, and used increasingly in rural and metropolitan regions (Figure 1).

Figure 1. A mesmerizing view of artificial lake/wetland with great recreational and environmental significance (Photo By Moharana Choudhury)

(xii) Tidal Flats and Salt Pans

Tidal flats and salt pans are coastal landforms in intertidal zones where seawater routinely covers and recedes from the surface (Warren & Warren, 2016). Supporting various marine life, including mollusks and crustaceans, tidal flats, also known as mudflats, formed by sediment deposition in estuaries, lagoons, and bays, are flat, muddy environments. These areas are vital for migratory birds and natural filters for coastal water systems, so they play important ecological functions.

(xiii) Riparian Wetlands

Found along rivers, streams, and other water bodies, riparian wetlands are habitats that are vital for preserving water quality, managing floods, and fostering biodiversity (Gumiero et al., 2013). Their grasses, shrubs, and trees help to stabilize riverbanks and control water flow, therefore maintaining the balance of an ecosystem. Furthermore, important stopping locations for migrating birds that help replenish groundwater are riparian wetlands. Maintaining aquatic and terrestrial life depends on the preservation of these wetlands.

As per the research conducted under the Ministry of Environment, Forests and Climate Change (MoEFCC), these organizations, the National Wetland Conservation Programme (NWCP) and National Lake Conservation Plan (NLCP), are conserving around 4.63 percent of the wetlands (MoEFCC, 2012). According to the Worldwide Fund for Nature India (WFNI), wetlands have been recorded as one of the most pressing threats to all ecological units of India (Anand et al., 2012).

After the convention of Ramsar in 1971, several laws were passed in India to preserve the ecosystem of wetlands (Prasad et al., 2002). MoEFCC has proclaimed 75 wetland sites that are covered under the Ramsar Convention (Table 2). In 2022, 26 new Ramsar sites were included, and some examples are Pichavaram Mangrove & Pallikaranai Marsh Reserve Forest from Tamil Nadu, Karikili Bird Sanctuary, Sakhya Sagar from Madhya Pradesh, Pala wetland from Mizoram, etc. Tamil Nadu has the greatest number of lakes (4369) in India, followed by Uttar Pradesh (3684) and West Bengal (1327) (Chandra, 2021). Kol wetland is the second-largest wetland in India and is situated in Kerala (Nisari & Sujatha, 2021). Pollution, land-use changes, and human growth pose a greater threat to these wetlands and surrounding ecosystems. Hence, maintaining and restoring urban wetlands is essential to reducing the effects of climate change in our country and implementing environmental sustainability measures in urban areas. The aim of this study was to measure the water quality indices (WQI) of the Surajpur Wetland in the year 2024 by using satellites such as NDWI, NDCI, NDSI, and NDTI. A diagrammatic representation of the overview of this study is presented in Figure 3.

Figure 2. The distribution of wetland areas in India

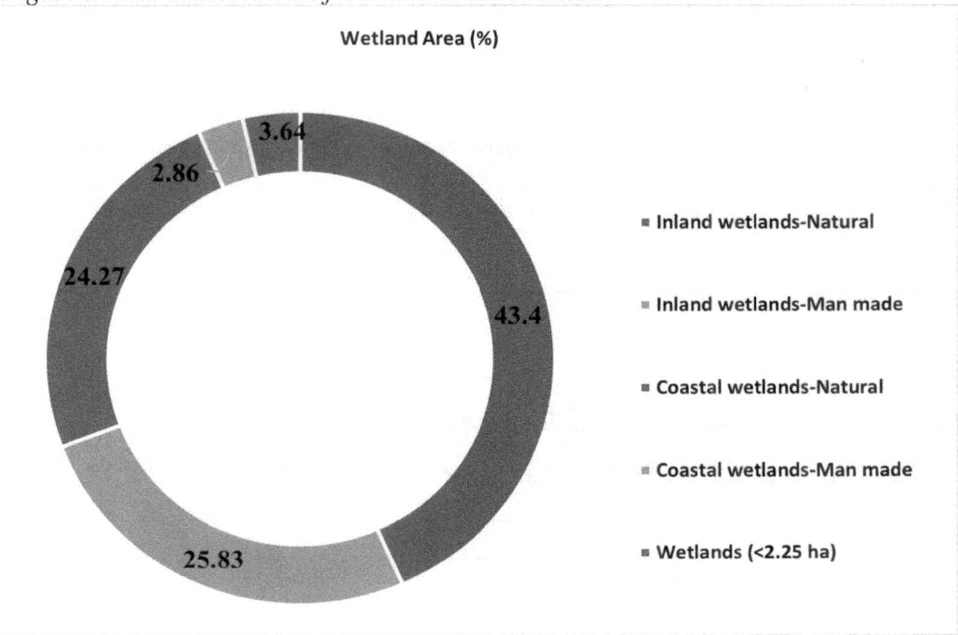

Figure 3. An overview of the study at Surajpur Wetland

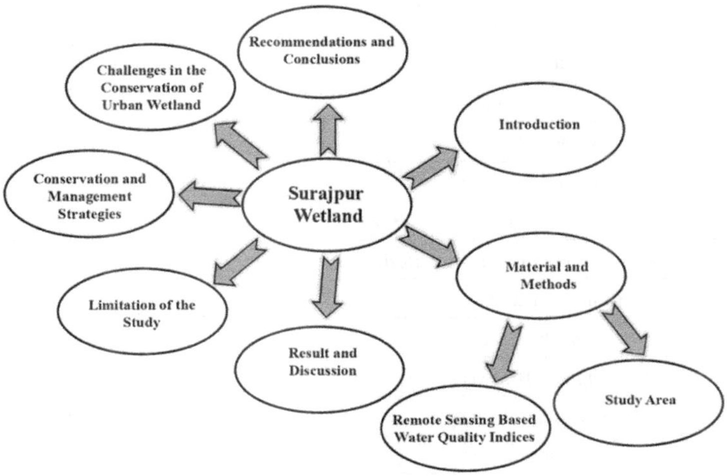

Table 1. Inland and coastal wetlands with their classes

S. No.	Wetland Category	Class	Cover Class
1	Inland wetlands	Natural	Lakes, Riverine wetlands, Ox-Blow lakes, Reservoirs/Barrages, Waterlogged, and River/Stream
		Man-made	Tank/Ponds, Waterlogged, Salt pans, Creeks, and Lagoons
2	Coastal wetlands	Natural	Sand beaches, Coral reefs, Salt marshes, and Mangroves
		Man-made	Salt pans and Aquaculture ponds

Table 2. List of Ramsar sites in India with their area

S. No.	Ramsar Wetland Sites	Location	Area (hectare)
1	Kolleru Lake	Andhra Pradesh	901
2	Deepor Beel	Assam	40
3	Kabartal Wetland	Bihar	26.2
4	Nanda Lake	Goa	0.42
5	Khijadia Wildlife Sanctuary	Gujarat	5.12
6	Nalsarovar Bird Sanctuary		120
7	Thol Lake Wildlife Sanctuary		6.99
8	Wadhvana Wetland		6.3
9	Bhindawas Wildlife Sanctuary	Haryana	4.12
10	Sultanpur National Park		1.43
11	Chandertal Wetland	Himachal Pradesh	0.49
12	Pong Dam Lake		156.62
13	Renuka Wetland		0.2
14	Hokera Wetland	Jammu and Kashmir	13.75
15	Hygam Wetland Conservation Reserve		8.02
16	Shallbugh Wetland Conservation Reserve		16.75
17	Surinsar Mansar Lakes		3.5
18	Wular Lake		189
19	Ranganathittu Bird Sanctuary	Karnataka	5.18
20	Asthamudi Wetland	Kerala	61.4
21	Sasthamkotta Lake		3.73
22	Vembanad Kol Wetland		1512.5

continued on following page

Table 2. Continued

S. No.	Ramsar Wetland Sites	Location	Area (hectare)
23	Tso Kar Wetland	Ladakh	95.77
24	Tsomoriri Lake		120
25	Bhoj Wetlands	Madhya Pradesh	32.01
26	Sakhya Sagar		2.48
27	Sirpur Wetland		1.61
28	Yashwant Sagar		8.23
29	Lonar Lake	Maharashtra	4.27
30	Nandur Madhameshwar		14.37
31	Thane Creek		65.21
32	Loktak Lake	Manipur	266
33	Pala Wetland	Mizoram	18.5
34	Ansupa Lake	Odisha	2.31
35	Bhitarkanika Mangroves		650
36	Chilka Lake		1165
37	Hirakud Reservoir		654
38	Satkosia Gorge		981.97
39	Tampara Lake		3
40	Beas Conservation Reserve	Punjab	64.29
41	Harike Lake		41
42	Kanjli Lake		1.83
43	Keshopur Miani Community Reserve		3.44
44	Nangal Wildlife Sanctuary		1.16
45	Ropar Lake		13.65
46	Keoladeo Ghana NP	Rajasthan	28.73
47	Sambhar Lake		240

continued on following page

Table 2. Continued

S. No.	Ramsar Wetland Sites	Location	Area (hectare)
48	Chitrangudi Bird Sanctuary	Tamil Nadu	2.6
49	Gulf of Mannar Marine Biosphere Reserve		526.72
50	Kanjirankulam Bird Sanctuary		0.97
51	Karikili Bird Sanctuary		0.58
52	Koonthankulam Bird Sanctuary		0.72
53	Pallikaranai Marsh Reserve Forest		12.48
54	Pichavaram Mangrove		14.79
55	Point Calimere Wildlife and Bird Sanctuary		385
56	Suchindram Theroor Wetland		0.94
57	Udhayamarthandapuram Bird Sanctuary		0.44
58	Vaduvur Bird Sanctuary		1.13
59	Vedanthangal Bird Sanctuary		0.4
60	Vellode Bird Sanctuary		0.77
61	Vembannur Wetland		0.2
62	Rudrasagar Lake	Tripura	2.4
63	Bakhira Wildlife Sanctuary	Uttar Pradesh	28.94
64	Haiderpur Wetland		69.08
65	Nawabganj Bird Sanctuary		2.25
66	Parvati Agra Bird Sanctuary		7.22
67	Saman Bird Sanctuary		5.26
68	Samaspur Bird Sanctuary		7.99
69	Sandi Bird Sanctuary		3.09
70	Sarsai Nawar Jheel		1.61
71	Sur Sarovar		4.31
72	Upper Ganga River		265.9
73	Asan Conservation Reserve		4.44
74	East Kolkata Wetlands	West Bengal	125
75	Sundarbans Wetland		4230

2. MATERIALS AND METHODS

2.1. Study Area

Surajpur Wetland ecosystem was the study area, and it is situated in Surajpur village in Greater Noida, Gautam Budh Nagar district of Uttar Pradesh in India. It was constructed in 2010. Uttar Pradesh Forest Department, Greater Noida Industrial Development Authority, and WWF–India play a major role in constructing this wetland. The area of this land is ~ 308 hectares. The latitude and longitude of the land are 28°31.425' N and 77°29.714'E. Different species of birds and plants are found in this region. Spot-billed Duck, Cotton Pygmy Goose, Greylag Goose, Lesser-whistling Duck, Northern Shoveler, Red-crested Pochard, Comb Duck, Common Teal, Ferruginous Pochard, Bar-headed Goose, and Gadwall are the different species of waterfowl and birds that are found in this region. The location map of Surajpur Wetland is shown in Figure 4.

Figure 4. The location map of Surajpur wetland

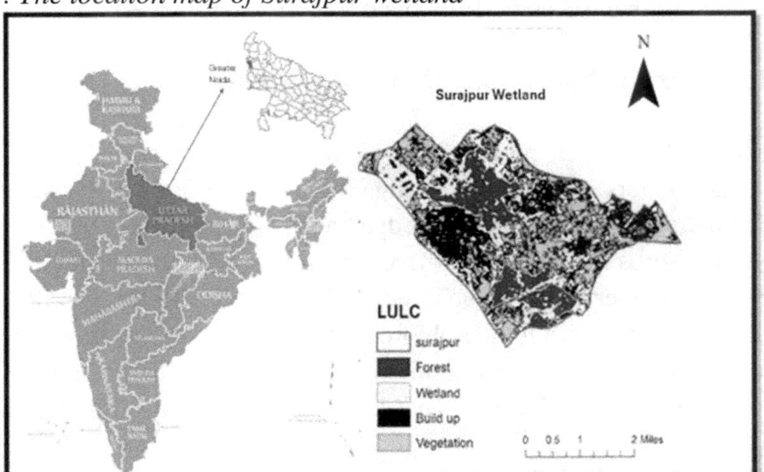

2.2. Remote Sensing-Based Water Quality Indices (WQI)

WQI produced from remote sensing can also be substituted for on-site measurements and laboratory testing of water samples drawn from surface water bodies. Such indices are derived from mathematical combinations of different spectral bands for estimating turbidity, chlorophyll concentration, and total suspended matter

(TSM) parameters. They leverage the spectral reflectance characteristics of water, as recorded by sensors. This study utilizes Landsat images to derive various WQIs for the year 2024.

2.2.1. Normalized Difference Water Index (NDWI)

NDWI extracts surface water bodies and estimates water content (Yang et al., 2017). Although NDWI does not directly measure WQI, it is a significant remote sensing tool that helps recognize water bodies and indirectly assess surface conditions. The indicator may indicate the presence of turbidity or polluted waters where sediment, algae, or pollutants reduce the water signal. A clean lake can have a high NDWI value because the clear water reflects more in the green band than the near-infrared (NIR) band (Yang et al., 2017). The NDWI is calculated using the following Landsat band combination formula.

NDWI = (Green + NIR) / (Green−NIR) Eq.(1)

Where Green is band 3 and NIR is band 8 in Landsat imagery.

2.2.2. Normalized Difference Chlorophyll Index (NDCI)

Superfluous NDCI detection has recognized high concentrations of chlorophyll in surface waters associated with algal bloom conditions in the water bodies (Mishra & Mishra, 2012). Red and red edge bands are employed to determine how chlorophyll-a absorbs radiation and reflects a strong value of radiant energy in the red edge band. A higher NDCI value indicates increased reflectance in the red edge band, which correlates with higher chlorophyll-α concentrations in water (Jang et al., 2024). Therefore, NDCI helps monitor and map eutrophication in water bodies caused by nutrient loading. The NDCI can be estimated using the following formula.

NDCI = (Red Edge + Red) / (Red Edge − Red) Eq.
(2)

Where Red Edge is band 5 and Red is band 4 in Sentinel 2 imagery.

2.2.3. Normalized Difference Salinity Index (NDSI)

It is utilized as an index of water quality for assessing salinity levels in an aquatic environment (Uddin et al., 2022). Salinity is essential for measuring water quality and terrestrial ecosystem conditions, affecting the environmental relationship between agriculture and drinking water resources (Bhateria & Jain, 2016). The salinity index uses the spectral reflectance characteristics of salinity in the water. Higher salinity levels in the water increase the amount of total dissolved solids (TDS), which scatter light and thus increase reflectance in the Shortwave Infrared (SWIR) band (Adjovu et al., 2023). On the other hand, reflectance in the red band may decrease due to absorption by organic matter or sediments associated with high salinity. The NDSI index, therefore, provides an indirect measure of salinity by capturing the contrast between the SWIR and red bands. The NDSI can be measured using the following equation.

NDCI = (Red Edge + Red) / (Red Edge – Red) Eq.
(3)

The high values of NDSI suggest higher salinity and vice versa. High salinity levels can harm aquatic organisms. Similarly, saline water is unsuitable for irrigation and human consumption.

2.2.4. Normalized Difference Turbidity Index (NDTI)

The NDTI evaluates the turbidity of water bodies based on the reflectance related to the number of suspended particles (Bid & Siddique, 2019). It estimates turbidity by analyzing the contrast between red and green reflectance. It has been studied that higher NDTI values indicate more turbidity. Higher turbidity, on the other hand, adversely affects aquatic plants since it hinders light penetration into water bodies (Jacobsen & Engström-Öst, 2018). Sediments, pollutants, and organic matter in water reduce the transparency, such as in the quality of water. Water bodies with heavy turbidity levels will considerably disturb their aquatic ecosystem and affect human health. For better quality of plant life in marine systems, high turbidity could hinder light penetration into the water bodies (Jacobsen and Engström-Öst, 2018).

The NDTI can be measured using the following formula.

NDTI = (Red + Green) / (Red – Green) Eq. (4)

Where the Green and Red bands are bands 3 and 4 in Sentinel 2, respectively.

3. RESULTS AND DISCUSSION

Over time, remote sensing has been utilized to map huge wetland regions (Rebelo et al., 2009). Moreover, remote sensing in aerial photography worked effectively to delineate, identify, and measure the spatial degree of wetlands. With repeated tracks of remote sensing instruments over a place, land evidence in the form of multispectral images and multi-date can be achieved within a persistent period. With the introduction of remote sensing satellites like the Landsat Series Thematic Mapper (LSTM) and Landsat Series with Multispectral Scanner (LSMSS), it has developed cost-effective and suitable to attain multi-date digital pictures over a larger array of spatial and temporal rulers than was probable with aerial photography (Kumar et al., 2023).

LSMSS has been effectively used to research big wetlands (Demarquet et al., 2023). However, its usage is very restricted for analyses of numerous aspects (e.g., species identification, species discrimination, etc.) of all categories of wetlands, specifically inland wetlands, which commonly have minor aerial degrees and complicated combinations of vegetation species. In early findings, the undeveloped limitations of using LSMSS data for wetland mapping records were geometric inaccuracy and inadequate spatial, spectral, and radiometric data outcomes. The accessibility of LSTM data explained this challenge of coarse solutions to some degree. With a spatial solution of 30 meters, it becomes feasible to research relatively smaller areas. Numerous researchers have worked with remotely sensed data for determining water quality and mapping lakes and reservoirs. LSMSS, SPOT satellite systems, and LSTM have been used to research wetlands (Kulawardhana et al., 2007; Dos Santos et al., 2019). Other findings have included JERS-1, AVHRR, SIR-C, IRS, RADARSAT, and ERS-1 (Sarkar & Jian, 2007; Jain et al., 2008; Joshi et al., 2016).

Remote Sensing-Based Water Quality Mapping

NDWI is used to determine the water quality of the Surajpur Wetland area. The high NDWI values are concentrated in a few packets within the wetland area, suggesting a reduced water quality level (Figure 5C). The highest value recorded for the wetland is 0.05, whereas the lowest value recorded is -0.38. These values suggest that the water quality of the Surajpur Wetland is decreasing due to increased sedimentation and that the water level is decreasing to a shallow level.

Figure 5. Diagrammatic Representation of Remote Sensing-Based Water Quality Indices (WQI) for the Surajpur Wetland Area (A) Normalised Difference Chlorophyll Index (NDCI), (B) Chlorophyll Index, (C) Normalized Difference Water Index (NDWI), (D) Normalized Difference Turbidity Index (NDTI), (E) Normalized Difference Salinity Index (NDSI).

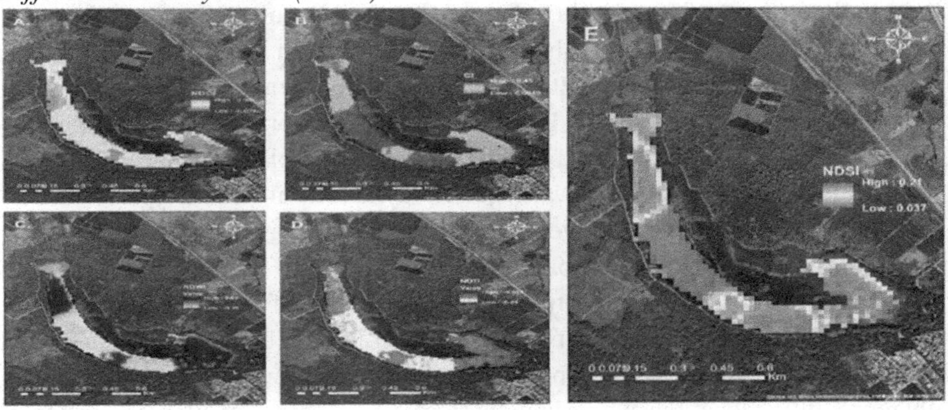

The NDCI is seen as high in Surajpur Wetland areas, where the NDWI values were recorded as the lowest within the wetland (Figure 5A). This suggests that increased sedimentation within the Surajpur Wetland has accelerated the growth of vegetation in that area and thus shows that the wetland is experiencing extensive algal growth. The index has recorded a high value of 0.0866 and a low value of -0.039. The spatial extent of the wetland is, therefore, showing a decreasing trend, causing considerable effects on the water quality of the wetland. A comparable pattern was noted in the Chlorophyll Index (see Figure 5B). The maximum value of the Chlorophyll Index was measured at 0.45, while the minimum value recorded was 0.0049. The NDSI is a crucial instrument for assessing water quality in wetlands, and its importance is notably significant in the Surajpur Wetland. As depicted in Figure 5E, there is a marked variation in the salinity levels across the wetland, with the highest recorded value reaching 0.21 and the lowest at 0.037. These values reflect significant spatial heterogeneity in salinity, suggesting that salinity is not uniformly distributed throughout the wetland. Instead, it exhibits patterns that appear to be closely linked with other critical water quality boundaries, such as chlorophyll concentration and water index values. High salinity values tend to align with areas of increased chlorophyll concentrations, which are often associated with higher levels of aquatic vegetation or biological activity. This correlation suggests that areas with elevated salinity may support more abundant or denser vegetation, potentially due to favorable nutrient conditions or other factors that promote plant growth. This connection, however, creates an issue for ecological balance; for example, elevated salinity levels may put pressure on specific aquatic organisms, diminishing their chances of survival or growth. Therefore, comprehending the distribution of salinity is crucial when evaluating the health of the wetland ecosystem and its ability to sustain various forms of life.

Furthermore, the water index affects the salinity distribution, which measures water quality in aspects like turbidity, dissolved materials, and contamination levels. Higher salinity levels are commonly observed in areas with low water index, which often indicates poorer water quality. This suggests a potential link between high salinity and reduced water quality. The issues involve evaporation, decreasing water levels, or pollution. The relationship among salinity, chlorophyll, and the water index illustrates the intricate nature of water quality dynamics in the Surajpur Wetland, highlighting the necessity for diligent monitoring and management to maintain the area's ecological health over the long term.

Regions with elevated salinity often coincide with areas of heightened chlorophyll index, suggesting they are zones of increased vegetation or biological productivity. Chlorophyll serves as the most commonly utilized indicator of plant activity, typically representing locations where aquatic flora, such as algae or submerged plants, are flourishing (Dubey et al., 2022; Farooq et al., 2024). The correlation of high salinity with elevated chlorophyll levels may indicate that these saline regions foster a greater diversity of plant life, likely due to nutrient availability or the influence of other favorable growth factors. Nonetheless, higher salinity can also be a stressor for some marine classes, potentially resulting in shifts in species composition or a decline in biodiversity over time.

In addition to the chlorophyll index, the water index, which often indicates water quality attributes such as turbidity, dissolved solids, or the occurrence of pollutants, also shows a role in spreading salinity. Lower water index values, indicative of poor water quality, coincide with areas with higher salinity. This pattern suggests that the increased salinity in these areas might be linked to reduced water levels, higher evaporation rates, or contamination sources that elevate the salinity. Such conditions can exacerbate the challenges faced by aquatic ecosystems, as higher salinity often leads to unfavorable conditions for many freshwater species. Therefore, the interplay between NDSI, chlorophyll, and water index in the Surajpur Wetland highlights complex dynamics between salinity and overall water quality. While higher chlorophyll levels may indicate healthy vegetation in some areas, the associated high salinity and low water index in certain zones could also point to underlying water quality issues, further emphasizing the need for a comprehensive assessment of the wetland's ecological health.

The salinity distribution throughout the wetland suggests it spans a large water body area, particularly in regions with minimal vegetation or very shallow water. Salinity levels in these areas could negatively affect the entire wetland ecosystem, as aquatic organisms may be impacted, leading to consequences for biodiversity. These salinity levels could have lasting effects on water quality, especially regarding how less suitable the water may become for various plant and animal species that thrive under neutral salinity conditions. The deeper sections of the wetland, which might initially be assumed to be less susceptible to water quality issues, are also not immune to contamination. These areas are impacted by turbidity, as indicated by the NDTI depicted in Figure 5D. Despite the presence of turbidity, the intensity levels measured in the water body do not appear to be alarmingly high, with the highest recorded NDTI value being -0.014 and the lowest at -0.09. The turbidity levels observed are moderate and not particularly harmful to the water. However, they do influence the clarity and visual appeal of the water. Additionally, these levels differ across the area, with the central portion of the wetland exhibiting the lowest turbidity values, which corresponds with regions of dense vegetation. This could indicate that these plants contribute to stabilizing water quality in those areas by minimizing sediment disturbance and enhancing water clarity through natural filtration mechanisms (Bhardwaj et al., 2023; Bhardwaj et al., 2025b).

Beyond the scientific investigation of the Surajpur Wetland via remote sensing techniques, it is mandatory to cast some light on the much larger social interface that, in different ways, affects the health and sustainability of the wetlands. Community participation is key for wetland conservation and resilience. When the local communities are engaged in citizen science, cleaning activities, or monitoring of biota, real-time information flows openly, while long-term stewardship backs that information. Such community-based monitoring committees greatly enhance monitoring, fostering stakeholder participation and ownership as well.

Policy support ranks next in importance. In a broad perspective, legislative enforcement would make conservation regimes sufficiently rigorous to be enforced and well-integrated into regional planning. Even though the Wetlands (Conservation and Management) Rules, 2017, provide an ample framework for regulation, the actual implementation, especially the cross-departmental coordination and enforcement to stop encroachment and pollution, should be strengthened. Mechanistically, following by formalizing wetlands into the Environmental Impact Assessment (EIA) process under land development projects for towns will give another layer of defence to the ecological zones. Wetlands like Surajpur are outlets for education and recreation from a social perspective. Awareness campaigns, eco-tourism, and academic fieldwork will instill greater appreciation and significance for the public.

Remote sensing has been proposed as a powerful tool to map and monitor these large wetland ecosystems and acts as an alternative and cost-efficient option for traditional aerial photography. Satellite systems like Landsat and SPOT have aided multi-temporal assessments of wetlands in the analysis of water quality, vegetation, and habitat degradation. However, free scientific data provided by these technologies can only be fully realized in conjunction with community-based monitoring and policy-led conservation. The local communities are important for the ground-truthing of remote sensing data, which adds to its accuracy and practical use. But it is true that coarse spatial resolution and atmospheric interference limit detailed assessments, especially for smaller one such complex wetland as Surajpur. This, therefore, calls for a hybrid approach that couples satellite data and participatory mapping with local stakeholders, thus increasing the frequency and reliability of monitoring.

Wetland aspects of Surajpur are time worrying in the quality of water, as values of NDWI lie between -0.38 and 0.05, suggesting increased sedimentation and decreased water levels (Figure 5C). High NDCI and high chlorophyll values (Figure 5 A to B) point to algal blooms, likely driven by nutrient pollution from agricultural runoff and urban wastewater. These findings highlight the consequences of weak municipal waste management policies. The social inferences of this deprivation are important, including public health dangers for societies relying on the wetland for irrigation and domestic usage, livelihood fatalities for fishermen and farmers, and inconsistent influences on marginalized groups who depend maximum on wetland properties. Strengthening policy implementation, such as the Wetlands (Conservation and Management) Rules, 2017, alongside community-led water property monitoring programs, could help alleviate these problems.

4. LIMITATIONS OF THE STUDY

While Landsat satellite imagery offers valuable spectral data for monitoring water quality, there are several limitations to its effectiveness in detailed water quality assessments. First, the spatial resolution of Landsat (10 to 60 meters) can be insufficient for smaller water bodies or heterogeneous environments where fine-scale variations in water quality exist. Additionally, Landsat lacks specific spectral bands needed to accurately capture certain water quality indicators like chlorophyll-α and dissolved organic carbon, which are often targeted by hyperspectral sensors. Atmospheric interference, such as clouds and aerosols, further limits data quality and availability. Lastly, the satellite's revisit time of five days may not capture rapid changes in water quality, limiting its effectiveness in areas where frequent monitoring is essential.

This research on Surajpur Wetland in Greater Noida, Uttar Pradesh, has significant restrictions, even if it offers insightful analysis of the function of urban wetlands in climate resistance. First, the study only uses a few field observations and secondary data sources, so it may not give a true picture of the wetland's biological and hydrological processes. Extreme weather events and seasonal fluctuations could greatly affect wetland operations, yet the study might not fairly represent long-term changes and trends due to temporal restrictions. Furthermore, the study lacks socioeconomic records on community interactions with the wetland, potentially impacting the evaluation of its broader resilience benefits. Second, although other important issues, such as governance challenges, policy frameworks, and land-use disputes, are not thoroughly investigated in the study, their main focus is climate resilience and environmental features. Greater Noida's fast urbanization seriously threatens Surajpur Wetland; yet, a more thorough investigation of land encroachment, pollution sources, and the implementation of conservation measures is outside the purview of this study. Moreover, the research does not particularly include indigenous knowledge or community-led conservation initiatives, which can offer insightful analysis of wetland viability.

Finally, data availability and accessibility restrictions naturally restrict the approach. Though valuable, remote sensing and GIS-based analysis might not reflect micro-level ecological changes completely. Continuous long-term investigations are necessary for water quality assessments and biodiversity monitoring; these were not possible within the scope of this work. Furthermore, the limited involvement of stakeholders, including local authorities and legislators, may impede the development of conservation plans. Future studies should fill these voids by combining several disciplines and conducting more comprehensive empirical investigations.

5. CONSERVATION AND MANAGEMENT STRATEGIES

By reducing floods, controlling local temperatures, and maintaining biodiversity, urban wetlands significantly contribute to increasing climate resilience (Alikhani et al., 2021; Choudhury et al., 2025). A shining example of an urban wetland that naturally offsets climate change and offers ecological and socioeconomic advantages is the Surajpur Wetland in Greater Noida, Uttar Pradesh. Still, pollution, rapid urbanization, and encroachment jeopardize its sustainability. For Surajpur Wetland, conservation and management techniques comprise community involvement, pollution control, and habitat restoration. Maintaining its natural equilibrium depends on efforts including reforestation, wetland restoration, and strict control of industrial effluents. Encouragement of wetland education projects and ecotourism will also help increase local involvement in environmental preservation efforts (Yaseen & Chalotra, 2022).

We need an integrated management approach incorporating scientific research, policy frameworks, and technological interventions to enhance climate resilience (Dixit et al., 2025). While sustainable urban design can avoid encroachment and guarantee buffer zones surrounding the wetland, remote sensing and GIS mapping can monitor wetland health. Effectively applying conservation measures depends on cooperation among local populations, non-governmental organizations, and governmental institutions. Surajpur Wetland preservation is not only essential for preserving regional biodiversity but also for guaranteeing climate resilience in the fast-growing metropolitan scene of Greater Noida.

The long-term preservation of the Surajpur Wetland depends critically on public involvement and policy support. Community-driven projects, such as citizen science programs, wetland clean-up campaigns, and afforestation campaigns, help local people develop stewardship. Furthermore, including wetland preservation in urban development programs through strict environmental impact assessments, incentives for sustainable behaviors, and legislative frameworks guarantees a balanced expansion. Further helping to maintain this important ecosystem will be strengthening legal safeguards under the Wetlands (Conservation and Management) Rules, 2017 and working with research facilities for ongoing ecological monitoring. Surajpur Wetland may remain a natural protection against environmental difficulties and support biodiversity and sustainable urban development by prioritizing wetland preservation as a basic component of climate resilience plans.

6. CHALLENGES IN THE CONSERVATION OF URBAN WETLAND

Rapid urbanization and land use modifications pose an ongoing threat to urban wetlands (Li et al., 2014; Anand & Oinam, 2020). The main obstacles are habitat loss and degradation brought on by encroachment, pollution, and infrastructure development (Sharma & Singh, 2021). Cities sometimes drain or fill wetlands to accommodate highways, buildings, and other developments, diminishing their ecological value. Furthermore, stormwater runoff from metropolitan areas, which carries chemicals, heavy metals, and oil, greatly affects water quality and aquatic life. The disturbance of natural hydrology presents another major obstacle. While drainage systems can direct water away from wetlands, resulting in desiccation, concrete surfaces stop water intrusion, producing faster runoff and erosion. By changing rainfall patterns and intensifying the frequency of extreme weather events, climate change aggravates these problems even more and makes wetlands more susceptible to floods and flooding (Bhardwaj et al., 2024).

Important obstacles to wetland preservation also include public knowledge and policy execution. Many city officials and citizens undervalue wetlands' ecological and financial advantages, which results in poor legal protection and insufficient conservation initiatives. Laws notwithstanding, enforcement is sometimes insufficient because of conflicting urban development interests and limited financing for conservation programs. Stronger legislation, community involvement, and sustainable urban design, including wetlands as vital green infrastructure rather than impediments to growth, are what effective wetland protection calls for. The long-term preservation of urban wetlands depends on community involvement and environmentally friendly living methods. Lack of public participation in many conservation projects results in failure since local populations cannot completely appreciate the biological value of these places. Teaching neighbors the advantages of wetlands, such as water filtering, flood control, and supporting biodiversity, can help inspire care. Incorporating nature-based solutions, such as built wetlands and green infrastructure, can also help to offset urban effects. In order to guarantee that conservation projects are sustainable and successful in the face of continuous urbanization, cooperation among local communities, environmental organizations, and government agencies is vital.

7. CASE STUDIES AND COMPARISON WITH SURAJPUR WETLAND

Case Study 1: Yamuna Biodiversity Park, Delhi and Comparison with Surajpur Wetland: In North Delhi, the Yamuna Biodiversity Park is located and has been restored. It was a waterlogged floodplain that scientists began restoring in 2000. Through the park's ecological restoration, more than 1500 kinds of animals and plants now flourish there; it also serves as an important carbon sink and groundwater recharge area. Its location on the city outskirts of Delhi is essential for fighting air pollution.

Yamuna Biodiversity Park and Surajpur Wetland are both close to urban areas and are important for biodiversity. Migratory species, in particular, attract special attention from birdwatchers convening at Surajpur Wetland. Yamuna Park, in contrast, is more concerned with broad biodiversity conservation aims and an educational programme as well for the local community on site at the park. In terms of governance, there is another difference: Yamuna Biodiversity Park is managed by the Delhi Development Authority, while Surajpur Wetland comes under the control of forest departments with support from NGOs and local people.

Case Study 2: Mangalajodi Wetlands, Odisha, and Comparison with Surajpur Wetland: Relatively more rural, but Mangalajodi Wetlands serve as a peri-urban wetland. Earlier, notorious for poaching, it was long ago transformed into a paradise for birdwatchers. Extending from the north bank of Chilika Lake, it's now chartered as a community-driven conservation zone. Over 300 species of birds make their home in the wetland, especially during the winter-rainy period. In the course of the ecotourism drive, this wetland has become a “home” with bamboo groves in which people can stay overnight.

Both of these wetlands are important habitats for birds. This is especially true for migratory species. But when it comes to environmental protection, Mangalajodi highlights community involvement together with eco-tourism, whereas Surajpur relies for the most part on government-led and NGO-supported partnerships. Mangalajodi is bigger and on a larger scale. Furthermore, it is linked with other lake systems, while Surajpur is a single offshore wetland confronting more urban problems since it is inside the National Capital Region.

Case Study 3: Rabindra Sarobar Lake, Kolkata, and Comparison with Surajpur Wetland: Rabindra Sarobar is an artificially created lake surrounded by parkland. It is a national lake conservation project site. Originally, it was used for amusement. It has become increasingly so, an urban wetland home to a number of water birds and water creatures. The lake is being threatened by urban development, industrial waste pollution, river effluent, and religious activities such as cremation, even though it has been taken over by the city, managed with citizen committees maintaining it.

Both wetlands are man-made. Rabindra Sarobar wetland is embedded deeply within the fabric of an urban area, part. Surajpur Wetland supports a rich avifauna, while Rabindra Sarobar wetland actually serves in practice as evidence of the impacts seen from cultural and recreational usage upon ecological health. Yet both wetlands show that involving citizens in urban wetland conservation is the most important thing.

8. RECOMMENDATIONS

Maintaining biodiversity, controlling water cycles, and avoiding natural events, including floods and erosion, depend on protecting wetlands. Strict conservation policies limiting activities, including land reclamation, too much water extraction, and pollution, are one major recommendation. Encouragement of sustainable land-use methods that strike a balance between development and preservation is still another crucial tactic. Promoting sensible travel and leisure activities can also increase understanding of wetlands' value and minimize human impact. More efficient monitoring and protection initiatives resulting from government, conservation, and local community cooperation also help guarantee that wetlands remain essential ecosystems for the next generation. The protection of wetlands also depends much on scientific study and technology. Modern monitoring techniques, including drone surveillance and satellite photography, can assist in tracking changes in wetland conditions and identifying hazards, including pollution or illegal development.

This study suggests several recommendations for the protection of these ecosystems.

Water hyacinths and other aquatic weeds should be removed from the wetland.
Indigenous fruiting varieties should be planted around the wetland area.
A visitor or tourist zone should be developed, and they should be trained properly before entering the wetland area.
Regular monitoring and cleaning of the wetland area should be done.
Strict laws should be made by the authorities for the cleaning of these ecosystems.
A cleaning or pollution-prohibiting signage should be pasted at the entrance of the wetland.
Plastic bottles, wrappers, and other food items should be prohibited inside the wetland area.
Dumping of any litter and other materials in proximity to the wetland should be strictly banned.
Awareness campaigns should inform local residents, legislators, and companies about the ecological and financial advantages of wetlands.
Sanitary landfills should be managed for final solid waste disposal to prevent acute pollution problems associated with discharging waste into wetlands.
The state government must formulate and accept a wetland policy and confirm its proper functioning.
Colleges and universities can incorporate wetland preservation into their courses, promoting environmental stewardship among the younger generation.
We can guarantee the long-term existence of these important ecosystems by combining contemporary technologies with conventional conservation initiatives.
Moreover, we should implement restoration initiatives to revive degraded wetlands, ensuring they continue to provide crucial biological functions.

9. CONCLUSIONS

Outdated wetland surveys are too time-consuming and costly, while the remote sensing imagery technique is a time-saving and cost-effective solution. Remote sensing has offered a great means to find various earth environments, including wetlands, by presenting cost- and time-effective information. Remote sensing proposes a cost-effective means for delineating wetlands over a substantial region at separate sites in time and can present useful data on wetland descriptions. Due to increasing population and urbanization, large-scale amendments in land use, burgeoning development projects, and inadequate usage of watersheds have all triggered a substantial decrease in wetland resources in the territory. Over-grazing and fishing actions have also resulted in the deprivation of wetlands.

Urban wetlands are important for enhancing climate resilience because they control regional hydrology, enhance biodiversity, and offer key ecosystem services like flood control, groundwater recharge, groundwater elevation, and air purification in the surrounding areas. Wetlands have appropriate productive soils, and the drainage of wetlands may not be without consequences, of generating acidic soils. Large wetland regions have been transformed in the past for the purpose of farming. The significance of these wetlands has long been acknowledged, and there is also a requirement to preserve and protect these wetlands. In addition, the Ramsar Convention recognizes fish and shrimp ponds, gravel pits, saltpans, reservoirs, etc., as wetlands. Worldwide Fund for Nature India (WWFNI) gave significant ecological protection inputs from time to time to extend this region as a valuable wildlife refuge. Greater Noida Industrial Development Authority and the Forest Department implement the inputs given by WWF India for the expansion and supervision of the Surajpur Wetland.

ACKNOWLEDGEMENT

The authors are grateful to their respective institutions for the support and motivation to carry out this research study.

REFERENCES

Acreman, M., & Holden, J. (2013). How wetlands affect floods. *Wetlands*, *33*(5), 773–786. DOI: 10.1007/s13157-013-0473-2

Adjovu, G. E., Stephen, H., James, D., & Ahmad, S. (2023). Measurement of total dissolved solids and total suspended solids in water systems: A review of the issues, conventional, and remote sensing techniques. *Remote Sensing (Basel)*, *15*(14), 3534. DOI: 10.3390/rs15143534

Alikhani, S., Nummi, P., & Ojala, A. (2021). Urban wetlands: A review on ecological and cultural values. *Water (Basel)*, *13*(22), 3301. DOI: 10.3390/w13223301

Alongi, D. M. (2012). Carbon sequestration in mangrove forests. *Carbon Management*, *3*(3), 313–322. DOI: 10.4155/cmt.12.20

Anand, A., Chandan, P., & Singh, R. B. (2012). Homestays at Korzok: Supplementing rural livelihoods and supporting green tourism in the Indian Himalayas. *Mountain Research and Development*, *32*(2), 126–136. DOI: 10.1659/MRD-JOURNAL-D-11-00109.1

Anand, V., & Oinam, B. (2020). Future land use land cover prediction with special emphasis on urbanization and wetlands. *Remote Sensing Letters*, *11*(3), 225–234. DOI: 10.1080/2150704X.2019.1704304

Bassi, N., Kumar, M. D., Sharma, A., & Pardha-Saradhi, P. (2014). Status of wetlands in India: A review of extent, ecosystem benefits, threats and management strategies. *Journal of Hydrology. Regional Studies*, *2*, 1–19. DOI: 10.1016/j.ejrh.2014.07.001

Bhardwaj, L. K., Kumar, D., & Kumar, A. (2023). Phytoremediation potential of Ocimum sanctum: A sustainable approach for remediation of heavy metals. In *Phytoremediation potential of medicinal and aromatic plants* (pp. 46–57). CRC Press. DOI: 10.1201/9781003290780-5

Bhardwaj, L. K., Rath, P., Jain, H., & Choudhury, M. (2025a). *A comprehensive review of wetlands in India and their conservation*. Nature-Based Technologies for Wastewater Treatment and Bioenergy Production., DOI: 10.2166/9781789064100_0043

Bhardwaj, L. K., Rath, P., Jain, H., Purohit, S., Yadav, P., & Singh, V. (2024). The Impact of Climate-Induced Livelihood, Health, and Migration on Women and Girls: A Review. *Global Insights on Women Empowerment and Leadership*, 100-118.

Bhardwaj, L. K., Upadhyay, D., Jain, H., Rath, P., Kumar, D., & Bajpai, S. (2025b). Recent trends of phytoremediation for the wastewater treatment and resource recovery. In *Biotechnologies for Wastewater Treatment and Resource Recovery* (pp. 135–146). Elsevier. DOI: 10.1016/B978-0-443-27376-6.00008-6

Bhateria, R., & Jain, D. (2016). Water quality assessment of lake water: A review. *Sustainable Water Resources Management*, *2*(2), 161–173. DOI: 10.1007/s40899-015-0014-7

Bid, S., & Siddique, G. (2019). Identification of seasonal variation of water turbidity using NDTI method in Panchet Hill Dam, India. *Modeling Earth Systems and Environment*, *5*(4), 1179–1200. DOI: 10.1007/s40808-019-00609-8

Boutwell, J. L., & Westra, J. V. (2016). The role of wetlands for mitigating economic damage from hurricanes. *Journal of the American Water Resources Association*, *52*(6), 1472–1481. DOI: 10.1111/1752-1688.12473

Butt, M. A., Zafar, M., Ahmed, M., Shaheen, S., Sultana, S., Butt, M. A., ... & Sultana, S. (2021). Types of Wetland and Wetland Plants. *Wetland Plants: A Source of Nutrition and Ethno-medicines*, 35-54.

Calhoun, A. J., Mushet, D. M., Bell, K. P., Boix, D., Fitzsimons, J. A., & Isselin-Nondedeu, F. (2017). Temporary wetlands: Challenges and solutions to conserving a 'disappearing'ecosystem. *Biological Conservation*, *211*, 3–11. DOI: 10.1016/j.biocon.2016.11.024

Chakraborty, S. K., Sanyal, P., & Ray, R. (2023a). Introductory and Basic Eco-biological Aspects of Wetlands. In *Wetlands Ecology: Eco-biological uniqueness of a Ramsar site (East Kolkata Wetlands, India)* (pp. 1–38). Springer International Publishing. DOI: 10.1007/978-3-031-09253-4_1

Chakraborty, S. K., Sanyal, P., & Ray, R. (2023b). Diversity and classification of wetlands in international and national perspectives. In *Wetlands Ecology: Eco-biological uniqueness of a Ramsar site (East Kolkata Wetlands, India)* (pp. 167–226). Springer International Publishing. DOI: 10.1007/978-3-031-09253-4_3

Chandra, K. (2021). *Faunal diversity in Ramsar wetlands of India*. Jointly published by Zoological Survey of India & Wetland Division, Ministry of Environment, Forest and Climate Change, Government of India.

Choudhury, M., Dixit, G., & Majumdar, S. (Eds.). (2025). *Sustainable Synergy: Harnessing Ecosystems for Climate Resilience*. Springer., DOI: 10.1007/978-3-031-77957-2

Choudhury, M., Fazli, P., Pramanick, P., Gobato, R., Zaman, S., & Mitra, A. (2019). Sensitivity of the Indian Sundarban mangrove ecosystem to local level climate change. *Science and Education*, *5*(3), 24–28.

Choudhury, M., Sharma, A., Singh, P., & Kumar, D. (2021). Impact of climate change on wetlands, concerning Son Beel, the largest wetland of North East, India. In *Global climate change* (pp. 393–414). Elsevier. DOI: 10.1016/B978-0-12-822928-6.00006-X

Costanza, R., Anderson, S. J., Sutton, P., Mulder, K., Mulder, O., Kubiszewski, I., Wang, X., Liu, X., Pérez-Maqueo, O., Luisa Martinez, M., Jarvis, D., & Dee, G. (2021). The global value of coastal wetlands for storm protection. *Global Environmental Change*, *70*, 102328. DOI: 10.1016/j.gloenvcha.2021.102328

Demarquet, Q., Rapinel, S., Dufour, S., & Hubert-Moy, L. (2023). Long-term wetland monitoring using the landsat archive: A review. *Remote Sensing (Basel)*, *15*(3), 820. DOI: 10.3390/rs15030820

Dixit, G., Majumdar, S., & Choudhury, M. (2025). Introduction to Ecosystem Services for Climate Resilience: A Theoretical Framework. In *Sustainable Synergy: Harnessing Ecosystems for Climate Resilience* (pp. 3-16). Cham: Springer Nature Switzerland. https://doi.org/DOI: 10.1007/978-3-031-77957-2_1

Dos Santos, G. M., Meléndez-Pastor, I., Navarro-Pedreño, J., & Lucas, I. G. (2019). A Review of Landsat TM/ETM based Vegetation Indices as Applied to Wetland Ecosystems. *Journal of Geographical Research/Ma alla Al-bu ū Al- uġrāfiyya* , *2*(1).

Dubey, D., Kumar, S., & Dutta, V. (2022). Algae and Macrophytes as Bioindicators of Freshwater Ecosystem. In *Environmental Studies and Climate Change* (pp. 441–467). CRC Press. DOI: 10.1201/9781003220824-33

Farooq, K., Shah, F. A., Qayoom, I., Qureshi, K., & Fayaz, I. (2024). Freshwater Floral Diversities as Pollution Indicators. In *Aquatic Pollution* (pp. 135–144). CRC Press. DOI: 10.1201/9781003503705-6

Fluet-Chouinard, E., Stocker, B. D., Zhang, Z., Malhotra, A., Melton, J. R., Poulter, B., Kaplan, J. O., Goldewijk, K. K., Siebert, S., Minayeva, T., Hugelius, G., Joosten, H., Barthelmes, A., Prigent, C., Aires, F., Hoyt, A. M., Davidson, N., Finlayson, C. M., Lehner, B., & McIntyre, P. B. (2023). Extensive global wetland loss over the past three centuries. *Nature*, *614*(7947), 281–286. DOI: 10.1038/s41586-022-05572-6 PMID: 36755174

Forcey, G. M. (2006). *Landscape-level influences on wetland birds in the Prairie Pothole Region of the United States and Canada*. North Dakota State University.

Ford, H., Garbutt, A., Ladd, C., Malarkey, J., & Skov, M. W. (2016). Soil stabilization linked to plant diversity and environmental context in coastal wetlands. *Journal of Vegetation Science*, *27*(2), 259–268. DOI: 10.1111/jvs.12367 PMID: 27867297

Gogoi, N., Choudhury, M., Ul Hasan, M. S., Changmai, B., Baruah, D., & Samanta, P. (2024). Soil organic carbon pool in diverse land utilization patterns in North-East India: An implication for carbon sequestration. *Environment, Development and Sustainability*, •••, 1–27. DOI: 10.1007/s10668-024-05037-y

Gopal, B. (2013). Future of wetlands in tropical and subtropical Asia, especially in the face of climate change. *Aquatic Sciences*, *75*(1), 39–61. DOI: 10.1007/s00027-011-0247-y

Gopal, B., & Ghosh, D. (2008). Natural Wetlands. In *Applications in Ecological Engineering*. Elsevier BV Amsterdam. DOI: 10.1016/B978-008045405-4.00067-7

Gumiero, B., Mant, J., Hein, T., Elso, J., & Boz, B. (2013). Linking the restoration of rivers and riparian zones/wetlands in Europe: Sharing knowledge through case studies. *Ecological Engineering*, *56*, 36–50. DOI: 10.1016/j.ecoleng.2012.12.103

Hopkinson, C. S., Wolanski, E., Cahoon, D. R., Perillo, G. M., & Brinson, M. M. (2019). Coastal wetlands: a synthesis. In *Coastal wetlands* (pp. 1–75). Elsevier. DOI: 10.1016/B978-0-444-63893-9.00001-0

Jacobsen, L., & Engström-Öst, J. (2018). Coping with environments; vegetation, turbidity and abiotics. In *Biology and ecology of pike* (pp. 32–61). CRC press. DOI: 10.1201/9781315119076-4

Jain, S. K., Sarkar, A., & Garg, V. (2008). Impact of declining trend of flow on Harike Wetland, India. *Water Resources Management*, *22*(4), 409–421. DOI: 10.1007/s11269-007-9169-9

Jang, W., Kim, J., Kim, J. H., Shin, J. K., Chon, K., Kang, E. T., Park, Y., & Kim, S. (2024). Evaluation of LandsatBased Chlorophyll-a Estimation in a Small-Scale Reservoir: Assessing Accuracy and Availability. *Remote Sensing (Basel)*, *16*(2), 315. DOI: 10.3390/rs16020315

Joshi, N., Baumann, M., Ehammer, A., Fensholt, R., Grogan, K., Hostert, P., Jepsen, M., Kuemmerle, T., Meyfroidt, P., Mitchard, E., Reiche, J., Ryan, C., & Waske, B. (2016). A review of the application of optical and radar remote sensing data fusion to land use mapping and monitoring. *Remote Sensing (Basel)*, *8*(1), 70. DOI: 10.3390/rs8010070

Kathiresan, K., & Bingham, B. L. (2001). Biology of mangroves and mangrove ecosystems.

Kulawardhana, R. W., Thenkabail, P. S., Vithanage, J. A. G. A. T. H., Biradar, C., Islam, A., Gunasinghe, S., & Alankara, R. (2007). Evaluation of the Wetland Mapping Methods using Landsat ETM+ and SRTM Data. *Journal of Spatial Hydrology*, *7*(2).

Kumar, A., Upadhyay, P., & Singh, U. (2023). *Multi-Sensor and Multi-Temporal Remote Sensing: Specific Single Class Mapping*. CRC Press. DOI: 10.1201/9781003373216

Kumar, D., & Choudhury, M. (2021). Recognizing economic values of wetland ecosystem services: A study of emerging role of monetary evaluation of Chandubi ecosystem and biodiversity. Wetlands Conservation: Current Challenges and Future Strategies, 87-110. https://doi.org/DOI: 10.1002/9781119692621.ch5

Kumar, D., Choudhury, M., & Rathore, A. (2020). Valuation of ecosystem services benefits of Son Beel wetland in Assam, India: a case study of natural solutions to climate change water. *Climate Change Conference*, Hamburg University, Germany.

Li, Y., Shi, Y., Zhu, X., Cao, H., & Yu, T. (2014). Coastal wetland loss and environmental change due to rapid urban expansion in Lianyungang, Jiangsu, China. *Regional Environmental Change*, *14*(3), 1175–1188. DOI: 10.1007/s10113-013-0552-1

Malaviya, P., & Singh, A. (2012). Constructed wetlands for management of urban stormwater runoff. *Critical Reviews in Environmental Science and Technology*, *42*(20), 2153–2214. DOI: 10.1080/10643389.2011.574107

Middleton, B., Grootjans, A. B., Jensen, K., Venterink, H. O., & Margóczi, K. (2006). Fen management and research perspectives: an overview. *Wetlands: functioning, biodiversity conservation, and restoration*, 247-268.

Ministry of Environment. Forests and Climate Change Annual report. (2012). MoEFCC, Government of India, New Delhi.

Mishra, S., & Mishra, D. R. (2012). Normalized difference chlorophyll index: A novel model for remote estimation of chlorophyll-a concentration in turbid productive waters. *Remote Sensing of Environment*, *117*, 394–406. DOI: 10.1016/j.rse.2011.10.016

Mitsch, W. J., Gosselink, J. G., Zhang, L., & Anderson, C. J. (2009). *Wetland ecosystems*. John Wiley & Sons.

Money, R. P., Wheeler, B. D., Baird, A. J., & Heathwaite, A. L. (2009). Replumbing Wetlands–managing water for the restoration of bogs and fens. *The wetlands handbook*, 755-779.

Nisari, A. R., & Sujatha, C. H. (2021). Assessment of trace metal contamination in the Kol wetland, a Ramsar site, Southwest coast of India. *Regional Studies in Marine Science*, *47*, 101953. DOI: 10.1016/j.rsma.2021.101953

Phukan, P., & Saikia, R. (2014). Wetland degradation and its conservation: A case study of some selected wetlands of Golaghat District, Assam, India. *Research Journal of Recent Sciences ISSN*, *2277*, 2502.

Pisani, O., Gao, M., Maie, N., Miyoshi, T., Childers, D. L., & Jaffé, R. (2018). Compositional aspects of herbaceous litter decomposition in the freshwater marshes of the Florida Everglades. *Plant and Soil*, *423*(1-2), 87–98. DOI: 10.1007/s11104-017-3495-3

Prasad, S. N., Ramachandra, T. V., Ahalya, N., Sengupta, T., Kumar, A., Tiwari, A. K., & Vijayan, L. (2002). Conservation of wetlands of India-a review. *Tropical Ecology*, *43*(1), 173–186.

Ramachandra, T. V. (2001). Restoration and management strategies of wetlands in developing countries. *Electronic Green Journal*, *1*(15). Advance online publication. DOI: 10.5070/G311510435

Rebelo, L. M., Finlayson, C. M., & Nagabhatla, N. (2009). Remote sensing and GIS for wetland inventory, mapping and change analysis. *Journal of Environmental Management*, *90*(7), 2144–2153. DOI: 10.1016/j.jenvman.2007.06.027 PMID: 18367311

Sarkar, A., & Jain, S. K. (2007, October). Using Remote Sensing data to study wetland dynamics–A case study of Harike Wetland. In *Proceedings of the Taal2007: The 12th World Lake Conference, Kusatsu, Japan* (Vol. 680, p. 684).

Sarkar, P., & Das, T. (2016). Wetland ecosystem services and its valuation with special reference to India—A review. *Biodiversity and Environmental Conservation*, 59-76.

Scuderi, L. A., Laudadio, C. K., & Fawcett, P. J. (2010). Monitoring playa lake inundation in the western United States: Modern analogues to late-Holocene lake level change. *Quaternary Research*, *73*(1), 48–58. DOI: 10.1016/j.yqres.2009.04.004

Sharma, S., & Singh, P. (Eds.). (2021). *Wetlands conservation: Current challenges and future strategies*. John Wiley & Sons. DOI: 10.1002/9781119692621

Stefanakis, A. I. (2019). The role of constructed wetlands as green infrastructure for sustainable urban water management. *Sustainability (Basel)*, *11*(24), 6981. DOI: 10.3390/su11246981

Subcommittee, W. (2013). *Classification of wetlands and deepwater habitats of the United States*. Federal Geographic Data Committee.

Uddin, M. G., Nash, S., Rahman, A., & Olbert, A. I. (2022). A comprehensive method for improvement of water quality index (WQI) models for coastal water quality assessment. *Water Research, 219*, 118532. DOI: 10.1016/j.watres.2022.118532 PMID: 35533623

Verma, M., Bakshi, N., & Nair, R. P. (2001). Economic valuation of Bhoj Wetland for sustainable use. *Unpublished project report for World Bank assistance to Government of India, Environmental Management Capacity-Building. Bhopal: Indian Institute of Forest Management, 35.*

Warren, J. K., & Warren, J. K. (2016). Sabkhas, saline mudflats and pans. *Evaporites: A Geological Compendium*, 207-301.

White, E., & Kaplan, D. (2017). Restore or retreat? Saltwater intrusion and water management in coastal wetlands. *Ecosystem Health and Sustainability, 3*(1), e01258. DOI: 10.1002/ehs2.1258

Xu, T., Weng, B., Yan, D., Wang, K., Li, X., Bi, W., Li, M., Cheng, X., & Liu, Y. (2019). Wetlands of international importance: Status, threats, and future protection. *International Journal of Environmental Research and Public Health, 16*(10), 1818. DOI: 10.3390/ijerph16101818 PMID: 31121932

Yamian, Z., Yifei, J., Shengwu, J., Qing, Z., Duoduo, F., Yumin, G., & Guangchun, L. (2012). Wuliangsuhai wetlands: A critical habitat for migratory water birds. *Journal of Resources and Ecology, 3*(4), 316–323. DOI: 10.5814/j.issn.1674-764x.2012.04.004

Yang, X., Zhao, S., Qin, X., Zhao, N., & Liang, L. (2017). Mapping of urban surface water bodies from LandsatMSI imagery at 10 m resolution via NDWI-based image sharpening. *Remote Sensing (Basel), 9*(6), 596. DOI: 10.3390/rs9060596

Yaseen, U., & Chalotra, E. S. C. E. S. (2022). Sustainable Management Strategies Urban Water Bodies and Wetlands in the Indian Context. *International Journal of Innovative Research in Engineering & Management, 9*(5), 257–265.

Zedler, J. B., & Kercher, S. (2005). Wetland resources: Status, trends, ecosystem services, and restorability. *Annual Review of Environment and Resources, 30*(1), 39–74. DOI: 10.1146/annurev.energy.30.050504.144248

Chapter 8
Responses of Carob Plants to Salt Stress:
Prospects for Sustainable Agriculture and the Bioeconomy

Yassine Mouniane
https://orcid.org/0000-0002-6037-5301
Faculty of Sciences, Ibn Tofail University, Kenitra, Morocco

Issam El-Khadir
https://orcid.org/0000-0002-3970-2865
Faculty of Sciences, Ibn Tofail University, Kenitra, Morocco

Ahmed Chriqui
Faculty of Sciences, Ibn Tofail University, Kenitra, Morocco

Abdelaati Soufiani
https://orcid.org/0000-0002-5164-7707
Faculty of Sciences, Ibn Tofail University, Kenitra, Morocco

Mahmoud Oudghiri
Faculty of Sciences, Ibn Tofail University, Kenitra, Morocco

Zineb Abbad
Faculty of Sciences, Ibn Tofail University, Kenitra, Morocco

Mohamed Haitam Zoubir
Faculty of Sciences, Ibn Tofail University, Kenitra, Morocco

Oumaima Ninich
https://orcid.org/0000-0003-1402-0590
Higher School of Technology of Sale, Mohammed V University in Rabat, Morocco

Driss Hmouni
https://orcid.org/0000-0001-7598-6204
Faculty of Sciences, Ibn Tofail University, Kenitra, Morocco

ABSTRACT

Increasing soil salinization threatens agriculture and biodiversity by limiting ecosys-

DOI: 10.4018/979-8-3373-1549-2.ch008

tem productivity and resilience. Carob (Ceratonia siliqua L.), native to Mediterranean regions, is notable for its drought tolerance and adaptability to degraded soils. This study examines the morphological, physiological, and biochemical responses of six Moroccan carob ecotypes to NaCl-induced salt stress under greenhouse conditions. Results reveal significant variability among ecotypes, emphasizing the role of genetic diversity in adaptation. Key parameters, such as proline and total sugar content, highlight promising adaptation mechanisms. Tolerant ecotypes could aid in saline soil restoration and support sustainable agriculture, positioning carob as a strategic resource for combating desertification and promoting resilient agroecosystems.

1. INTRODUCTION

Environmental degradation, exacerbated by climate change and human activities, poses a growing threat to the sustainability of ecosystems and global food security (Nguyen et al., 2023). Among the major challenges, soil salinization represents a significant obstacle to agricultural productivity, reducing land fertility and limiting the choice of adapted crops (Poveda, 2022; Bello and Ahmed, 2024). According to the FAO, over 20% of irrigated land is now affected by salinity, approximately 1125 million hectares of land are affected by salinity, and 1.5 million hectares are lost to salinity each year (Sanower, 2019). Salinity due to osmotic imbalance through ionic toxicity by sodium (Na^+) and chloride (Cl^-) ions, which lead to oxidative stress and cellular damage (Sapre et al., 2022), also reduced has an impact on the balancing of enzyme functions and various physiological processes such as photosynthesis, respiration and hormone regulation, which hampers growth and yield (Dlamini, 2021; Singh et al., 2022). As a result, salt stress significantly diminishes crop quality and is responsible for economic losses estimated at billions of dollars every year (Zörb et al., 2018). In this context, the identification and valorization of salt stress tolerant species appears to be a promising solution to mitigate these impacts and promote more resilient agriculture.

The carob tree (*Ceratonia siliqua* L.), a cultivar of the Fabaceae family, is largely grown in Mediterranean regions, particularly Morocco, Spain, Italy and Tunisia. Carob trees are distinguished by their evergreen foliage and capacity to adapt to arid and semi-arid environments, making them suitable for addressing soil erosion and desertification (Batlle & Tous, 1997; Ramon-Laca & Mabberley, 2004; Mouniane et al., 2024a). The fruit of the carob tree is called the "pod", and it possesses significant nutritional value, composed of sugars, fiber and bioactive compounds (Ayaz et al., 2009; Brassesco et al., 2021). Its use extends beyond its applications as a human nutritional supplement; it has many therapeutic effects and is a common ingredient in the pharmaceutical industry, particularly as a treatment option for diabetes and several infectious diseases (Rtibi et al, 2017; Stavrou et al., 2018; Richane et al., 2022).

The current research examines the effects of sodium chloride-induced salt stress on carob trees sourced from six Moroccan provinces. The response of carob trees to salt stress was assessed by measuring various parameters, including electrical conductivity, morphological parameters, physiological parameters, and biochemicals. These comprehensive studies aim to better understand the salt-stress tolerance and adaptation mechanisms of carob trees, which may help identify the most tolerant provinces for potential incorporation into sustainable agriculture in salt-affected areas.

2. LITERATURE REVIEW

2.1. Introduction to Soil Salinization as an Environmental Crisis

Soil salinization has gained prominence as one of the most significant aspects of environmental degradation, particularly in arid and semi-arid areas. With salinization affecting 20% of the irrigation lands globally, it is almost exclusively responsible for declining soil fertility, decreasing crop production and threatening food security (Han et al., 2024). Natural processes, including the weathering of parent materials, do contribute to soils salinization, but it is the anthropogenic activities such as poor irrigation methods, use of saline water, poor drainage, and over-fertilization that pose the most serious risk (El-Ramady et al., 2024). Climatic effects, such as rising sea levels and changes in the rainfall patters, can and have increased the capillary rise of salts in soil and worsened salinization (Maslin et al., 2025). Salinized soils are usually characterized by high electrical conductivity (EC), limited water availability due to osmotic stress, and toxic concentrations of sodium (Na^+) and chloride (Cl^-) ions, which disrupt cellular metabolism, affect enzyme activity and induce oxidative stress. Overall, these physiological disturbances reduce germination, impede plant growth and suppress reproduction success. Consequently, the long-term effects of salinized soils endanger agricultural ecosystems by either leading to land abandonment or a decreased supply of arable land.

2.2. Plant Responses to Salinity Stress

Saline conditions challenge plants at many levels, but they typically respond through a series of morphological, physiological and molecular changes to minimize damage and adapt to stress. These changes can be organized into 3 principal strategies, namely, osmotic adjustment, ion homeostasis, and antioxidant defense.

Osmotic Adjustment

Osmotic stress is the immediate response to high salinity, because high salt concentrations in the rhizosphere lead to lower soil water potentials, and thereby, restrict water uptake. To cope with osmotic stress, plants synthesize a number of osmolytes such as proline, soluble sugars, polyols, and glycine betaine. Osmolytes, also referred to as compatible solutes, are chemical compounds that function as solutes without interfering with the biochemical processes of living systems. They help maintain cell turgor, enzyme function and membrane integrity by maintaining normal concentrations. Proline is perhaps one of the most well documented osmoprotectants that also stabilizes sub-cellular structures and scavenges free radicals (Kang et al., 2024).

Ion Homeostasis and Compartmentalization

Elevated levels of Na+ and Cl- disrupt the balance of the nutrients, particularly the all-important K+/Na+ ratio necessary for enzyme activity and the maintenance of the membrane potential. Plants utilize several processes to help regulate ion concentration, including selective ion uptake; restricting translocation to the shoot; and sequestering toxic ions into vacuoles, through tonoplast transporters (e.g. NHX1). The Salt Overly Sensitive (SOS) pathway, especially SOS1 for Na+ extrusion, is well-defined in model organisms such as Arabidopsis (Zhang et al., 2025).

Antioxidant Defense

Salt stress generally leads to an excess in reactive oxygen species (ROS), such as hydrogen peroxide (H_2O_2), and free radicals like superoxide (O_2^-) and hydroxyl radicals (OH^-), which can produce oxidative effects on lipids, proteins and nucleic acids. Plants induction of enzymatic antioxidants such as superoxide dismutase, catalase, and peroxidases, as well as non-enzymatic antioxidants like ascorbate, glutathione, and phenolic compounds, help guard against ROS. These mechanisms reduce oxidative stress and are critical mechanisms during the other cellular stresses associated with salinity for restoring cellular homeostasis (Ahmed et al., 2024).

2.3. Carob Tree (*Ceratonia siliqua* L.) as a Resilient Species

Botanical and Ecological Characteristics

Ceratonia siliqua L, or carob tree, is an ever green leguminous tree that is native to the Mediterranean region. It can be found in semi-arid environments and grows in calcareous, rocky, and degraded soils. It has a deep root system that allows the tree to get water via groundwater and exhibits excellent drought and marginal soil tolerance. The carob tree plays a key role in protecting soil by alleviating erosion, contributes to soil structure via litter, and expands biodiversity by providing habitat. The carob trees versatility make it an ideal species for agroforestry systems and reforestation programs under degraded landscapes (Martins-Loução et al., 2024).

Nutritional and Economic Potential

They also produce high concentrations of natural sugars dietary fiber, and bioactive compounds (i.e., tannins, flavonoids, and phenolic acids) in their pods. A natural sweetener for the food and beverage industry and a mineral less substitute for cocoa in powder and syrup form can be made from the carob pulp. Carob seeds are used as locust bean gum, which is commonly used as a thicker in the food industry. The carob plants pharmacological function includes antioxidant, anti-inflammatory, anti-diabetic, and anti-diarrheal functions. These attributes make carob plants attractive to pharmaceutical and nutraceutical industries. The multipurpose use of carob is consistent with the circular bioeconomy concept that provides opportunities for creating value chains and improving rural livelihoods, and finally contributing to Sustainable Development Goals (SDGs) (Buzzanca et al., 2025).

2.4. Previous Studies on Salinity Tolerance in Carob

Research on carob's tolerance to salinity stress isn't as developed as literature developed for major staple crops but existing studies have identified several important adaptive traits of this species. Studies show that young carob can grow (and photosynthesize) in moderate salt if they are given a chance, likely because this tree is more adept at compartmentalizing excess sodium and chloride ions and starting osmotic adjustment to maintain water and cell integrity. Differences in responses to salinity appear to be mediated by genetic traits inherent to each ecotype. Comparing populations has led us to discover differences in drought and salt tolerance traits indicating that they may have physiological advantages to stress conditions. Variability can be seen in germination rate success and seedling vigor in salt and water-stressed treatments showcasing regional adaptedness strategies of carob cultivation. Important physiological characteristics such as retention of chlorophyll, ability to retain cellular water content, and increased production of proline can all serve as good predictors of the ability to tolerate salt stress. However, so far, most studies generally focus on a single trait or single stage of development. There is an apparent lack of integrative studies assessing morphological, physiological, and biochemical responses as a function of many ecotypes, meaning the entire range of carob's adaptive potential in relation to salinity remains largely unlooked at present and requires further and related study to begin to fully understand.

2.5. Research Gaps and Rationale for the Present Study

Though the existing literature indicates the potential of the carob tree as a resilient species, there are still some gaps. First, much of the literature is focused on seedlings or only limited physiological traits in seedlings, typically not examining long-term growth performance in the field at relevant stress levels. Second, there is very little work on, and understanding of, the genetic basis of tolerance mechanisms, and little research investigating intraspecific variation in ecotypes within a single country.

In Morocco, carob growing has ecological and economic viability, and an understanding of how the local ecotypes respond variably to salinity is important information when developing breeding programs, reforestation programs, and/or sustainable land management. The current study addresses these gaps by confronting six Moroccan carob ecotypes with a controlled salt stress and performance was evaluated using a multi-parametric approach including electrical conductivity, morphological growth parameters, water content, pigment content, proline and sugar content, and variance and correlation statistics.

2.6. Linking Carob Tolerance to Sustainable Agriculture and the Bioeconomy

The inclusion of salt-tolerant species such as carob in degraded agricultural systems offers a route to sustainable agriculture. Carob's low input requirements, multiple uses, and ecological functions make it an appropriate species for agroforestry systems in saline-affected areas. In the climate change context, and the goals of land degradation neutrality, carob assuming a cultivation role supports ecological restoration with economic motivation. Its valorization through food, pharmaceutical, and cosmetic industries is utilized to develop circular value chains and rural entrepreneurship. Additionally, recognizing and conserving the genetic diversity of carob ecotypes supports global goals to adapt agriculture for resilience, invest in food security, and sustainably use natural resources. The present study brings together empirical work on stress tolerance with society wide environmental and economic goals to support decision making in adaptive agroecosystem planning of carob.

3. MATERIALS AND METHODS

3.1. Plant Material

Carob seeds are obtained from mature pods collected during the optimum ripening period to guarantee seed quality and viability. This study selected six harvesting sites based on their ecology, which allowed a grab of the different climate and soil types of the country (figure 1).

Figure 1. Collection sites for carob pods (Mouniane et al., 2024b)

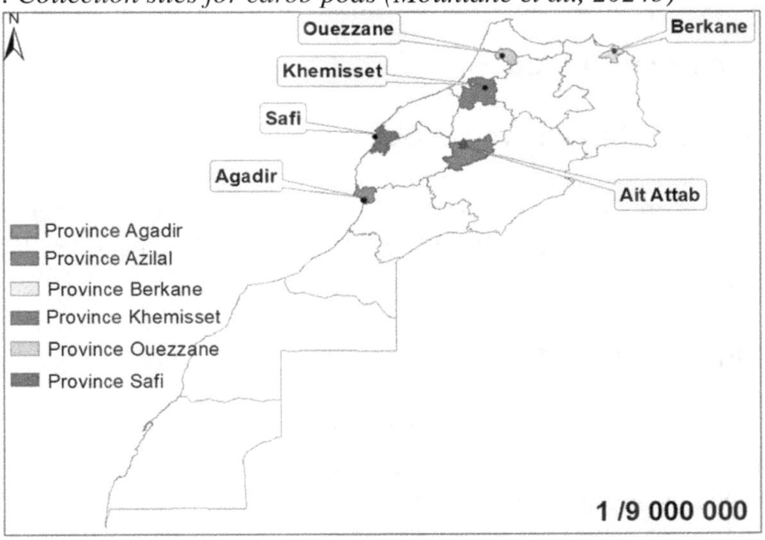

3.2. Growing Conditions

The study was carried out at University Ibn Tofail, Faculty of Sciences, Kenitra, Morocco, in an experimental greenhouse during July and August 2023. First, the seeds of each province underwent a scarification treatment by soaking in sulfuric acid for 20 minutes, followed by soaking in water for 24 hours to optimize germination rates. The germinated carob seedlings are then transplanted in plastic pots with a mixture of Maamoura soil, compost and peat in a ratio (2:1:1) for 6 months.

3.3. Salt Stress Treatments Applied

Three NaCl concentration solutions, and a control group with no added salt (Table 1), were used to simulate salt stress. Plants were irrigated with these solutions every 48 hours for a period of 60 days, to ensure continuous exposition to salt stress.

Table 1. NaCl treatment used to induce salt stress

Concentration	NaCl (g/L)	NaCl (mM/L)	CE (Ms/cm)
C_0	0	0	0.61
C_1	5	85	8.30
C_2	10	171	15.38
C_3	15	257	22.2

3.4. Parameters Studied

3.4.1. Electrical Conductivity of Soil

Electrical conductivity was determined by using a conductivity meter on a solution prepared from a mixture of 20 g of soil in 100 ml of distilled water (Derouiche et al., 2023).

3.4.2. Morphological Parameters

Three morphological parameters were assessed: biomass was determined using an electrical balance, and the height of the aerial parts and the length of the roots were measured using a graduated ruler.

3.4.3. Physiological Parameters

Relative water content was determined according to the method described by Barrs (1966). Chlorophyll content was assessed using the protocol described by Arnon (1949) and carotenoids were assessed using the following procedure used by (Wellburn, 1994).

3.4.4. Biochemical Parameters

Proline concentration was assessed using the protocol described by Bates et al. (1973), and total sugars were quantified using the protocol of Dubois et al. (1956).

3.5. Statistical Analysis

Results are presented as mean values ± standard deviations, and to identify significant variations, ANOVA was performed using IBM SPSS Statistics version 23 software. In addition, correlation analysis was also carried out.

4. RESULTS

4.1. Variation in Electrical Conductivity

Analysis of electrical conductivity showed a significant increase with NaCl concentration in all the ecotypes studied. Ouazzane province showed a high increase of 641.87%, followed by Berkane with 442.05%, Ait Attab with 361%, Safi with 354.96% and Khémissat with 344.18%, while Agadir province showed the lowest increase of 230.70%.

The ANOVA analysis confirms that NaCl, seed province and their interaction significantly influence the electrical conductivity of carob plants under salt stress.

Table 2. Variation in electrical conductivity as a function of salt stress in the six feedstocks studied

Provinces	**Electrical conductivity (mS/cm)**			
	0 g/L NaCl	**5 g/L NaCl**	**10 g/L NaCl**	**15 g/L NaCl**
Agadir	1,14 ± 0,02	1,47 ± 0,02	2,24 ± 0,03	3,77 ± 0,02
Aït Attab	0,78 ± 0,02	1,55 ± 0,03	2,76 ± 0,01	3,62 ± 0,02
Berkane	0,65 ± 0,02	1,45 ± 0,02	2,62 ± 0,02	3,52 ± 0,02
Khémissat	0,48 ± 0,02	1,80 ± 0,02	1,86 ± 0,02	2,14 ± 0,03
Safi	0,74 ± 0,02	1,36 ± 0,02	2,86 ± 0,02	3,36 ± 0,02
Ouazzane	0,55 ± 0,01	1,35 ± 0,02	2,54 ± 0,02	4,13 ± 0,02
Source of variation	***p*-Value**			
NaCl (A)	<0.001***			

continued on following page

Table 2. Continued

Provinces	Electrical conductivity (mS/cm)			
	0 g/L NaCl	5 g/L NaCl	10 g/L NaCl	15 g/L NaCl
Provinces of seeds (B)	<0.001***			
A x B	<0.001***			

(not significant; * p < 0.05; ** p < 0.01; *** p < 0.001).

4.2. Effect of Salt Stress on Morphological Parameters

Analysis of figure 2 showed a significant reduction in the morphological parameters studied, such as biomass, length of aerial part and roots, in carob plants from all provinces at all salinity levels (5, 10 and 15 g/L NaCl) compared with the control (0 g/L NaCl).

For biomass, the smallest reduction was in Berkane province with a 62.05% reduction, followed by Aït Attab province with a 64.94% reduction, Ouazzane with a 65.51% reduction, Safi with a 70.51% reduction, and Khemissat with an 82.45% reduction. Finally, the province of Agadir shows the greatest reduction, with 83.06%.

As regards the length of the aerial part, the percentages of reduction were 44.07%, 50.15%, 54.95%, 55.35%, 55.63% and 58.35% for the provinces of Safi, Khemissat, Ait Attab, Agadir, Berkane and Ouazzane respectively. In terms of roots, the smallest reduction was in Agadir province, with a fall of 38.29%, while Khemisset showed the biggest reduction, with a fall of 58.23%. For the other provinces, the reduction in roots varied between 48.38% and 54.23%.

Figure 2. Variation in morphological parameters of six carob provinces, a: biomass; b: length of aerial part; c: length of roots.

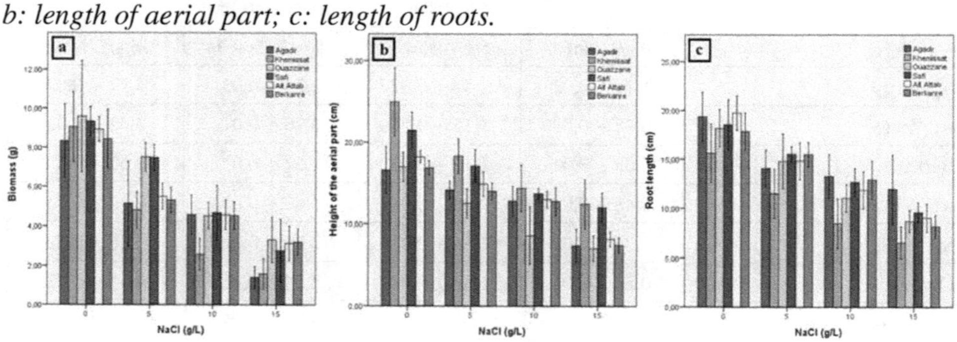

ANOVA analysis (Table 3) shows a significant effect of NaCl indicating that salt stress significantly affects morphological parameters in all provinces. In addition, the effect of seed province is also highly significant, reflecting local differences in response to salt stress. The interaction between NaCl and province is significant, suggesting that the impact of NaCl varies from province to province, some being more resistant or more sensitive to salt stress than others.

Table 3. Analysis of variance (ANOVA) of morphological parameters

Morphological parameters	**Biomass**	**Height of the aerial part**	**Root length**
Source of variation	*p*-Value	*p*-Value	*p*-Value
NaCl (A)	<0.001***	<0.001***	<0.001***
Provinces of seeds (B)	<0.001***	<0.001***	<0.001***
A x B	<0.001***	<0.001***	<0.01**

(not significant; * $p < 0.05$; ** $p < 0.01$; *** $p < 0.001$).

4.3. Effect of Salt Stress on Physiological Parameters

Salt stress negatively affects relative water content (Figure 3.a) and chlorophyll content (Figure 3.b). The Agadir province shows a 35.34% decrease in relative water content (RWC) and a 50.22% decrease in chlorophyll content. Khémissat showed a decrease of 47.27% in RWC and 45.69% in chlorophyll. For Ouazzane RWC dropped by 40.42% and its chlorophyll content also showed a significant reduction, down 52.40%. Safi saw a 49% reduction in RWC, and a 42.45% drop in chlorophyll content. Aït Attab saw a 38.61% drop in RWC and a 40.76% reduction in chlorophyll. Finally, Berkane recorded a 44.24% drop in RWC and 44.62% in chlorophyll.

Figure 3. Variation in physiological parameters of six carob provinces, a: relative water content; b: chlorophyll content.

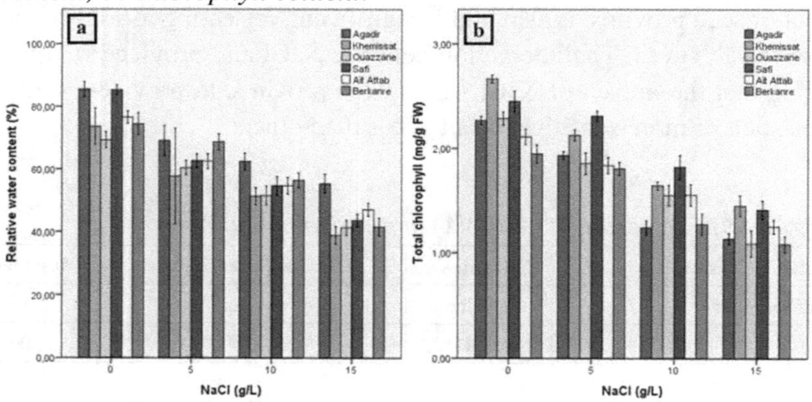

ANOVA analysis (Table 4) showed that carob seed province, salt stress, and their interaction have significant effects on the relative water content and chlorophyll concentration of carob plants.

Table 4. Analysis of variance (ANOVA) of physiological parameters

Morphological parameters	**Relative water content**	**Total chlorophyll**
Source of variation	*p*-Value	*p*-Value
NaCl (A)	<0.001***	<0.001***
Provinces of seeds (B)	<0.001***	<0.001***
A x B	<0.001***	<0.001***

(not significant; * $p < 0.05$; ** $p < 0.01$; *** $p < 0.001$).

4.4. Effect of Salt Stress on Biochemical Parameters

In general, proline (Figure 4.a) and sugar (Figure 4.b) content increased significantly for all provinces under salt stress.

For the Agadir ecotype, proline rose from 0.83 mg/g to 1.65 mg/g, representing an increase of 97.6%. In Khemissat, proline rose from 0.67 mg/g to 1.61 mg/g, the biggest increase of 138.4%. For Ouazzane, proline content rose from 0.75 mg/g to 1.58 mg/g, corresponding to an increase of 110.2%. Safi shows an increase from 0.84 mg/g to 1.84 mg/g, i.e. a 118% secretion. Aït Attab saw its proline content rise from 0.95 mg/g to 1.58 mg/g, an increase of 66.6%. Finally, Berkane recorded an increase from 0.87 mg/g to 1.74 mg/g, i.e. an increase of 100.7%.

For the Agadir ecotype, sugar content rose from 8.99 mg/g to 15.48 mg/g, representing an increase of 72.2%. In Khemissat, it rose from 13.96 mg/g to 20.89 mg/g, an increase of 49.6%. Ouazzane shows a very marked increase, rising from 10.59 mg/g to 29.07 mg/g, an increase of 174.5%. The Safi ecotype saw its sugar content rise from 9.59 mg/g to 17.69 mg/g (+84.4%). Aït Attab also saw a significant increase, from 8.40 mg/g to 18.44 mg/g, corresponding to an increase of 119.4%. Finally, Berkane saw an increase from 9.41 mg/g to 19.76 mg/g, a rise of 110%.

Figure 4. Variation in physiological parameters of six carob provinces, a: relative water content; b: chlorophyll content.

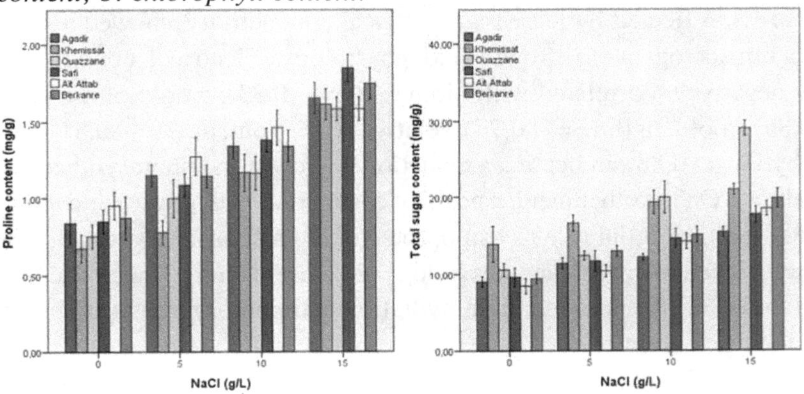

Analysis of variance (ANOVA) was used to assess the impact of two factors (seed province and salt stress) and their interaction on proline and total sugar content (Table 5). The effect of province on proline and total sugar content was significant, with marked variations between ecotypes. Similarly, NaCl concentration has a strong influence on proline and sugar accumulation. The interaction between these two factors is also significant.

Table 5. Analysis of variance (ANOVA) of physiological parameters

Morphological parameters	**Proline content**	**Total sugar content**
Source of variation	*p*-Value	*p*-Value
NaCl (A)	<0.001***	<0.001***
Provinces of seeds (B)	<0.001***	<0.001***
A x B	<0.001***	<0.001***

(Different letters in the column indicate significance at the 5% level. ns: not significant; * $p < 0.05$; ** $p < 0.01$; *** $p < 0.001$).

4.5. Correlation Analysis

Salt stress, reflected by increased electrical conductivity, showed a significant negative impact on carob growth and physiology. Electrical conductivity was strongly negatively correlated with biomass (r = - 0,83), height of the aerial part (r = - 0,85), root length (r = - 0,77), relative water content (r = - 0,81), and total chlorophyll (r = - 0,88), indicating a reduction in growth and photosynthetic activity under salinity. On the other hand, a positive correlation with proline and total sugars underlines their essential role as osmoprotectors in response to stress. In addition, parameters linked to growth and physiology are closely interconnected, confirming that the ability of plants to maintain hydration and photosynthesis is essential for their resilience to salinity.

Figure 5. Correlations between electrical conductivity and morphological, physiological and biochemical parameters of carob trees under salt stress

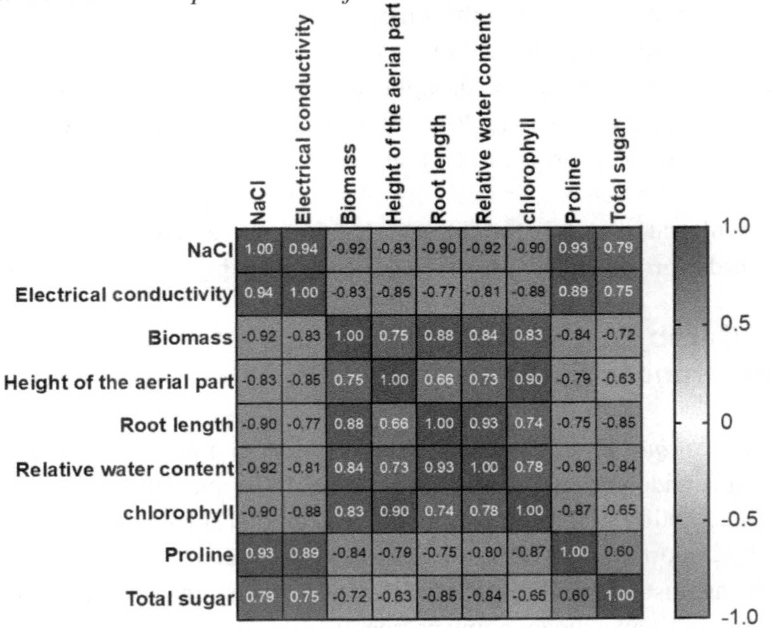

5. DISCUSSION

Salinity is a critical constraint to agricultural productivity, in both naturally saline soils and irrigated soils (Elloumi et al., 2024). In the Mediterranean Basin the low quality of water available to grow olives is another obstacle to growing olives and has detrimental impacts on yields (Chartzoulakhis, 2005; Achraf and Foolad, 2007). Of all forms of environmental degradation, salinization of soils is causing considerable concern due to land productivity of agricultural land, ecosystems (biodiversity), and global food security. In the face of these threats, there has been a renewed focus on the study of plant tolerance to salinity in the pursuit of preparing for sustainable agricultural management and increased resilience of socio-ecological systems. The aim of this work was to evaluate the response of various Moroccan ecotypes of carob (*Ceratonia siliqua* L.) to sodium chloride (NaCl) induced salt stress, in order to propose innovative and integrated solutions to contemporary environmental environmental and ecological challenges.

5.1. Salinity, A Major Factor in Farmland Degradation

Soil salinization is caused by both human activity (bad farming practices, excessive irrigation, inadequate management of water resources), and climate change (rising sea levels, prolonged periods of drought, increased rates of evapotranspiration). Currently, it threatens more than 20% of the world's irrigated land; approximately 1,125 million hectares, and is on the rise. By affecting the physical and chemical properties of soil (structure, permeability, plant nutrients), salinization can severely reduce agricultural productivity. The sustainable management of this land requires a shift towards agro-ecological practices, such as growing salt-tolerant crops.

5.2. The Carob Tree, A Strategic Species for Mediterranean Ecosystems

Ceratonia siliqua L. (the carob tree), is considered a primarily Mediterranean species, and is widely grown in Morocco, Spain, Italy, and Tunisia. Its robustness, longevity, and ability to thrive in poor and/or degraded soils make it a particularly ecologically important and economically promising species. It is also an important component against desertification, soil fixation, moisture retention, and carbon capture/storage. From a bioeconomy perspective, its fruits are rich in sugars, fibres, and bioactive ingredients, which can be used in agri-food, pharmacy, and cosmetic sectors. Our study demonstrates that the species exhibits exceptional tolerance capacities to salt stress, and that salt tolerance will usually depend on the origin of the seeds. As a result, these findings may create opportunities for reforestation, phytoremediation, and valorization in marginal areas (Dadach et al., 2024).

5.3. Impact of Salt Stress on Electrical Conductivity and Morphological Parameters

In our study, salt stress led to an increase in electrical conductivity, as expected. Morphologically, carob growth was inhibited with increasing NaCl concentration, as evidenced by decreases in biomass, aerial part length and root length. The results achieved indicate a marked decrease in biomass, height of aerial part and root length in all ecotypes suffering from salt stress. This inhibition of growth reflects the classical impact of osmotic stress that limits the uptake of water and nutrients and leads to ionic imbalance and the accumulation of toxic free radicals. However, not all ecotypes responded in the same way. The Berkane ecotype showed the lowest reduction in biomass (62%) which clearly reflects a certain morphological plasticity. The Agadir ecotype in fact showed an 83% reduction which indicates greater sensitivity. These results highlight the probable existence of a local genetic variability that could be used as source material to select resistant genotypes adapted to saline soil.

5.4. Physiological Responses: Hydric and Phototrophic Adjustments

Physiologically, relative leaf water content and chlorophyll concentration also decreased significantly in all ecotypes. High levels of salinity impose a disproportionate decrease of relative leaf water content (RWC), a key component of plant water balance. The reduction in RWC indicates the problems plants must overcome to remain turgid given the underlying osmotic gradient. At high levels of salinity, chlorophyll degrades as it has a direct downstream effect on limiting photosynthetic activity. This loss of functionality following salinity stress is well documented in the literature. During salt stress, stomata close in turn reducing the amount of CO_2 fixed, limiting its overall uptake, and chlorophyll pigment degrades as part of its adaptive capacity to mitigate light related oxidative damage (photo-inhibition). Our results support what many have documented while providing interesting variation between inter-ecotype plant dynamics. For example, the Aït Attab ecotype has a higher relative chlorophyll which may indicate its comparative phototrophic adaptive capacity.

5.5. Accumulation of Osmoprotectants: Proline and Sugars

At the same time, an increase in proline and total sugars was observed, indicating mechanisms of adaptation and tolerance to salt stress. One of the fundamental ways that plants are able to cope with salt stress is through the accumulation of compatible solutes, or osmoprotectants, in the form of proline and soluble sugars. These solutes are important in stabilizing cell membranes, providing protection to proteins and enzymes, and preserving osmotic pressure. In the present study, all ecotypes had a significant increase in proline from 66% to 138%, and total sugar content increased for each ecotype with the Ouazzane ecotype (+174%) having the most significant increase. It is expected for halophytic or tolerant plants to produce this metabolic response, which is an efficient method of survival in challenging environments.

Based on particular physiological and biochemical mechanisms, such as the accumulation of osmoprotectants (proline, soluble sugars), and regulation of photosynthetic activity, the ability is granted (Correia et al., 2010). Due to its various responses, it can still grow and produce in limited situations, making it an important crop in farming adaptability plans (Gadoum et al., 2019). Carob is an important tree that is used to fight against the degradation of devastated ecosystems. Carob has great economic potential due to its diverse uses in various sectors including agri-food, pharmaceuticals, and cosmetics. The use of carob is not limited to its mitigating effects on ecosystem degradation. The flour that comes from carob is full of fiber and useful chemical compounds and fits sustainably in a bioeconomy approach (Laaraj et al., 2023).

5.6. Significant Correlations and Interactions

Correlation analysis revealed strong negative associations between soil electrical conductivity (EC) and morphological and physiological parameters, which reflect the detrimental effects of salts on growth. Positive correlations were observed with osmoprotective compounds. These results corroborate the critical role of metabolic plasticity in adaptation to stress. The strong interaction between NaCl concentration and seed provenance, indicated by ANOVA, emphasises the need for a ecotypic approach to breeding strategies. This confirms the requirement of in situ and ex situ conservation programs for local genetic resources. These results are in line with previous studies highlighting the negative impact of salinity on plant growth, while also highlighting the accumulation of osmoprotectants as an adaptive response (Wu et al., 2023; Hamouda et al., 2023; Arif et al., 2024; Tian et al., 2024). Certainly, plants employ various defence responses against salt stress such as reduction in leaf area and development of deeper root system for better uptake of water and nutrients (Mehta and Vyas, 2023). Proline, sugars, and other metabolites act as osmoprotectants to keep osmotic balance and help stabilise membranes (Acosta-Motos et al., 2017; Zarat et al., 2020).

5.7. Ecological and Agronomic Implications

Utilizing salt-tolerant ecotypes of carob trees can provide an extensive ecological and agronomic opportunity to address the increasing salinization of soils (Dadach et al., 2023). These are plants that can acclimatize under adverse conditions and directly aid in rehabilitating degraded land. They are effective in the stabilization of soil and limiting erosion, and are setting into recovery important and essential ecological functions often found in marginal habit that are implored to conventional crop production as a result of desiccation and degradation (Chalise et al., 2019). Their use in more deliberative targeted reforestation projects creates an opportunity to also strategically maximize and optimize the goals of environmental protection and sustainable agricultural production. As an adaptation of a crop, particularly in Mediterranean agroforestry systems, the carob tree is a potentially advantageous choice because it can produce under stress from both water and salt and profoundly limit erosion. Creating and maintaining intra-specific diversity within carob populations and communities will recreate variety and certainly improve the species' genetic resilience against both physical and possible stresses that expand beyond conventional farming management while also supporting local plant genetic resources. Human activity and climate change are putting more pressure on agricultural and natural ecosystems that are causing the degradation of soils and salinization of agricultural lands. As a result, adapting crops to extreme abiotic conditions which include salt stress is emerging as a major challenge for food security and sustainability of agricultural systems (Dagar et al., 2016). The carob tree (*Ceratonia siliqua* L.) is an important and commonly cultivated Mediterranean tree species in Morocco that grows well in dry and saline habitats (Gharnit and Ennabili, 2016; Dadar et al., 2024).

5.8. Towards a Circular Bio-Economic Approach

Apart from its environmentally sustainable characteristics, the carob tree contributes to a circular bioeconomy. The fruits of carob trees are high in dietary fiber, natural sugars, tannins and polyphenols and can produce high-value products for the food, pharmaceutical and cosmetics industries. Processing them into flours, syrups and natural additives can provide major economic opportunities to rural areas, many of which are marginal. While generating direct income for local actors through added-value processing of carobs, it also encourages actors to build regional value chains from a renewable, low-input resource. It helps to reduce the need for raw materials that are produced through intensive agriculture (often monocultures) or synthetic chemistries. The development of the carob tree inevitably eases the pressure on natural resources and also supports a greener, environmental economy. Therefore, addition of the carob tree to national policies for rural development, food security and the bioeconomy is not only a technical decision; it is a strategic choice in attempting to address the impacts of climate change and sustainability issues.

Starting carob planting and varietal improvement programs in salinization-threatened areas is a promising strategy. This could help improve soils and biodiversity while boosting the local economy through the valorisation of endogenous resources. However, looking at how the Moroccan carob tree responds to salt stress, it leaves the door open for agriculture that can still be sustainable, based on the fact that climate change and environmental degradation are challenges that the future holds. Conservation of this species in agricultural and environmental policies can enhance ecosystem resilience while supporting the shift towards a circular and sustainable bioeconomy (Palaiogianni et al., 2022).

5.9. Scientific and Strategic Perspectives

While the results achieved in this study are promising, they also reveal the necessity for further research in a number of complementary areas. In particular, it would be useful to obtain a clearer understanding of the molecular mechanisms of salt tolerance in the different ecotypes (i.e., identification of genes involved, signalling pathways triggered and differential expression profiles). In addition, the resilience of these ecotypes should be assessed over appropriate time periods under natural field conditions in order to validate the stability of the traits observed in controlled conditions. The development of a cross-breeding program between the best-performing ecotypes could be used to develop improved varieties that are better adapted to saline zones. Lastly, a seed bank of tolerant ecotypes could be a strategic asset for establishing future planting and ecological restoration activities. However, in order to make this dynamic truly effective, it will need to be bolstered by strong public policies that promote the cultivation of carob trees in marginal areas and include the integration of small farmers into production, processing, and marketing circuits.

6. CONCLUSION

In conclusion, our study indicates that carob plants show a gene-based ability to tolerate salt stress. Selected resistance ecotypes of carob demonstrate great potential for use in reforestation programs and the rehabilitation of saline-affected marginal soils. Yet, further studies particularly at the molecular level are needed to better understand the mechanisms involved and help reinforce conservation and enhancement strategies of this valuable species in the future. With the proper utilization of local genetic resources, the carob tree is a key strategic species for sustainable agriculture and the bio-economy. Its potential for restoring saline soils and agri-food production makes it a viable solution for mitigating the effects of environmental degradation and promoting more resilient agricultural systems in the face of climate change.

REFERENCES

Achraf, M., & Foolad, M. R. (2007). Roles of glycine betaine and proline in improving plant abiotic stress resistance. *Environmental and Experimental Botany*, *59*(2), 206–216. DOI: 10.1016/j.envexpbot.2005.12.006

Acosta-Motos, J. R., Ortuño, M. F., Bernal-Vicente, A., Diaz-Vivancos, P., Sanchez-Blanco, M. J., & Hernandez, J. A. (2017). Plant Responses to Salt Stress: Adaptive Mechanisms. *Agronomy (Basel)*, *7*(1), 18. DOI: 10.3390/agronomy7010018

Ahmed, M., Tóth, Z., & Decsi, K. (2024). The impact of salinity on crop yields and the confrontational behavior of transcriptional regulators, nanoparticles, and antioxidant defensive mechanisms under stressful conditions: A review. *International Journal of Molecular Sciences*, *25*(5), 2654. DOI: 10.3390/ijms25052654 PMID: 38473901

Arif, M., Khurshid, H., Rabbani, M. A., Jahanzaib, M., Khattak, S. H., Ullah, I., Farah, M. A., Singh, R., & Li, L. (2024). Morpho-physiological traits and stress indices reveal divergence in early-stage salt stress response in elite soybean germplasm. *Journal of King Saud University. Science*, *36*(11), 103522. DOI: 10.1016/j.jksus.2024.103522

Arnon, D. I. (1949). Copper enzymes in isolated chloroplasts. polyphenoloxidase in beta vulgaris. *Plant Physiology*, *24*(1), 1–15. DOI: 10.1104/pp.24.1.1 PMID: 16654194

Ayaz, F. A., Torun, H., Glew, R. H., Bak, Z. D., Chuang, L. T., Presley, J. M., & Andrews, R. (2009). Nutrient content of carob pod (*Ceratonia siliqua* L.) flour prepared commercially and domestically. *Plant Foods for Human Nutrition (Dordrecht, Netherlands)*, *64*(4), 286–292. DOI: 10.1007/s11130-009-0130-3 PMID: 19763833

Barrs, H. D. (1966). Root pressure and leaf water potential. *Science*, *152*(3726), 1266–1268. DOI: 10.1126/science.152.3726.1266 PMID: 17769543

Bates, L. S., Waldren, R. P., & Teare, I. D. (1973). Rapid determination of free proline for water studies. *Plant and Soil*, *39*(1), 205–208. DOI: 10.1007/BF00018060

Batlle, I., & Tous, J. (1997). *Carob tree Ceratonia siliqua L. Promoting the conservation and use of underutilized and neglected crops. 17* (Vol. 17). Bioversity International.

Bello, A. S., & Ahmed, T. (2024). Moringa leaf extract alleviates salt stress in tomato (*Solanum lycopersicum* L.) by activating antioxidant defenses, reducing osmolyte accumulation, improving water status, and enhancing yield. *Plant Stress*, *14*, 100640. DOI: 10.1016/j.stress.2024.100640

Brassesco, M. E., Brandao, T. R., Silva, C. L., & Pintado, M. (2021). Carob bean (*Ceratonia siliqua* L.): A new perspective for functional food. *Trends in Food Science & Technology, 114*, 310–322. DOI: 10.1016/j.tifs.2021.05.037

Buzzanca, C., D'Amico, A., Pistorio, E., Di Stefano, V., & Melilli, M. G. (2025). Carob-based functional beverages: Nutritional value and health properties. *Beverages, 11*(1), 1. DOI: 10.3390/beverages11010001

Chalise, D., Kumar, L., & Kristiansen, P. (2019). Land degradation by soil erosion in Nepal: A review. *Soil Systems, 3*(1), 12. DOI: 10.3390/soilsystems3010012

Chartzoulakhis, K. S. (2005). Salinity and olive: Growth, salt tolerance, photosynthesis and yield. *Agricultural Water Management, 78*(1-2), 108–121. DOI: 10.1016/j.agwat.2005.04.025

Correia, P. J., Gama, F., Pestana, M., & Martins-Loução, M. A. (2010). Tolerance of young (*Ceratonia siliqua* L.) carob rootstock to NaCl. *Agricultural Water Management, 97*(6), 910–916. DOI: 10.1016/j.agwat.2010.01.022

Dadach, M., Bhatt, A., Gairola, S., Elnaggar, A., & El-Keblawy, A. (2024). Germination responses of carob (*Ceratonia siliqua* L.) seeds to salinity and water stresses: Implications for the restoration of Mediterranean marginal lands. *Botany, 103*, 1–13. DOI: 10.1139/cjb-2024-0070

Dadach, M., Gairola, S., Hameed, A., & El-Keblawy, A. (2023). Effects of different chloride salts on seed germination and early seedling performance of Ceratonia siliqua, a multipurpose tree of the Mediterranean region. *South African Journal of Botany, 158*, 8–17. DOI: 10.1016/j.sajb.2023.04.044

Dagar, J. C., Sharma, P. C., Chaudhari, S. K., Jat, H. S., & Ahamad, S. (2016). Climate Change vis-a-vis Saline Agriculture: Impact and Adaptation Strategies. In Dagar, J., Sharma, P., Sharma, D., & Singh, A. (Eds.), *Innovative Saline Agriculture*. Springer., DOI: 10.1007/978-81-322-2770-0_2

Derouiche, M., Mzabri, I., Ouahhoud, S., Dehmani, I., Benabess, R., Addi, M., Hano, C., Boukroute, A., Berrichi, A., & Kouddane, N. (2023). The effect of salt stress on the growth and development of three *Aloe* species in eastern Morocco. *Plant Stress, 9*, 100187. DOI: 10.1016/j.stress.2023.100187

Dlamini, P. J. (2021). Drought stress tolerance mechanisms and breeding effort in sugarcane: A review of progress and constraints in South Africa. *Plant Stress, 2*, 100027. DOI: 10.1016/j.stress.2021.100027

Dubois, M. K., Gils, J. K., Hanniton, P. A., & Smith, F. (1956). Use of phenol reagent for the determination of total sugar. *Analytical Chemistry*, *28*, 350–356. DOI: 10.1021/ac60111a017

El-Ramady, H., Prokisch, J., Mansour, H., Bayoumi, Y. A., Shalaby, T. A., Veres, S., & Brevik, E. C. (2024). Review of crop response to soil salinity stress: Possible approaches from leaching to nano-management. *Soil Systems*, *8*(1), 11. DOI: 10.3390/soilsystems8010011

Elloumi, O., Amar, F. B., Trigui, A., Gargouri, K., & Ghrab, M. (2024). Early evaluation of salt-stress tolerance of new released olive cultivars based on physiological and biomass allocation indicators. *Journal of Agriculture and Food Research*, *18*, 101503. DOI: 10.1016/j.jafr.2024.101503

Gadoum, A., Adda, A., Sahnoune, M., & Aid, F. (2019). Physiological and biochemical responses of three ecotypes of carob (Ceratonia siliqua L.) against drought stress in Algeria. *Applied Ecology and Environmental Research*, *17*(2), 1929–1945. DOI: 10.15666/aeer/1702_19291945

Gharnit, N., & Ennabili, A. (2016). Categories of Carob Tree (*Ceratonia siliqua* L.) from Morocco. *International Journal of Fruit Science*, *16*(3), 259–274. DOI: 10.1080/15538362.2015.1102674

Hamouda, M. M., Badr, A., Ali, S. S., Adham, A. M., Ahmed, H. I. S., & Saad-Allah, K. M. (2023). Growth, physiological, and molecular responses of three phaeophyte extracts on salt-stressed pea (*Pisum sativum* L.) seedlings. *Journal of Genetic Engineering and Biotechnology*, *21*(1), 32. DOI: 10.1186/s43141-023-00483-z PMID: 36929363

Han, J., Kim, M., Mammadov, Z., Lee, S., Elzinga, E. J., Mammadov, G., Hwang, W., & Ro, H. M. (2024). Synergistic effect of climate change and water management: Historical and future soil salinity in the Kur-Araz lowland, Azerbaijan. *The Science of the Total Environment*, *907*, 167720. DOI: 10.1016/j.scitotenv.2023.167720 PMID: 37827308

Kang, J., Wang, X.-L., Yan, S.-J., Guo, H., & Cui, Y.-N. (2024). Comparative physiological and gene expression analyses provide insights into ion transports and osmotic adjustment of sweet sorghum under salt stress. *Agronomy (Basel)*, *14*(8), 1849. DOI: 10.3390/agronomy14081849

Laaraj, S., Salmaoui, S., Addi, M., El-rhouttais, C., Tikent, A., Elbouzidi, A., Taibi, M., Hano, C., Noutfia, Y., & Elfazazi, K. (2023). Carob (*Ceratonia siliqua* L.) Seed Constituents: A Comprehensive Review of Composition, Chemical Profile, and Diverse Applications. *Journal of Food Quality*, *2023*, 1–14. DOI: 10.1155/2023/3438179

Martins-Loução, M. A., Correia, P. J., & Romano, A. (2024). Carob: A Mediterranean resource for the future. *Plants, 13*(9), 1188. DOI: 10.3390/plants13091188 PMID: 38732403

Maslin, M., Ramnath, R. D., Welsh, G. I., & Sisodiya, S. M. (2025). Climate health emergency: Understanding the health impacts of the climate crisis. *Future Healthcare Journal, 12*(1), 100240. DOI: 10.1016/j.fhj.2025.100240 PMID: 40236934

Mehta, D., & Vyas, S. (2023). Comparative bio-accumulation of osmoprotectants in saline stress tolerating plants: A review. *Plant Stress, 9*, 100177. DOI: 10.1016/j.stress.2023.100177

Mouniane, Y., Chriqui, A., El-Khadir, I., Bensaid, A., Arabi, R., Zoubir, M. H., Mabrouki, J., & Hmouni, D. (2024a). Investigation on the Phytosociology and Soil Characteristics of Carob Tree (*Ceratonia siliqua* L.) in Northeast Morocco: Implications for Biodiversity Conservation and Ecosystem Restoration. In Mabrouki, J., & Azrour, M. (Eds.), *Integrated Solutions for Smart and Sustainable Environmental Conservation. Studies in Systems, Decision and Control* (Vol. 527). Springer., DOI: 10.1007/978-3-031-55787-3_10

Mouniane, Y., Chriqui, A., El-Khadir, I., Hbyaj, K., Rochd, A., Aitouhanni, I., Kouighat, M., & Hmouni, D. (2024b). Assessing Morphological Trait Variability in Moroccan Carob (*Ceratonia siliqua* L.) Ecotypes for Adaptive Breeding in Response to Climate Change and Sustainable Food Security. In BIO Web of Conferences. 109:01040. https://doi.org/DOI: 10.1051/bioconf/202410901040

Nguyen, T. T., Grote, U., Neubacher, F., Do, M. H., & Paudel, G. P. (2023). Security risks from climate change and environmental degradation: Implications for sustainable land use transformation in the Global South. *Current Opinion in Environmental Sustainability, 63*, 101322. DOI: 10.1016/j.cosust.2023.101322

Palaiogianni, A., Stylianou, M., Sarris, D., & Agapiou, A. (2022). Carob-Agro-Industrial Waste and Potential Uses in the Circular Economy. In Ramadan, M. F., & Farag, M. A. (Eds.), *Mediterranean Fruits Bio-wastes*. Springer., DOI: 10.1007/978-3-030-84436-3_33

Poveda, J. (2022). Effect of volatile and non-volatile metabolites from Leptosphaeria maculans on tomato calli under abiotic stresses. *Plant Stress, 3*, 100054. DOI: 10.1016/j.stress.2021.100054

Ramon-Laca, L., & Mabberley, D. J. (2004). The ecological status of the carob-tree (*Ceratonia siliqua* Leguminosae) in the Mediterranean. *Botanical Journal of the Linnean Society, 144*(4), 431–436. DOI: 10.1111/j.1095-8339.2003.00254.x

Richane, A., Rim, B. M., Wided, M., Riadh, K., Khaoula, A., Nizar, M., & Hanen, B. I. (2022). Variability of phenolic compounds and antioxidant activities of ten *Ceratonia siliqua* L. provenances. *Biochemical Systematics and Ecology*, *104*, 104486. DOI: 10.1016/j.bse.2022.104486

Rtibi, K., Selmi, S., Grami, D., Amri, M., Eto, B., El-Benna, J., & Marzouki, L. (2017). Chemical constituents and pharmacological actions of carob pods and leaves (*Ceratonia siliqua* L.) on the gastrointestinal tract: A review. *Biomedicine and Pharmacotherapy*, *93*, 522–528. DOI: 10.1016/j.biopha.2017.06.088 PMID: 28686965

Sanower, H. (2019). Present scenario of global salt affected soils, its management and importance of salinity research. *International Research Journal of Biological Sciences*, *1*, 1–3.

Sapre, S., Gontia-Mishra, I., & Tiwari, S. (2022). Plant growth-promoting rhizobacteria ameliorates salinity stress in pea (*Pisum sativum*). *Journal of Plant Growth Regulation*, *41*(2), 647–656. DOI: 10.1007/s00344-021-10329-y

Singh, P., Choudhary, K. K., Chaudhary, N., Gupta, S., Sahu, M., Tejaswini, B., & Sarkar, S. (2022). Salt stress resilience in plants mediated through osmolyte accumulation and its crosstalk mechanism with phytohormones. *Frontiers in Plant Science*, *13*, 1006617. DOI: 10.3389/fpls.2022.1006617 PMID: 36237504

Stavrou, I. J., Christou, A., & Kapnissi-Christodoulou, C. P. (2018). Polyphenols in carobs: A review on their composition, antioxidant capacity and cytotoxic effects, and health impact. *Food Chemistry*, *269*, 355–374. DOI: 10.1016/j.foodchem.2018.06.152 PMID: 30100447

Tian, X., Chang, Z., Wang, R., Li, G., & Qi, Y. (2024). The abiotic stress gene (Asg) family member Asg2 as a modulator of plant responses to salt stress. *Plant Stress*, *14*, 100596. DOI: 10.1016/j.stress.2024.100596

Wellburn, A. R. (1994). The spectral determination of chlorophylls a and b, as well as total carotenoids, using various solvents with spectrophotometers of different resolution. *Journal of Plant Physiology*, *144*(3), 307–313. DOI: 10.1016/S0176-1617(11)81192-2

Wu, C., Zhang, M., Liang, Y., Zhang, L., & Diao, X. (2023). Salt stress responses in foxtail millet: Physiological and molecular regulation. *The Crop Journal*, *11*(4), 1011–1021. DOI: 10.1016/j.cj.2023.06.001

Zarat, L., Alfaiz, C., Chiadmi, N., Kallida, R., Gaboun, F., & Ibriz, M. (2020). Plant growth, physiological response and osmotic adjustment of wild *bituminaria bituminosa* (L.) stirton accessions under salt stress. *Plant Cell Biotechnology and Molecular Biology*, *21*(1&2), 59–69.

Zhang, M., Ma, C., Qiao, S., Li, H., Zhao, W., & Liu, B. (2025). Compensatory growth and ion balance adaptation mechanisms of Salix matsudana Koidz under heterogeneous salinity stress. *BMC Plant Biology*, *25*(1), 231. DOI: 10.1186/s12870-025-06252-z PMID: 39979843

Zörb, C., Geilfus, C. M., & Dietz, K. J. (2019). Salinity and crop yield. *Plant Biology*, *21*(S1), 31–38. DOI: 10.1111/plb.12884 PMID: 30059606

Chapter 9
Environmental Awareness for a Sustainable Future: A Study on University Students

Faruk Ay
Sivas Cumhuriyet University, Turkey

Nazan Yalçın Erik
Sivas Cumhuriyet University, Turkey

Ferda Alper Ay
https://orcid.org/0000-0002-6170-1578
Sivas Cumhuriyet University, Turkey

ABSTRACT

Today's modern life makes it increasingly difficult for individuals to live in a healthy environment. In this context, sustainability plays a critical role in solving environmental problems. The purpose of this research is to determine the environmental awareness and knowledge levels of university students. The sample consists of 384 students studying at Sivas Cumhuriyet University. The data were analyzed using the independent sample t-test, ANOVA, and Tukey test in the SPSS 23.0 program. The findings showed high environmental awareness of students receiving education on climate and environmental problems, family environment, 2nd-grade and natural students, and vocational school students. The results reveal the effect of education on more sensitivity to the environment and provide an important roadmap for educators and policymakers in developing strategies.

DOI: 10.4018/979-8-3373-1549-2.ch009

1. INTRODUCTION

Energy is an indispensable need that powers almost all aspects of human life, and since prehistoric times, the social and cultural development of societies has been achieved thanks to the efficient use of energy and technological advances that support the industrial field. From past to present, fossil fuels such as oil and coal have been the most important sources that meet primary energy demand and their intensive use has increased global warming and climate change (Casalo & Escario, 2018). Human-induced carbon dioxide emissions are added to emissions from natural sources. This means that the capacity of carbon dioxide to absorb infrared radiation is already saturated; that is, the stronger absorption lines have reached the saturation point. Therefore, some gases, such as anesthetic gases, are more effective greenhouse gases than carbon dioxide, because they act on weaker absorption lines (Slingo & Slingo, 2024). The global warming trend observed since the mid-20th century is the "greenhouse effect" that occurs when the atmosphere traps heat radiating from the Earth into space, and agreements are being made to prevent or reduce the future dangers of this situation, aiming for the participation of all countries in the world (such as the Kyoto and Paris agreements). The ecological effects of global change include various problems caused by climate change and global warming; these include melting glaciers, sea level rise, floods, drought and loss of biodiversity. These ecological changes can also cause negative effects on the social and economic structure, bringing problems such as unemployment, economic stagnation and mass migration, which can result in global crises and possible conflicts (Ay, 2021). The effects of climate change on human health are increasingly understood; problems such as increasing temperature, depletion of water resources and hunger lead to negative health consequences on people, especially infants and the elderly (Başıbüyük et al., 2017). The potential impacts of climate change are given in Table 1 (Ay, 2021).

Table 1. Potential impacts of global climate change

Category	Main effects
Sea Level Rise and Coastal Regions	Coastal Erosion
	Floods and Inundations
	Costs of Protecting Coastal Communities
Human Health	Climate-Related Deaths
	Epidemics
	Decrease in Air Quality

continued on following page

Table 1. Continued

Category	Main effects
Energy	Change in Energy Policies
	Change in Energy Consumption
	Change in Energy Costs
Water Resources	Water Supply Decline
	Water Quality Decline
	Competition for Water Resources
Agriculture	Crop Losses
	Irrigation Problems
	Change in Agricultural Areas
Natural Environment and Species	Losses in Natural Habitats
	Reduction in Species Diversity
Forests	Forest Composition
	Changes in Geographic Distribution of Forests
	Decreases in Forest Health and Productivity

Environmental awareness training started in Europe in the 1960s and 1970s, and the development of environmental awareness started to increase in academic and scientific circles after this process. Environmental education that should be given to gain environmental awareness and sensitivity is defined by UNESCO as "Developing environmental awareness in all segments of society, gaining environmentally sensitive, permanent and positive behavioral changes and protecting natural, historical, cultural and socio-aesthetic values, ensuring active participation and taking part in solving problems" (UNESCO-UNEP, 1991). Although the mentioned trainings are becoming widespread, as Sanera (1998) also stated, the environmental education process is complex, and in-depth examinations are required at affective, cognitive, behavioral, and metacognitive levels for its results to be beneficial. As Jeronen and Kaikkonen (2002) stated, education on environmental problems and solutions must be comprehensive and solution-oriented, focusing on developing environmental awareness about broader sustainability issues such as ecosystem change, pollution (air, water, soil) and biodiversity loss.

Environmental studies generally address issues such as individuals' environmental sensitivity, attitudes toward the environment, and environmental awareness. For example, Ajzen and Fishbein (1975) stated that environmental awareness and sensitivity are affected by beliefs, attitudes, and social pressures, while Jaus (1982) similarly stated that there is a positive relationship between environmental education and positive attitudes toward the environment.

Various studies conducted on environmental knowledge have revealed that this knowledge is significantly related to the individual's gender, grade level, parents' education and knowledge level, and perception of environmental problems (Dillon & Gayford, 1997; Ma & Bateson, 1999; Zimmermann, 1996; Arı & Yılmaz, 2017; Gökdere, 2005; Koçak & Akış, 2021; Ek et al., 2009; Çetin et al., 2020). These findings show that environmental awareness is nourished not only by individual interest but also by the socio-demographic context.

Studies conducted in the context of gender indicate contradictory results. While some studies suggest that men are more conscious and sensitive to environmental issues than women (Arcury et al., 1986; Hes-Quimbita & Pavel, 1996; Worsley & Skrzypiec, 1998), others have reversed this finding and found that women exhibit stronger environmental sensitivity (Davidson & Freudenburg, 1996; Gardos & Dodd, 1995; Tikka et al., 2000; Arı & Yılmaz, 2017; Uzun & Sağlam, 2006; Yılmaz & Arslan, 2011). On the other hand, Shobeiri et al. (2007) and Eagles & Demare (1999) did not find a statistically significant relationship between gender and environmental sensitivity. However, they stated that women adopt more ethical approaches towards the environment.

Freudenburg and McGinn (1989) examined the relationship between environmental sensitivity and social structure. They showed that agricultural workers have a higher level of environmental awareness than other individuals living in the same rural community. Uyeki and Holland (2000) showed that individuals with lower socioeconomic status exhibit more protective and sensitive attitudes toward the environment.

Researchers such as Musser and Diamond (1999), who focus on the effect of income level, have stated that university students from higher-income families develop weaker attitudes toward the environment than students from lower-income families. These findings show that environmental sensitivity depends on individual preferences and the value world shaped by social class.

Reducing the effects of greenhouse gases that cause environmental problems will control population growth and prevent global warming to some extent. People need to be made aware of these issues, and measures must be taken globally. In this context, the leading solutions are Increasing energy efficiency, reducing fossil fuel-based vehicle transportation, increasing the use of renewable energy sources, phasing out fossil fuels, managing forests and agriculture, exploring nuclear energy, developing and distributing new low and zero-carbon technologies, ensuring sustainable development, and giving importance to activities such as adapting to ongoing changes. In addition, one of the most important is to increase awareness through education. Education is important for increasing people's, businesses, and even countries' knowledge of the importance of global warming and climate change. This will increase awareness and make it easier for them to adapt to the necessary measures. Since there is no other world to go to, measures should be taken as soon as possible to not disappear in this universe, live healthily, and win the battle for survival (Ay, 2021).

Organizations also have significant responsibilities regarding environmental awareness and sustainability. Organizations and managers must consider external environmental factors for motivation and develop strategies to increase motivation and efficiency (Alper Ay, 2018; 2025). In this context, global environmental effects such as climate change can significantly impact employees' psychological state and job satisfaction. For example, increasing temperatures, natural disasters, and environmental uncertainties can increase employees' stress levels and negatively affect their motivation. Therefore, managers' development of flexible working conditions, supportive programs, and sustainable practices, considering not only workplace dynamics but also environmental changes and their adverse effects on employees, will play an important role in increasing employees' motivation. The concept of green organizational behavior has become popular in recent years. It is suggested that breakthroughs should be made to reinvent resource consumption approaches so people can learn to live sustainably. While climate change has profound effects on current and future generations, many factors shape individuals' behavior. Although solutions such as energy saving and recycling are insufficient, taking small steps for sustainable living and green organizational behavior has become important (Alavosius & Houmanfar, 2020). In this context, green organizational behavior provides a framework that enables organizations to achieve their sustainability goals by covering various strategies and approaches related to climate change and environmental problems (Reşat Cica & Karabulut, 2024). In addition, evaluating the needs of employees from a holistic perspective for effective management requires a structured understanding of corporate social responsibility (Alper Ay & Yılmaz, 2024) and creating environmental awareness for a more livable and sustainable future.

However, in recent years, studies on "future literacy," a rich subject that includes climate and environmental issues, have gained importance for a sustainable future. These studies focus on sustainable futures, such as designing the future, business development, and education. Numerous disasters, such as climate change and global pandemics, cast a shadow on our dreams for the future. Future literacy, an interdisciplinary subject with significant contributions from Miller (2007) with the support of UNESCO, is critical in terms of predicting possible scenarios and taking precautions (Gladwin et al., 2022; Mangnus et al., 2021; Alper Ay et al., 2024). Studies on this subject emphasize the importance of incorporating future literacy into the curriculum in education (Jónasson, 2016), and models are being developed for "future thinking" that include teachers and students (Häggström & Schmidt, 2021). Evaluating students' pro-environmental attitudes and behaviors before and after environmental education is important for a sustainable future. Because students are both the primary culprits of environmental problems today and in the future and the target of their consequences. They must embrace this issue as future managers, policymakers, and educators. Continuing the education process that begins in preschool in the university environment provides an environment for them to learn and discuss environmental issues while also encouraging positive environmental behaviors in society (Vicente-Molina et al., 2018; Yu et al., 2017; de Leeuw et al., 2015).

Such developments are crucial for developing policies and taking precautions on climate change issues. However, it is much more important for individuals to be involved by increasing their awareness.

The most important step to be taken in this context is to teach this issue to children and young people at every stage of education to increase their sensitivity to this global problem and to ensure that they become a part of the solution to environmental problems (Hayta, 2006; Bilgili, 2008). Environmental awareness, which is also the focus of our study, can be briefly expressed as the sensitivity of individuals who form societies to environmental problems and their interest in the subject (Bilgili, 2008). Environmental awareness and sensitivity encompass psychology, sociology, environmental science, and other disciplines (Ham et al., 2016). Some academics think environmental awareness is a part of environmental literacy, which can be defined as the combination of motivation, knowledge, and skills (Umuhire & Fang, 2016).

Evaluating the relationship between environmental education and environmental awareness/sensitivity, Edsand and Broich (2020) and Haryono et al. (2014) determined that environmental education at school significantly affects students' knowledge but not students' environmental attitudes and awareness. A similar study was conducted by Rakotomamonjy et al. (2015) in rural Madagascar and indicated that children in villages that received environmental education had higher environmental knowledge and more positive environmental attitudes than those in villages that did not. Dono et al. (2010) determined a statistically significant relationship between environmental attitude and environmental behavior. In contrast, Steg and Vlek (2009) argue that environmental attitudes should be the starting point of every study aiming to explain pro-environmental behavior. A similar result was also determined by Aydın and Kaya (2011), who stated that the magnitude of the positive relationship between environmental education and environmental awareness decreased and that students' demographic factors were no longer statistically significant. This result has also been expressed in some studies. It has been stated that environmental awareness does not have a direct effect on pro-environmental behaviors, but people with pro-environmental attitudes are more likely to exhibit pro-environmental behaviors and purchase environmentally friendly products (Tuncer et al., 2005; Berberoğlu & Tosunoğlu, 1995; Yılmaz et al., 2009). A study conducted with students determined that the most common sources of information on global warming and climate change were the Internet, scientific studies, and TV and radio. The student's views on the main causes of climate change were determined to be "air pollution, deforestation, and ozone layer depletion," respectively (Ay & Erik, 2020). This situation reveals that the media is important in information and educational activities.

This study aims to determine the environmental awareness level of university students and whether there are significant differences between their levels according to some sociodemographic and socioeconomic characteristics (family income level, education level, occupation of mother and father, etc.). Therefore, this study is based on a sample of students at Sivas Cumhuriyet University in Turkey.

Depending on the purpose of the study, the hypotheses of the research were formed as follows:

H1: There is a significant difference between the environmental awareness levels of students and gender.
H2: There is a significant difference between the environmental awareness levels of students and age.
H3: There is a significant difference between the environmental awareness levels of students and marital status.
H4: There is a significant difference between the environmental awareness levels of students and income.
H5: There is a significant difference between the environmental awareness levels of students and school type.
H6: There is a significant difference between the environmental awareness levels of students and the natural field.
H7: There is a significant difference between the environmental awareness levels of students and grades.
H8: There is a significant difference between students' environmental awareness levels and their mothers' educational status.
H9: There is a significant difference between students' environmental awareness levels and their fathers' educational status.
H10: There is a significant difference between students' environmental awareness levels and mothers' occupations.
H11: There is a significant difference between students' environmental awareness levels and their fathers' occupations.
H12: There is a significant difference between students' environmental awareness levels and the place where the family lives.
H13: There is a significant difference between students' environmental awareness levels and their taking courses on climate or the environment.

2. METHODS

This study aims to examine the knowledge levels of students studying at Sivas Cumhuriyet University regarding environmental awareness and environmental problems.

This research is a descriptive and cross-sectional study. The study universe consists of Sivas Cumhuriyet University students studying in the 2022-2023 academic year. The universe of the study consisted of 48296 students. According to the 95% confidence interval in the study, 382 students were found to be statistically suitable for the sample (Altunışık et al., 2005). A face-to-face survey was applied to 384 university students, constituting the study sample in the spring of 2023. In order to implement the study, an Ethics Committee Decision dated 27.12.2022 numbered 245559 was obtained from Sivas Cumhuriyet University.

Data Collection Tool: A questionnaire developed by Çabuk and Karacaoğlu (2003) and with content validity was used to determine students' knowledge levels at Sivas Cumhuriyet University about environmental awareness. The reliability of the environmental awareness questionnaire is 0.810 and consists of 24 items. The questionnaire consists of questions about air pollution (1-4), water pollution (5-8), soil pollution (9-15), ecological balance (16-18), participation in environmental studies (19-20) and receiving environmental education (21-24). In this study, the environmental awareness questionnaire's Cronbach's Alpha reliability coefficient was 0.861.

Evaluation of Data: The data showed a normal distribution and parametric tests were used. In evaluating the data, the independent sample t-test was used for two-category variables, and the One-Way ANOVA test and Tukey test were used for more than two categories through the SPSS 23.0 program.

3. FINDINGS

Introductory information of the participants participating in the study is shown in Table 2.

Table 2. Distribution of participants according to their sociodemographic characteristics (n=387)

Gender	f	%	*Age*	f	%
Woman	205	53.4	17-20	139	36.2
Male	179	46.6	21-25	229	59.6
Marital status			26-30	9	2.3
Married	13	3.4	31 and over	7	1.8
Single	371	96.6	***Science area***		
Income status			Social Sciences	107	27.9
Low	105	27.3	Health Sciences	92	24.0
Middle	263	68.5	Educational Sciences	20	5.2
High	16	4.2	Natural Sciences	165	43.0
School type			***Class***		
Faculty	263	68.5	1	108	28.1
Vocational School	121	31.5	2	160	41.7
Mother's education level			3	73	19.0
Primary school	203	52.9	4	25	6.5
Middle school	78	20.3	5	10	2.6
High school	56	14.6	6	8	2.1
University	38	9.9	**Mother's profession**		
Postgraduate	9	2.3	Housewife	302	78.6
Father's education level			Civil servant	25	6.5
Primary school	110	28.6	Employee	31	8.1
Middle school	95	24.7	Self-employment	26	6.8
High school	98	25.5	**Father's Profession**		
University	68	17.7	Not working	44	11.5
Postgraduate	13	3.4	Civil servant	73	19.0
Where your family lives			Employee	110	28.6
City	303	78.9	Self-employment	124	32.3
Rural	81	21.1	Farmer	33	8.6
Have you taken courses on climate or the environment?					
Yes	173	45.1			
No	211	54.9			

As shown in Table 2, 53.4% of the students participating in the study are female, 59.6% are between the ages of 21 and 25, and 96.6% are single. 68.5% of the students have a medium-level monthly income. 27.9% of the students study social sciences, 68.5% study at the faculty, and 41.7% are in their second year. 52.9% of the students' mothers and 28.6% of their fathers were primary school graduates. 78.6% of their mothers were homemakers, and 32.3% of their fathers were self-employed. 78.9% of the students' families lived in urban areas. 54.9% of the students had not taken a course on climate or the environment.

Table 3 shows the students' responses regarding air pollution sensitivity.

Table 3. Students' responses to air pollution sensitivity

Environmentally friendly substances	Always		Sometimes		Never	
I. Items related to air pollution sensitivity	f	%	f	%	f	%
1. Are you careful not to use consumer goods (deodorant and other sprays) that contain substances harmful to the ozone layer?	77	20.1	251	65.4	56	14.6
2. Even if you have your vehicle, do you use public transportation, taking into account not causing air pollution?	91	23.7	215	56.0	78	20.3
3. Do you make sure that other people are not affected while talking and using various tools?	272	70.8	97	25.3	15	3.9
4. Would you warn people to be sensitive about air pollution?	126	32.8	225	58.6	33	8.6

The majority of the students agreed with the ozone layer at 65.4%, used public transportation at 56%, and warned people about air pollution at 58.6% "sometimes." They were "always" careful at 70.8% not to create noise pollution while talking and using various vehicles (Table 3).

Table 4 shows the students' responses to water pollution sensitivities.

Table 4. Students' responses to water pollution sensitivity

Environmentally friendly substances	Always		Sometimes		Never	
II. Items regarding water pollution sensitivity	f	%	f	%	f	%
5. Do you buy cleaning products by paying attention to whether they contain harmful chemicals?	91	23.7	212	55.2	81	21.1
6. Are you economical in water use under all circumstances?	213	55.5	154	40.1	17	4.4
7. Do you take care to prevent harmful chemicals such as engine oil and paint from entering the sewer system?	206	53.6	144	37.5	34	8.9
8. Would you warn people to be sensitive about water pollution?	193	50.3	171	44.5	20	5.2

In the responses given by the majority of the students, it was determined that they pay attention to chemicals in cleaning materials at the level of "sometimes 55.2%". It was determined that they agree with the issues of being economical in water use at the level of 55.5%, not mixing chemicals into the sewage system at the level of 53.6%, and warning people about water use at the level of "always" at the level of 50.3% (Table 4).

Table 5 shows the students' responses to soil pollution sensitivities.

Table 5. Students' responses to soil pollution sensitivity

Environmentally friendly substances	**Always**		**Sometimes**		**Never**	
III. Substances related to soil pollution sensitivity	**f**	**%**	**f**	**%**	**f**	**%**
9. Do you take care to use both sides of the paper you write on?	284	74.0	89	23.2	11	2.9
10. Are you economical in using paper napkins under all circumstances?	209	54.4	151	39.3	24	6.3
11. Would you plant saplings, taking into account the appropriate conditions for them to grow?	131	34.1	194	50.5	59	15.4
12. Do you take care to ensure that the waste reaches the trash bin?	279	72.7	97	25.3	8	2.1
13. Do you put waste in appropriate recycling bins so that they can be reused?	220	57.3	155	40.4	9	2.3
14. Do you classify garbage when throwing it away?	115	29.9	221	57.6	48	12.5
15. Would you warn the people around you to be sensitive about soil pollution?	133	34.6	212	55.2	39	10.2

In the responses given by the majority of the students, it was determined that they always agreed with the use of paper 74.0%, the use of napkins 54.4%, taking waste to the trash can 72.7%, and throwing waste into recycling bins 57.3%. It was determined that they sometimes agreed with the planting of saplings 50.5%, classifying and disposing of garbage 57.6%, and warning people about soil pollution 55.2% (Table 5).

Table 6 shows the students' responses regarding their sensitivity to ecological balance.

Table 6. Students' responses to ecological balance sensitivity

Environmentally friendly substances	Always		Sometimes		Never	
IV. Articles related to Ecological Balance Sensitivities	**f**	**%**	**f**	**%**	**F**	**%**
16. If you were/are married, would you pay attention to population planning considering ecological balance?	187	48.7	144	37.5	53	13.8
17. Do you think it is appropriate for humanity to conduct all kinds of experiments on humans and animals?	47	12.2	97	25.3	240	62.5
18. Would you warn the people around you to be sensitive about preserving ecological balance?	121	31.5	226	58.9	37	9.6

The majority of the students agreed with the responses given to the following: paying attention to population planning "always 48.7%", finding all kinds of experiments on humans and animals appropriate "never 62.7%", and warning others to protect the ecological balance "sometimes 58.9%" (Table 6).

Table 7 shows the students' responses regarding their participation in environmental studies.

Table 7. Students' responses to participation in environmental studies

Environmentally friendly substances	Always		Sometimes		Never	
V. Articles regarding participation in environmental studies	**F**	**%**	**f**	**%**	**F**	**%**
19. Do you participate in scientific studies such as seminars, panels, and conferences on the environment?	60	15.6	223	58.1	101	26.3
20. Do you participate in the work of voluntary organizations working on the environment?	59	15.4	194	50.5	131	34.1

The majority of the students responded that they participated in scientific activities related to the environment at a rate of 58.1% and in helping voluntary organizations related to the environment at a rate of 50.5% "sometimes." (Table 7).

Table 8 shows the students' answers regarding the environmental education they received in formal education institutions.

Table 8. Students' answers about receiving environmental education

Environmentally friendly substances	Always		Sometimes		Never	
VI. Articles related to receiving environmental education	**f**	%	**f**	%	**f**	%
21. Do you believe that you have received sufficient training to raise awareness about air pollution?	99	25.8	160	41.7	125	32.6
22. Do you believe that you have received sufficient training to raise awareness about water pollution?	110	28.6	163	42.4	111	28.9
23. Do you believe that you have received sufficient training to raise awareness about soil pollution?	107	27.9	152	39.6	125	32.6
24. Do you believe that you have received sufficient education to raise awareness about ecological balance?	90	23.4	165	43.0	129	33.6

The majority of the students agreed with the level of "sometimes" about air pollution (41.7%), water pollution (42.4%), soil pollution (39.6%), and environmental balance (43%) (Table 8).

Table 9 shows the results of the independent sample t-test analysis regarding whether there are significant differences in environmental awareness levels according to the students' sociodemographic characteristics.

Table 9. Students' environmental awareness according to sociodemographic characteristics (Independent sample t-test findings)

Variables	Category	Number	Average	Standard deviation	t	P
Gender	Woman	205	2.20	0.30	- 0.142	0.887
	Male	179	2.21	0.32		
Marital status	Married	13	2.12	0.39	- 0.990	0.323
	Single	371	2.21	0.31		
Where your family lives?	City	303	2.19	0.32	- 1.868	0.062
	Rural	81	2.26	0.27		
Have you taken courses on climate or the environment?	Yes	173	2.28	0.33	4.272	0.000
	No	211	2.14	0.28		
School type	Faculty	263	2.16	0.30	- 4.634	0.000
	Vocational School	121	2.31	0.31		

No significant difference was found between the environmental awareness levels of the students, their gender, marital status, and the place where their families live (P>0.05). (Table 9). Accordingly, hypotheses H1, H3, and H12 were rejected. A statistically significant difference was found between the status of taking courses on climate or environment, the type of school, and the environmental awareness levels. The environmental awareness levels of those who took courses on climate or environment were higher than those who did not take courses on this subject ($P<0.05$). In addition, the environmental awareness levels of those who studied at the Vocational School were higher than those who studied at the Faculty regarding school type ($P<0.05$). Accordingly, hypotheses H5 and H13 were accepted.

Table 10 shows the One-way ANOVA test analysis results regarding whether there is a significant difference in environmental awareness levels according to the students' sociodemographic characteristics.

Table 10. Students' environmental awareness according to sociodemographic characteristics (One-way ANOVA test findings)

Variables	Category	Number	Average	Standard deviation	f	p	Tukey test results
Age	17-20	139	2.23	0.32	1.105	0.347	-
	21-25	229	2.19	0.31			
	26-30	9	2.06	0.25			
	31 and over	7	2.26	0.18			
Income	Low	105	2.18	0.34	1.622	0.199	-
	Middle	263	2.21	0.30			
	High	16	2.33	0.30			
Science area	Social Sciences	107	2.22	0.30	8.270	0.000	3<1,2,4 4>2,3
	Health Sciences	92	2.14	0.29			
	Educational Sciences	20	1.94	0.30			
	Natural	165	2.26	0.31			
Class	1	108	2.23	0.28	2.513	0.030	4<2
	2	160	2.24	0.34			
	3	73	2.15	0.25			
	4	25	2.04	0.35			
	5	10	2.23	0.27			
	6	8	2.14	0.20			

continued on following page

Table 10. Continued

Variables	Category	Number	Average	Standard deviation	f	p	Tukey test results
Mother's education level	Primary school	203	2.24	0.28	3.358	0.010	1>4
	Middle school	78	2.20	0.31			
	High school	56	2.20	0.31			
	University	38	2.05	0.36			
	Postgraduate	9	2.08	0.48			
Father's education level	Primary school	110	2.28	0.29	4.411	0.002	1>4 3>4
	Middle school	95	2.18	0.31			
	High school	98	2.23	0.25			
	University	68	2.09	0.33			
	Postgraduate	13	2.11	0.54			
Mother's profession	Housewife	302	2.21	0.29	1.612	0.186	-
	Civil servant	25	2.08	0.37			
	Employee	31	2.25	0.33			
	Self-employment	26	2.21	0.47			
Father's profession	Not working	44	2.22	0.36	1.658	0.159	-
	Civil servant	73	2.13	0.33			
	Employee	110	2.21	0.29			
	Self-employment	124	2.22	0.31			
	Farmer	33	2.29	0.22			

No significant difference was found between the environmental awareness levels of the students and the variables of age, monthly income, mother's profession, and father's profession (P> 0.05)(Table 10). Accordingly, H2, H4, H10 and H11 were rejected.

A statistically significant difference was found between the students' Natural field, grade, mother's education level, father's education level, and their environmental awareness levels. The environmental awareness levels of the students studying in the field of educational sciences were found to be lower than those studying in the fields of social, health, and natural. The environmental awareness levels of natural students were statistically significantly higher than those of health and educational sciences. The environmental awareness levels of the 4th-grade students were lower than those of the 2nd-grade students (5th and 6th-grade students are students of pharmacy, veterinary, and medical faculties). The environmental awareness levels of the students whose mothers have a university education level were found to be lower than those who graduated from primary school. The environmental awareness levels of the students whose fathers have a university education level were found to be lower than those who graduated from primary school and high school. Accordingly, the hypotheses H6, H7, H8, and H9 were accepted (Table 10).

4. DISCUSSION

The primary purpose of this study is to determine the environmental awareness level of university students and its relationship with some sociodemographic and socioeconomic characteristics such as family income level, education level, and occupation of mother and father. For this purpose, a survey was applied to 384 students, and analyses were made. The results of the findings obtained in the study are explained below in the following items.

Environmental awareness level: In general, it is stated that the observation of pro-environmental behaviors among university students is moderately high and relatively positive (Hemayatkhah Jahromi et al., 2017; Shafiei & Maleksaeidi, 2020). This study determined that Sivas Cumhuriyet University students have relatively positive environmental behaviors, and their energy consumption behaviors are more favorable. This finding is consistent with the findings of Pazokinejad and Salehi (2014), Salehi et al. (2016), Bakhshi et al. (2017), and Shafiei and Maleksaeidi, 2020.

Environmental education: This study determined that the environmental awareness levels of those who took courses on climate or environment were higher than those who did not. A study that supports our findings determined that students who took courses on climate and environmental conditions had higher knowledge about the causes, effects, and ways to combat climate change (Ay & Erik, 2020). While providing environmental awareness training to children that starts in the family and continues for a long time afterward significantly supports environmental awareness (Coertjens et al., 2010; Lin & Shi, 2014; Edsand & Broich, 2020), it can also be said that education is not a magic wand for developing environmental awareness or awareness. Environmental education should consider the relationship between multidisciplinary scientific approaches, technology, society, and environmental knowledge. According to research, in the face of the increasing frequency and impact of global disasters, the higher awareness levels of individuals who receive disaster education reveal the importance of disseminating and diversifying this training in a way everyone can access (Alper Ay et al., 2023).

Family: Education level of mother and father. Since families play an important role in shaping students' attitudes and views on environmental problems, the occupation and education level of the parents were also controlled. This factor, which determines the quality of communication in the family, also determines the living space and social environment. According to Gambro and Switzky (1999), Kuhlemeier et al. (1999), and Vlahov and Treagust (1988) determined that environmental knowledge is significantly related to the gender of the students and the education level of their parents. Similar studies by Işıldar (2008) revealed that the education process significantly differs in forming and clarifying environmental approaches. It was also found in the studies of Rideout et al. (2005), Günden and Miran (2008), Işıldar (2008), Alnıaçık and Koç (2009), Alnıaçık (2010). This result was consistent with the Edsand and Broich (2020) and Aydın (2017) studies.

Research has identified that, in general, mothers are more concerned about the well-being and health of the family (concerns more closely related to the quality of environmental conditions such as water and air quality and waste). At the same time, fathers are more concerned with the family's financial and economic situation. In studies conducted in this context, it has been stated that the level of education of mothers has an effect on environmental literacy levels and that the high level of education of students' mothers is associated with the high level of knowledge of students about the environment (George & Southwell, 1986; Dietz et al., 1998; Arı & Yılmaz, 2017; Erkal et al. (2012). In this study, no significant difference was found in terms of profession; however, the environmental awareness levels of students whose mothers have a university education level were lower than those of primary school graduates. The environmental awareness levels of students whose fathers have a university education level were lower than those of primary school and high school graduates. It can be said that the higher environmental awareness of those whose mothers and fathers have a low level of education stems from the student's awareness of their families living in a natural/rural environment and their awareness of protecting their living and livelihood areas. A similar result was also determined in the study of Gürsoy (2008).

The place where the family lives (living area); The place where students live affects their attitudes towards the environment due to socioeconomic factors. However, studies on this subject have not yielded the same results. For example, Ek et al. (2009) stated that the attitudes of students living in urban areas towards environmental problems are more positive than those living in rural areas, while Alnıaçık (2010) determined that the type of settlement where the students' families live does not have a significant effect on the students' environmental orientation, but as the settlement grows (for example, as one moves from a village to a big city), the environmentalist worldview also increases. In the study of Demirel et al. (2009), no significant relationship was found between the settlement where the students lived the longest and their sensitivity towards the environment. Similarly, in this study, no significant difference was found in environmental sensitivity in terms of "the place where the students' families live". This result is probably due to the fact that the students generally have similar socioeconomic status and family life.

Gender: Although many studies suggest possible relationships between gender and environmental concerns, it is difficult to draw clear boundaries for this relationship. When the literature on this subject is examined, it is seen that gender is effective in some studies, while others do not find an effect of gender. For example, in the study of Çetin et al. (2020), it was seen that male students were closer to the idea of the possibility of environmental disaster and gave more importance to it than female students. A study by McEvoy (1972) showed that men were more concerned about the environment than women due to their higher education levels and participation in communities and political issues. Some studies show that women are more concerned about the environment than men because men are more concerned about economic growth and stability and see the environment as a constraint on economic growth (Chang, 2015; Hoşgör et al., 2015). Similarly, Xiao and Hong (2018), Alnıaçık (2010), Sam et al. (2010), and Erkal et al. (2012) also emphasized that women are more sensitive than men in terms of nature-centeredness. In this context, the advanced analysis conducted by Mohai (Stern, Dietz, & Kalof, 1993) suggests that women tend to be more interested in local environmental problems than men. However, this difference is more minor in terms of national or global problems. In the studies of Işıldar (2008), Demirel et al. (2009), Shobeiri et al. (2007), Pazokinejad and Salehi (2014), Edsand and Broich (2020), it was stated that gender is not an effective variable on environmental approaches and behaviors. Similarly, no significant difference was found in gender-related environmental sensitivity in our study.

Age: It has been stated that the younger generation generally tends to be more concerned about environmental quality than the older generation (Van Liere & Dunlap, 1980; Lowe & Pinhey, 1982; Mohai & Twight, 1987). In contrast, studies conducted by Harris (1991) and Furman (1998) have shown only a weak relationship between environmental attitude and age. Similarly, in this study conducted with university students, no significant difference was found in environmental awareness due to the small age range.

Marital status: In our study, no significant difference was found in environmental awareness regarding marital status. This finding is consistent with the results of studies conducted by Shafiei and Maleksaeidi (2020) and Pazokinejad and Salehi (2014).

School type, field, and class: Many studies have observed significant differences among students in terms of the faculty, fields of study, and classes they study in (Shafiei & Maleksaeidi, 2020; Alnıaçık & Koç, 2009). This study found a statistically significant difference between the type of school, field of natural, class, and environmental awareness levels of students. The environmental awareness levels of students studying in the field of educational sciences were found to be lower than those studying in the fields of social, health, and natural. The environmental awareness levels of natural students were statistically significantly higher than those of health and educational sciences students. This may be because the curriculum in the field of natural includes more learning about the environment and nature. It can be said that students' awareness and perception levels, especially about global climate change and events happening in the world, have increased in this way and that these topics are not given effectively academically during the education process in educational sciences. A study conducted to support our findings determined that students studying in social sciences have lower knowledge levels about the effects of climate change and ways to combat it compared to natural fields (Ay & Erik, 2020).

Some studies indicate that individuals become more concerned about the environment as they become more educated (Kohut & Shriver, 1989; Vining & Ebreo, 1990; Mainieri et al., 1997). In contrast, this study found the opposite, and 4th-grade students had lower environmental awareness levels than 2nd-grade students. These findings are similar to the results of the studies conducted by Müderrisoğlu and Altanlar (2011), Aydın and Kaya (2011), Sarvestani (2012), and Ahmadvand and Nooripoor (2011) but different from the study conducted by Çabuk and Karacaoğlu (2003) and Gürsoy et al. (2022). In addition, in our study, the environmental awareness of vocational school students was higher than that of faculty students. These results may be due to the fact that the number of non-field elective courses has increased in recent years and the students in the 2nd year have taken elective courses on climate and environment, while the students in the 4th year have not received education on this subject.

Income status: The socioeconomic status of students with family-based income affects their attitudes towards the environment. The general idea is that wealthier people have more positive attitudes toward environmental problems than low-income and average-income students. Worsley and Skrzypiec (1998) stated that male students with lower socioeconomic status are more pessimistic than female students but support environmental development and scientific solutions more. Edsand and Broich (2020) pointed out that the higher a student's socioeconomic status, the more equal other conditions are, and the more interested he/she is in environmental problems. This result is supported by Erkal et al. (2012), while Alnıaçık (2010) showed that family income does not significantly affect environmentally friendly behavior in general, and students from low-income families give more importance to anthropocentrism. The results of this study are also consistent with the studies conducted in Colombia and the United States of America (Edsand & Broich, 2020; Lin & Shi, 2014). However, no significant difference was found in our study in income-related environmental awareness. This can be expressed as the result of the study on students with similar economic status.

Since only measurable things can be managed, measuring environmental awareness based on scientific criteria is becoming increasingly interesting for scientists working in different disciplines (Yılmaz & Ertekin, 2019). To continue this progress, it is necessary to ensure its manageability in all relevant aspects and areas by constantly working with a multidisciplinary perspective. This is achieved by providing environmental awareness with a determined and internalized understanding from the beginning to the last stage of education and by gaining environmental sensitivity as a behavioral pattern.

5. CONCLUSIONS AND FUTURE IMPLICATIONS

Education is an indispensable element that enables societies to progress academically and gain and maintain their social development and economic prosperity. Due to the precise relationship between the socioeconomic development levels of societies and energy use, energy demand continues to increase despite the striking effects of global warming and climate change. Although it is impossible to stop climate change and global warming, it is impossible to stop energy demand or completely switch to clean energy sources (such as wind, solar, and hydro) in a short time for measures that will slow down its development. Therefore, a livable world for future generations is only possible with the effective use of existing energy sources, accelerating the transition to clean energy sources and energy conservation. This can be achieved by continuous education for a sustainable future by providing environmental awareness with a determined and internalized understanding and by gaining environmental sensitivity as a behavioral pattern.

This study has presented results similar to those of scientific research conducted in many fields and groups with similar scopes. It is important to determine the errors and deficiencies in effective environmental education required for the development of environmental sensitivity and environmental awareness. Therefore, continuous work on developing environmental sensitivity and awareness will significantly contribute to taking the necessary precautions. Providing training at regular intervals during the education period or ensuring that students take courses for at least two semesters can provide significant developments in this regard. It may be recommended to develop programs that will provide students with training on climate and the environment, primarily in the first year (to increase knowledge and awareness) and in the last year (to transform knowledge and attitudes into behavior, to ensure reinforcement and continuity, and to be able to follow current changes).

It is also crucial that this study be conducted with university students from diverse regions, as regional characteristics can significantly influence the degree of environmental sensitivity, perception, and application method. Therefore, it is important for researchers from different universities to come together and evaluate the subject with scientific collaborations. This sense of unity and shared responsibility is key to taking an active role in the processes required for solution-oriented actions. This study is currently limited to the perspectives of Sivas Cumhuriyet University students. However, for future studies, comparisons can be made with larger samples or students from different universities. To further enhance the knowledge and attitudes of university students regarding the causes and consequences of environmental and climate change, and to increase their awareness and knowledge levels not only in the field of natural but also in different fields, education should be emphasized. For effective environmental education, an interdisciplinary approach should be adopted, and education must be designed as a continuous process. The environmental education curriculum should be strengthened in courses in the educational sciences, and environmental awareness-raising content should be added. Campaigns and community-based projects should be organized to increase environmental awareness among young people. Socio-demographic factors should be taken into account in education policies. Strengthening communication within the family can help develop a more conscious and active attitude towards environmental issues. Studies can be conducted to evaluate the attitudes of individuals from different cultures and income groups toward environmental problems.

REFERENCES

Ahmadvand, M., & Nooripoor, M. (2011). Agricultural students' environmental attitudes of Yasouj University: A gender analysis. *Ulum-i Tarvij va Amuzish-i Kishavarzi*, *6*(2), 1–14.

Ajzen, I., & Fishbein, M. (1975). A Bayesian analysis of attribution processes. *Psychological Bulletin*, *82*(2), 261–277. DOI: 10.1037/h0076477

Alavosius, M. P., & Houmanfar, R. A. (2020). Global warming: Analysis of behavior and organizational practices as climate impacts increase. In Cihon, T. M., & Mattaini, M. A. (Eds.), *Behavior science perspectives on culture and community* (pp. 109–126). Springer., DOI: 10.1007/978-3-030-45421-0_10

Alnıaçık, Ü. (2010). Çevreci yönelim, çevre dostu davraniş ve demografik özellikler: Üniversite öğrencileri üzerinde bir araştirma. *Sosyal Ekonomik Araştırmalar Dergisi*, *10*(20), 507–532.

Alper Ay, F. (2018). *Örgütsel davranışta ve insan kaynakları yönetimi uygulamalarında örgütsel vatandaşlık davranışı*. Beta Yayınevi.

Alper Ay, F. (2025). The relationship between work motivation and productivity: Bibliometric analysis of articles from 1953 to 2024. *Human Systems Management*, *44*(1), 187–205. Advance online publication. DOI: 10.3233/HSM-240020

Alper Ay F. Demir, G., & Karakaya, A. (2024). Mapping futures literacy: Bibliometric analysis and future research directions. *Journal of Humanities and Tourism Research*, *14*(3), 414–437.

Alper Ay F. Yılmaz, K. (2024). Corporate social responsibility and digital transformation interaction: Determining strategic orientations with bibliometric analysis. *Journal of Management and Economic Studies*, *6*(3), 329–348.

Altunışık, R., Coşkun, R., Bayraktaroğlu, S., & Yıldırım, E. (2005). *Sosyal Bilimlerde Araştırma Yöntemleri (SPSS Uygulamalı)*. Sakarya Kitabevi, Sakarya.

Arcury, T. A., Johnson, T. P., & Scollay, S. J. (1986). Ecological worldview and environmental knowledge: The "new environmental paradigm. *The Journal of Environmental Education*, *17*(4), 35–40. DOI: 10.1080/00958964.1986.9941424

Arı, E., & Yılmaz, V. (2017). Effects of environmental illiteracy and environmental awareness among middle school students on environmental behavior. *Environment, Development and Sustainability*, *19*(5), 1779–1793. DOI: 10.1007/s10668-016-9826-3

Ay, F. (2021). *Jeoloji tarihindeki iklim, küresel ısınma ve etkileri*. Akademisyen Kitabevi. DOI: 10.37609/akya.257

Ay, F., & Erik, N. Y. (2020). Üniversite öğrencilerinin küresel ısınma ve iklim değişikliğine yönelik bilgi ve algı düzeyleri. *Cumhuriyet Üniversitesi Fen-Edebiyat Fakültesi Sosyal Bilimler Dergisi, 44*(2), 1–18.

Aydın, F. (2017). Determining the level of knowledge of university students on global warming. *Journal of Social Sciences and Humanities, 1*(1), 118–132.

Aydın, F., & Kaya, H. (2011). Evaluation of Social Sciences High school Students' Sensitivity towards Environment. *Marmara Geographical Review, 24*(2), 229–257.

Bakhshi, M. R., Pirdadeh Bayranvand, K., & Falaki, M. (2017). Investigating the relationship between social capital and environmental behaviors of students in university of birjand. *Social Capital Management, 4*(3), 451–473.

Başıbüyük, G. Ö., & Ay F. Acar, S. (2017). Küresel ısınma ve küresel ısınmanın yaşlılar üzerindeki etkisi. *Türkiye Sosyal Araştırmalar Dergisi, 21*(1), 203–218.

Berberoglu, G., & Tosunoglu, C. (1995). Exploratory and confirmatory factor analyses of an environmental attitude scale (EAS) for Turkish university students. *The Journal of Environmental Education, 26*(3), 40–43. DOI: 10.1080/00958964.1995.9941444

Bilgili, S. (2008). *İlköğretim 7. sınıf fen ve teknoloji dersinde çevre konularının öğretiminde, yapılandırmacı yaklaşıma dayalı işbirlikli öğrenmenin öğrencilerin erişine etkisi*. Yayımlanmış Yüksek Lisans Tezi, Gazi Üniversitesi.

Çabuk, B., & Karacaoğlu, C. (2003). Üniversiteli öğrencilerinin çevre duyarlılıklarının incelenmesi. [JFES]. *Ankara University Journal of Faculty of Educational Sciences, 36*(1), 189–198. DOI: 10.1501/Egifak_0000000079

Casaló, L. V., & Escario, J. J. (2018). Heterogeneity in the association between environmental attitudes and pro-environmental behavior: A multilevel regression approach. *Journal of Cleaner Production, 175*, 155–163. DOI: 10.1016/j.jclepro.2017.11.237

Çetin, T., Turan, M. E., Aytekin, P., & Yurdusev, M. A. (2020). Üniversite öğrencilerinin çevresel yaklaşımlarını ölçmeye yönelik bir araştırma: Manisa Celal Bayar Üniversitesi örneği. *Doğal Afetler ve Çevre Dergisi, 6*(1), 12–23. DOI: 10.21324/dacd.546369

Chang, C. H. (2015). Teaching climate change–a fad or a necessity? *International Research in Geographical and Environmental Education, 24*(3), 181–183. DOI: 10.1080/10382046.2015.1043763

Coertjens, L., Boeve-de Pauw, J., De Maeyer, S., & Van Petegem, P. (2010). Do Schools Make a difference in their students' environmental attitudes and awareness? evidence from pisa 2006. *International Journal of Science and Mathematics Education*, *8*(3), 497–522. DOI: 10.1007/s10763-010-9200-0

Davidson, D. J., & Freudenburg, W. R. (1996). Gender and environmental risk concerns: A review and analysis of available research. *Environment and Behavior*, *28*(3), 302–339. DOI: 10.1177/0013916596283003

De Leeuw, A., Valois, P., Ajzen, I., & Schmidt, P. (2015). Using the theory of planned behavior to identify key beliefs underlying pro-environmental behavior in high-school students: Implications for educational interventions. *Journal of Environmental Psychology*, *42*, 128–138. DOI: 10.1016/j.jenvp.2015.03.005

Demirel, M., Gürbüz, B., & Karaküçük, S. (2009). Rekreasyonel aktivitelere katilimin çevreye yönelik tutum üzerindeki etkisi ve yeni ekolojik paradigma ölçeği'nin geçerliği ve güvenirliği. *Spormetre Beden Eğitimi ve Spor Bilimleri Dergisi*, *7*(2), 47–50.

Dietz, T., Stern, P. C., & Guagnano, G. A. (1998). Social structural and social psychological bases of environmental concern. *Environment and Behavior*, *30*(4), 450–471. DOI: 10.1177/001391659803000402

Dillon, P. J., & Gayford, C. G. (1997). A psychometric approach to investigating the environmental beliefs, intentions and behaviours of pre-service teachers. *Environmental Education Research*, *3*(3), 283–297. DOI: 10.1080/1350462970030303

Dono, J., Webb, J., & Richardson, B. (2010). The relationship between environmental activism, pro- environmental behaviour and social identity. *Journal of Environmental Psychology*, *30*(2), 178–186. DOI: 10.1016/j.jenvp.2009.11.006

Eagles, P. F., & Demare, R. (1999). Factors influencing children's environmental attitudes. *The Journal of Environmental Education*, *30*(4), 33–37. DOI: 10.1080/00958969909601882

Edsand, H. E., & Broich, T. (2020). The impact of environmental education on environmental and renewable energy technology awareness: Empirical evidence from Colombia. *International Journal of Science and Mathematics Education*, *18*(4), 611–634. DOI: 10.1007/s10763-019-09988-x

Ek, H. N., Kılıç, N., Öğdüm, P., Düzgün, G., & Peker, S. (2009). Adnan Menderes Üniversitesinin farklı akademik alanlarında öğrenim gören ılk ve son sınıf öğrencilerinin çevre sorunlarına yönelik tutumları ve duyarlılıkları. *Kastamonu Eğitim Dergisi*, *17*(1), 125–136.

Erkal, S., Kilic, I., & Sahin, H. (2012). Comparison of environmental attitudes of university students determined via the new environmental paradigm scale according to the students' personal characteristics. *Eurasian Journal of Educational Research, 49*, 21–39.

Freudenburg, W. R. (1991). Rural-urban differences in environmental concern: A closer look. *Sociological Inquiry, 61*(2), 167–198. DOI: 10.1111/j.1475-682X.1991.tb00274.x

Freudenburg, W. R., & Mcginn, B. (1989). Rural Urban Differences. In *Environmental Concern: A Closer Look*. Eric Document Reproduction Service N.

Furman, A. (1998). A note on environmental concern in a developing country: Results from an Istanbul survey. *Environment and Behavior, 30*(4), 520–534. DOI: 10.1177/001391659803000406

Gambro, J. S., & Switzky, H. N. (1999). Variables associated with American high school students' knowledge of environmental issues related to energy and pollution. *The Journal of Environmental Education, 30*(2), 15–22. DOI: 10.1080/00958969909601866

Gardos, V. T., & Dodd, D. K. (1995). An immediate response to environmentally disturbing news and the environmental attitudes of college students. *Psychological Reports, 77*(3), 1121–1122. DOI: 10.2466/pr0.1995.77.3f.1121

George, D. L., & Southwell, P. L. (1986). Opinion on the Diablo Canyon nuclear power plant: The effects of situation and socialization. *Social Science Quarterly, 67*(4), 722.

Gladwin, D., Horst, R., James, K., & Sameshima, P. (2022). Imagining futures literacies: A collaborative practice. *Journal of Higher Education Theory and Practice, 22*(7), 27–40.

Gökdere, M. (2005). A study on environmental knowledge level of primary students in Turkey. In *Asia-Pacific Forum on Science Learning and Teaching,* 6(2), 1-13.

Günden C. Miran B. (2008). Yeni çevresel paradigma ölçeğiyle çiftçilerin çevre tutumunun belirlenmesi: İzmir ili Torbalı ilçesi örneği. *Ekoloji dergisi*, 18(69),41-50.

Gürsoy, S., Hocaoğlu Bahadır, N., & Yalçın, E. C. (2022). Çevre bilinci dinamikleri çevresel tutumda bir farklılık yaratır mı? Kırklareli üniversitesi örneği. *Ankara Üniversitesi SBF Dergisi, 77*(4), 779–799. DOI: 10.33630/ausbf.1081113

Gürsoy, T. (2008). *Öğretmen adaylarının öğrenme stillerinin çeşitli değişkenler açısından incelenmesi.* Master's thesis, Adnan Menderes Üniversitesi, Sosyal Bilimler Enstitüsü, Manisa.

Häggström, M., & Schmidt, C. (2021). Futures literacy–to belong, participate and act!: An educational perspective. *Futures, 132*, 102813. DOI: 10.1016/j.futures.2021.102813

Ham, M., Mrčela, D., & Horvat, M. (2016). Insights for measuring environmental awareness. *Ekonomski vjesnik. Review of Contemporary Entrepreneurship, Business, and Economic Issues*, *29*(1), 159–176.

Harris, C. (1991). Environmental education in England and Wales: A brief review. *European Journal of Education*, *26*(4), 287–290. DOI: 10.2307/1503129

Haryono A. Soemarno Djati, M. S., & Setyoleksono, A. (2014). Learning attitude and awareness against students in cultured environmental success in Probolinggo. *Journal of Environment and Earth Science*, *4*(16), 72–81.

Hayta, A. B. (2006). Çevre kirliliğinin önlenmesinde ailenin yeri ve önemi. *Ahi Evran Üniversitesi Kırşehir Eğitim Fakültesi Dergisi*, *7*(2), 359–376.

Hemayatkhah Jahromi, M., Ershad F. Danesh, P., & Ghorbani, M. (2017). Explore environmental behavior based on the model of TPB. [Previously Human Development]. *Quarterly Journal of Social Development*, *11*(3), 31–66.

Hes-Quimbita, G., & Pavel, M. (1996) *Assessing an Environmental Attitude Development Model: Factors Influencing the Environmental Attitudes of College Students. Eric Document. American Education Research Association Conference*, April 8-12, New York.

Hoşgör, H., Hoşgör, D. G., & Tosun, N. K. (2015). Sağlık bilimleri fakültesi öğrencilerinin çevreye yönelik tutumlarının belirlenmesi: Kıyaslamalı bir analiz. *Journal of Health Sciences and Professions*, *2*(2), 198–208.

Işildar, G. Y. (2008). Meslek yüksekokulları boyutunda" çevre eğitimi" nin çevreci yaklaşımlar ve davranışlar üzerindeki etkilerinin değerlendirilmesi. *Türk Eğitim Bilimleri Dergisi*, *6*(4), 759–778.

Jaus, H. H. (1982). The effect of environmental education ınstruction on children's attitudes toward the environment. *Science Education*, *66*(5), 689–692. DOI: 10.1002/sce.3730660504

Jeronen, E., & Kaikkonen, M. (2002). Thoughts of children and adults about the environment and environmental education. *International Research in Geographical and Environmental Education, 11*(4), 341–353. DOI: 10.1080/10382040208667501

Jónasson, J. T. (2016). Educational change, inertia and potential futures: Why is it difficult to change the content of education? *European Journal of Futures Research, 4*(1), 1–14. DOI: 10.1007/s40309-016-0087-z

Koçak B. Akış A. (2021). Sosyal bilgiler öğretmen adaylarının çevre eğitimi hakkında görüşleri. *Turkish Studies-Educational Sciences*, 16(1).

Kohut A. Shriver J. (1989). *Environment regaining a foothold on the national agenda*. Gallup Report, no. 285.

Kuhlemeier, H., Van Den Bergh, H., & Lagerweij, N. (1999). Environmental knowledge, attitudes, and behavior in Dutch secondary education. *The Journal of Environmental Education, 30*(2), 4–14. DOI: 10.1080/00958969909601864

Lin, E., & Shi, Q. (2014). Exploring individual and school-related factors and environmental literacy: Comparing U.S. and Canada using PISA 2006. *International Journal of Science and Mathematics Education, 12*(1), 73–97. DOI: 10.1007/s10763-012-9396-2

Lowe, G. D., & Pinhey, T. K. (1982). Rural-urban differences in support for environmental protection. *Rural Sociology, 47*(1), 114.

Ma, X., & Bateson, D. J. (1999). A multivariate analysis of the relationship between attitude toward science and attitude toward the environment. *The Journal of Environmental Education, 31*(1), 27–32. DOI: 10.1080/00958969909598629

Mainieri, T., Barnett, E. G., Valdero, T. R., Unipan, J. B., & Oskamp, S. (1997). Green buying: The influence of environmental concern on consumer behavior. *The Journal of Social Psychology, 137*(2), 189–204. DOI: 10.1080/00224549709595430

Mangnus, A. C., Oomen, J., Vervoort, J. M., & Hajer, M. A. (2021). Futures literacy and the diversity of the future. *Futures, 132*, 102793. DOI: 10.1016/j.futures.2021.102793

McEvoy J. (1972). The American concern with the environment. *Social behavior, natural resources and the environment*, 214-236.

Miller, D. L. (2007). The seeds of learning: Young children develop important skills through their gardening activities at a midwestern early education program. *Applied Environmental Education and Communication, 6*(1), 49–66. DOI: 10.1080/15330150701318828

Mohai, P., & Twight, B. W. (1987). Age and environmentalism: An elaboration of the Buttel model using national survey evidence. *Social Science Quarterly*, *68*(4), 798.

Müderrisoğlu, H., & Altanlar, A. (2011). Attitudes and behaviors of undergraduate students toward environmental ıssues. *International Journal of Environmental Science and Technology*, *8*(1), 159–168. DOI: 10.1007/BF03326205

Musser, L. M., & Diamond, K. E. (1999). The children's attitudes toward the environment scale for preschool children. *The Journal of Environmental Education*, *30*(2), 23–30. DOI: 10.1080/00958969909601867

Pazokinejad, Z., & Salehi, S. (2014). An analysis of social factors influencing studentsâ environmental attitudes and performance. *Journal of Applied Sociology*, *25*(3), 71–88.

Rakotomamonjy, S. N., Jones, J. P. G., Razafimanahaka, J. H., Ramamonjisoa, B., & Williams, S. J. (2015). The effects of environmental education on children's and parents' knowledge and attitudes towards lemurs in rural Madagascar. *Animal Conservation*, *18*(2), 157–166. DOI: 10.1111/acv.12153

Reşat Cica, H., & Karabulut, T. (2024). Çalışanların örgütsel bağlılık ile örgütsel özdeşleşme davranışının yeşil örgütsel davranış üzerindeki etkisinde çevre tutkusunun aracı rolü. *Gaziantep Üniversitesi Sosyal Bilimler Dergisi*, *23*(1), 304–326. DOI: 10.21547/jss.1312993

Rideout, B. E., Hushen, K., McGinty, D., Perkins, S., & Tate, J. (2005). Endorsement of the new ecological paradigm in systematic and e-mail samples of college students. *The Journal of Environmental Education*, *36*(2), 15–23. DOI: 10.3200/JOEE.36.2.15-23

Salehi, S., Nejad, Z. P., Mahmoudi, H., & Burkart, S. (2016). Knowledge of global climate change: View of Iranian university students. *International Research in Geographical and Environmental Education*, *25*(3), 226–243. DOI: 10.1080/10382046.2016.1155322

Sam, N., Gürsakal, S., & Sam, R. (2010). Üniversite öğrencilerinin çevresel risk algısı ve çevresel tutumlarının belirlenmesi. *Akademik Bakış Dergisi*, *20*, 1–16.

Sanera, M. (1998). Environmental education: Promise and performance. [CJEE]. *Canadian Journal of Environmental Education*, •••, 9–26.

Sarvestani, A. A. (2012). Environmental attitude and behavior of students of Gorgan University of Agricultural Sciences and Natural Resources. *Ulum-i Tarvij va Amuzish-i Kishavarzi*, *7*(2), 77–92.

Shafiei, A., & Maleksaeidi, H. (2020). Pro-environmental behavior of university students: Application of protection motivation theory. *Global Ecology and Conservation, 22*, e00908. DOI: 10.1016/j.gecco.2020.e00908

Shobeiri, S. M., Omidvar, B., & Prohallada, N. N. (2007). A comparative study of environmental awareness among secondary school students in Iran and India. *International Journal of Environmental Research, 1*(1), 28–34.

Slingo, J. M., & Slingo, M. E. (2024). The science of climate change and the effect of anaesthetic gas emissions. *Anaesthesia, 79*(3), 252–260. DOI: 10.1111/anae.16189 PMID: 38205585

Steg, L., & Vlek, C. (2009). Encouraging pro-environmental behaviour: An integrative review and research agenda. *Journal of Environmental Psychology, 29*(3), 309–317. DOI: 10.1016/j.jenvp.2008.10.004

Stern, P., Dietz, T., & Kalof, L. (1993). Value orientations, gender, and environmental concern. *Environment and Behavior*, •••, 25.

Tikka, P. M., Kuitunen, M. T., & Tynys, S. M. (2000). Effect of educational background on students' attitude, activity levels, and knowledge concerning the environment. *The Journal of Environmental Education, 31*(3), 12–19. DOI: 10.1080/00958960009598640

Tuncer, G., Ertepinar, H., Tekkaya, C., & Sungur, S. (2005). Environmental attitudes of young people in Turkey: Effect of school type and gender. *Environmental Education Research, 11*(2), 215–233. DOI: 10.1080/1350462042000338379

Umuhire, M. L., & Fang, Q. (2016). Method and application of ocean environmental awareness measurement: Lessons learnt from university students of China. *Marine Pollution Bulletin, 102*(2), 289–294. DOI: 10.1016/j.marpolbul.2015.07.067 PMID: 26242972

UNESCO–UNEP. (1991). United Nations environment programme: report of the governing council on the work of its 16th session, 20-31 May 1991. UNEP. Governing Council (16th sess.: 1991: Nairobi)

Uyeki, E. S., & Holland, L. J. (2000). Diffusion of pro-environment attitudes? *The American Behavioral Scientist, 43*(4), 646–662. DOI: 10.1177/00027640021955478

Uzun, N., & Sağlam, N. (2007). The environmental education in secondary school and teachers' opinions on environmental education programs. *Eurasian Journal of Educational Research, 26*, 176–187.

Van Liere, K. D., & Dunlap, R. E. (1981). Environmental concern: Does it make a difference how it's measured? *Environment and Behavior*, *13*(6), 651–676. DOI: 10.1177/0013916581136001

Vicente-Molina, M. A., Fernández-Sainz, A., & Izagirre-Olaizola, J. (2018). Does gender make a difference in pro-environmental behavior? The case of the Basque Country University students. *Journal of Cleaner Production*, *176*, 89–98. DOI: 10.1016/j.jclepro.2017.12.079

Vining, J. R., & Ebro, A. (1990). What makes a recycler? A comparison of recyclers and non-recyclers. *Environment and Behavior*, *22*(1), 56–73. DOI: 10.1177/0013916590221003

Vlahov, S. J., & Treagust, D. F. (1988). Students' knowledge of energy and attitudes to energy conservation. *School Science and Mathematics*, *88*(6), 452–458. DOI: 10.1111/j.1949-8594.1988.tb11837.x

Worsley, A., & Skrzypiec, G. (1998). Environmental attitudes of senior secondary school students in South Australia. *Global Environmental Change*, *8*(3), 209–225. DOI: 10.1016/S0959-3780(98)00016-8

Xiao, C., & Hong, D. (2018). Gender differences in environmental behaviors among the Chinese public: Model of mediation and moderation. *Environment and Behavior*, *50*(9), 975–996. DOI: 10.1177/0013916517723126

Yılmaz S. Ertekin Ö. (2019). Bilimsel Kriterlere Dayalı Çevre Bilinci Ölçümü: Bir Literatür Taraması. *İşletme Bilimi Dergisi*, 7(1), 155-181.

Yılmaz, V., & Arslan, T. (2011). Üniversite öğrencilerinin çevre koruma vaatleri ve çevre dostu tüketim davranışlarının incelenmesi. *Anadolu University Journal of Social Sciences*, *11*(3), 1–10.

Yılmaz, V., Çelik, E., & Yağızer, C. (2009). Investigating the effects of environmental sensitivity and environmental behavior on ecological product buying behavior through structural equation model. *Anadolu University Journal of Social Science*, *9*(2), 1–14.

Yu, W., Ramanathan, R., & Nath, P. (2017). Environmental pressures and performance: An analysis of the roles of environmental innovation strategy and marketing capability. *Technological Forecasting and Social Change*, *117*, 160–169. DOI: 10.1016/j.techfore.2016.12.005

Zimmermann, L. K. (1996). Knowledge, affect, and the environment: 15 years of research (1979–1993). *The Journal of Environmental Education*, *27*(3), 41–44. DOI: 10.1080/00958964.1996.9941466

Compilation of References

Abosedra, S., & Baghestani, H. (1989). New evidence on the causal relationship between United States energy consumption and gross national product. *The Journal of Energy and Development*, •••, 285–292.

Acaravci, A. (2010). Structural breaks, electricity consumption and economic growth: Evidence from Turkey. *Journal for Economic Forecasting*, *2*, 140–154.

Achraf, M., & Foolad, M. R. (2007). Roles of glycine betaine and proline in improving plant abiotic stress resistance. *Environmental and Experimental Botany*, *59*(2), 206–216. DOI: 10.1016/j.envexpbot.2005.12.006

Acosta-Motos, J. R., Ortuño, M. F., Bernal-Vicente, A., Diaz-Vivancos, P., Sanchez-Blanco, M. J., & Hernandez, J. A. (2017). Plant Responses to Salt Stress: Adaptive Mechanisms. *Agronomy (Basel)*, *7*(1), 18. DOI: 10.3390/agronomy7010018

Acreman, M., & Holden, J. (2013). How wetlands affect floods. *Wetlands*, *33*(5), 773–786. DOI: 10.1007/s13157-013-0473-2

Adamantiades, A., & Kessides, I. (2009). Nuclear power for sustainable development: Current status and future prospects. *Energy Policy*, *37*(12), 5149–5166. DOI: 10.1016/j.enpol.2009.07.052

Adjovu, G. E., Stephen, H., James, D., & Ahmad, S. (2023). Measurement of total dissolved solids and total suspended solids in water systems: A review of the issues, conventional, and remote sensing techniques. *Remote Sensing (Basel)*, *15*(14), 3534. DOI: 10.3390/rs15143534

Adukaite, A., & Cantoni, L. (2016). Raising awareness and promoting informal learning on World Heritage in Southern Africa: The case of WHACY, a gamified ICT-enhanced tool. *International Journal of Education and Development using ICT*, *12*(2). https://www.learntechlib.org/p/173447/

African Development Bank (AfDB). (2022). *Renewable Energy in Africa: Trends and Opportunities.* Retrieved from https://www.afdb.org/en/topics-and-sectors/initiatives-partnerships/africa-renewable-energy-initiative

African Renewable Energy Initiative (AREI). (2023). *African Energy Transition Report.* Retrieved from https://www.arei.org/energy-transition

Agena, T. (2007). The relationship between economic activities and environmental degradation in Africa. *Journal of the Historical Society of Nigeria, 17*, 28–40. https://www.jstor.org/stable/41857144

Ahanger, M. A., Akram, N. A., Ashraf, M., Alyemeni, M. N., Wijaya, L., & Ahmad, P. (2017). Plant responses to environmental stresses—From gene to biotechnology. *AoB Plants, 9*(4), 9. DOI: 10.1093/aobpla/plx025 PMID: 28775828

Ahmad, P., & Prasad, M. N. V. (Eds.). (2011). *Environmental adaptations and stress tolerance of plants in the era of climate change*. Springer Science & Business Media.

Ahmadvand, M., & Nooripoor, M. (2011). Agricultural students' environmental attitudes of Yasouj University: A gender analysis. *Ulum-i Tarvij va Amuzish-i Kishavarzi, 6*(2), 1–14.

Ahmed, M., Tóth, Z., & Decsi, K. (2024). The impact of salinity on crop yields and the confrontational behavior of transcriptional regulators, nanoparticles, and antioxidant defensive mechanisms under stressful conditions: A review. *International Journal of Molecular Sciences, 25*(5), 2654. DOI: 10.3390/ijms25052654 PMID: 38473901

Ahmed, Y. A. E., Yue, B., Gu, Z., & Yang, J. (2023). An overview: Big data analysis by deep learning and image processing. *International Journal of Quantum Information, 21*(7), 2340009. DOI: 10.1142/S0219749923400099

Ajzen, I., & Fishbein, M. (1975). A Bayesian analysis of attribution processes. *Psychological Bulletin, 82*(2), 261–277. DOI: 10.1037/h0076477

Akhmat, G., & Zaman, K. (2013). Nuclear energy consumption, commercial energy consumption and economic growth in South Asia: Bootstrap panel causality test. *Renewable & Sustainable Energy Reviews, 25*, 552–559. DOI: 10.1016/j.rser.2013.05.019

Akkaya, Ş., & Bakkal, U. (2005). Çevre Vergileri ve Çifte Yarar. *İstanbul Üniversitesi İktisat Fakültesi Mecmuası, 55*(2), 1–22.

Aktar, W., Sengupta, D., & Chowdhury, A. (2009). Impact of pesticides use in agriculture: Their benefits and hazards. *Interdisciplinary Toxicology, 2*(1), 1–12. DOI: 10.2478/v10102-009-0001-7 PMID: 21217838

Alabi, O., Ackah, I., & Lartey, A. (2017). Re-visiting the renewable energy–economic growth nexus: Empirical evidence from African OPEC countries. *International Journal of Energy Sector Management, 11*(3), 387–403. DOI: 10.1108/IJESM-07-2016-0002

Alam, A., Upadhyay, S., Murthy, C. H., Reddy, M. J. B., Jana, K. C., & Mohanta, D. K. (2012, May). *Reliability evaluation of solar photovoltaic microgrid. In 2012 11th International Conference on Environment and Electrical Engineering*. IEEE., https://ieeexplore.ieee.org/document/6221427

Alavosius, M. P., & Houmanfar, R. A. (2020). Global warming: Analysis of behavior and organizational practices as climate impacts increase. In Cihon, T. M., & Mattaini, M. A. (Eds.), *Behavior science perspectives on culture and community* (pp. 109–126). Springer., DOI: 10.1007/978-3-030-45421-0_10

Alikhani, S., Nummi, P., & Ojala, A. (2021). Urban wetlands: A review on ecological and cultural values. *Water (Basel), 13*(22), 3301. DOI: 10.3390/w13223301

Allen, C. D., Macalady, A. K., Chenchouni, H., Bachelet, D., McDowell, N., Vennetier, M., & Cobb, N. (2010). A global overview of drought and heat-induced tree mortality reveals emerging climate change risks for forests. *Forest Ecology and Management, 259*(4), 660–684. DOI: 10.1016/j.foreco.2009.09.001

Al-Mulali, U., Saboori, B., & Ozturk, I. (2015). Investigating the environmental Kuznets curve hypothesis in Vietnam. *Energy Policy, 76*, 123–131. DOI: 10.1016/j.enpol.2014.11.019

Alnıaçık, Ü. (2010). Çevreci yönelim, çevre dostu davraniş ve demografik özellikler: Üniversite öğrencileri üzerinde bir araştirma. *Sosyal Ekonomik Araştırmalar Dergisi, 10*(20), 507–532.

Alongi, D. M. (2012). Carbon sequestration in mangrove forests. *Carbon Management, 3*(3), 313–322. DOI: 10.4155/cmt.12.20

Alonso, A., Dal Lmeier, F., Granek, E., & Raven, P. (2001). Biodiversity connecting with the tapestry of life, Smithsonian Institution / Monitoring and assessment of biodiversity Program and President's committee of Advisors on Science and Technology. Washington DC, United States of America.

Alper Ay F. Demir, G., & Karakaya, A. (2024). Mapping futures literacy: Bibliometric analysis and future research directions. *Journal of Humanities and Tourism Research, 14*(3), 414–437.

Alper Ay F. Yılmaz, K. (2024). Corporate social responsibility and digital transformation interaction: Determining strategic orientations with bibliometric analysis. *Journal of Management and Economic Studies*, *6*(3), 329–348.

Alper Ay, F. (2018). *Örgütsel davranışta ve insan kaynakları yönetimi uygulamalarında örgütsel vatandaşlık davranışı*. Beta Yayınevi.

Alper Ay, F. (2025). The relationship between work motivation and productivity: Bibliometric analysis of articles from 1953 to 2024. *Human Systems Management*, *44*(1), 187–205. Advance online publication. DOI: 10.3233/HSM-240020

Alper, A., & Oguz, O. (2016). The role of renewable energy consumption in economic growth: Evidence from asymmetric causality. *Renewable & Sustainable Energy Reviews*, *60*, 953–959. DOI: 10.1016/j.rser.2016.01.123

Altunışık, R., Coşkun, R., Bayraktaroğlu, S., & Yıldırım, E. (2005). *Sosyal Bilimlerde Araştırma Yöntemleri (SPSS Uygulamalı)*. Sakarya Kitabevi, Sakarya.

Anand, A., Chandan, P., & Singh, R. B. (2012). Homestays at Korzok: Supplementing rural livelihoods and supporting green tourism in the Indian Himalayas. *Mountain Research and Development*, *32*(2), 126–136. DOI: 10.1659/MRD-JOURNAL-D-11-00109.1

Anand, V., & Oinam, B. (2020). Future land use land cover prediction with special emphasis on urbanization and wetlands. *Remote Sensing Letters*, *11*(3), 225–234. DOI: 10.1080/2150704X.2019.1704304

Anderson, M. E. (2020). From 'exorcism' to engagement: The private sphere of perpetratorship in twenty-first century South African documentary film. *Continuum (Perth)*, *34*(2), 210–225. DOI: 10.1080/10304312.2020.1737433

Andrews, N. (2014). Driving factors for national competitiveness in Africa as measured by GDP per capita. https://repository.up.ac.za/bitstream/handle/2263/43965/Andrews_Driving_2014.pdf?sequence=1&isAllowed=y

Apergis, N., Payne, J., Menyah, K., & Wolde-Rufael, Y. (2010). On the causal dynamics between emissions, nuclear energy, renewable energy, and economic growth. *Ecological Economics*, *69*(11), 2255–2260. DOI: 10.1016/j.ecolecon.2010.06.014

Aravind, N. A., Dinesh, R. A. O., Ganeshaiah, K. N., Uma Shaanker, R., & Poulsen, J. G. (2010). Impact of the invasive plant, *Lantana camara*, on bird assemblages at Male Mahadeshwara Reserve Forest, South India. *Tropical Ecology*, *51*(2S), 325–338.

Aravind, N. A., Rao, D., Vanaraj, G., Ganeshaiah, K. N., Shaanker, R. U., & Poulsen, J. G. (2006). *Impact of Lantana camara on plant communities at Malé Mahadeshwara reserve forest, South India. Invasive Alien Species and Biodiversity in India*. Banaras Hindu University.

Arcury, T. A., Johnson, T. P., & Scollay, S. J. (1986). Ecological worldview and environmental knowledge: The "new environmental paradigm. *The Journal of Environmental Education*, *17*(4), 35–40. DOI: 10.1080/00958964.1986.9941424

Arı, E., & Yılmaz, V. (2017). Effects of environmental illiteracy and environmental awareness among middle school students on environmental behavior. *Environment, Development and Sustainability*, *19*(5), 1779–1793. DOI: 10.1007/s10668-016-9826-3

Ariffin, N. A., Yunus, A. M., & Kadir, I. K. (2021). The role of big data in the healthcare industry. *Journal of Islamic*, *6*(36), 235–245.

Arif, M., Khurshid, H., Rabbani, M. A., Jahanzaib, M., Khattak, S. H., Ullah, I., Farah, M. A., Singh, R., & Li, L. (2024). Morpho-physiological traits and stress indices reveal divergence in early-stage salt stress response in elite soybean germplasm. *Journal of King Saud University. Science*, *36*(11), 103522. DOI: 10.1016/j.jksus.2024.103522

Arnon, D. I. (1949). Copper enzymes in isolated chloroplasts. polyphenoloxidase in beta vulgaris. *Plant Physiology*, *24*(1), 1–15. DOI: 10.1104/pp.24.1.1 PMID: 16654194

Aroca, R., Vernieri, P., & Ruiz-Lozano, J. M. (2008). Mycorrhizal and non-mycorrhizal Lactuca sativa plants exhibit contrasting responses to exogenous ABA during drought stress and recovery. *Journal of Experimental Botany*, *59*(8), 2029–2041. DOI: 10.1093/jxb/ern057 PMID: 18469324

Atkinson, N. J., & Urwin, P. E. (2012). The interaction of plant biotic and abiotic stresses: From genes to the field. *Journal of Experimental Botany*, *63*(10), 3523–3543. DOI: 10.1093/jxb/ers100 PMID: 22467407

Ayaz, F. A., Torun, H., Glew, R. H., Bak, Z. D., Chuang, L. T., Presley, J. M., & Andrews, R. (2009). Nutrient content of carob pod (*Ceratonia siliqua* L.) flour prepared commercially and domestically. *Plant Foods for Human Nutrition (Dordrecht, Netherlands)*, *64*(4), 286–292. DOI: 10.1007/s11130-009-0130-3 PMID: 19763833

Aydın, F. (2017). Determining the level of knowledge of university students on global warming. *Journal of Social Sciences and Humanities*, *1*(1), 118–132.

Aydın, F., & Kaya, H. (2011). Evaluation of Social Sciences High school Students' Sensitivity towards Environment. *Marmara Geographical Review*, *24*(2), 229–257.

Ay, F. (2021). *Jeoloji tarihindeki iklim, küresel ısınma ve etkileri*. Akademisyen Kitabevi. DOI: 10.37609/akya.257

Ay, F., & Erik, N. Y. (2020). Üniversite öğrencilerinin küresel ısınma ve iklim değişikliğine yönelik bilgi ve algı düzeyleri. *Cumhuriyet Üniversitesi Fen-Edebiyat Fakültesi Sosyal Bilimler Dergisi*, *44*(2), 1–18.

Bakhshi, M. R., Pirdadeh Bayranvand, K., & Falaki, M. (2017). Investigating the relationship between social capital and environmental behaviors of students in university of birjand. *Social Capital Management*, *4*(3), 451–473.

Balestrini, R., Chitarra, W., Ghirardo, A., Nardini, A., & Nerva, L. (2022). A stressful life: How plants cope with multiple biotic and abiotic adverse factors. *Plant Stress*, *5*, 5. DOI: 10.1016/j.stress.2022.100095

Baloyi, E. (2015). The Christian church's role in the escalating mob justice system in our black townships-An African pastoral view. *HTS Theological Studies, 71*(2), 01-07. https://www.tandfonline.com/doi/abs/10.1300/J118v04n04_12

Banday, U., & Aneja, R. (2020). Renewable and non-renewable energy consumption, economic growth and carbon emission in BRICS: Evidence from bootstrap panel causality. *International Journal of Energy Sector Management*, *14*(1), 248–260. DOI: 10.1108/IJESM-02-2019-0007

Baninla, Y., Sharifi, A., Allam, Z., Tume, S. J. P., Gangtar, N. N., & George, N. (2022). An overview of climate change adaptation and mitigation research in Africa. *Frontiers in Climate*, *4*, 976427. https://www.frontiersin.org/journals/climate/articles/10.3389/fclim.2022.976427/full. DOI: 10.3389/fclim.2022.976427

Bao, Y., Tang, Z., Li, H., & Zhang, Y. (2019). Computer vision and deep learning–based data anomaly detection method for structural health monitoring. *Structural Health Monitoring*, *18*(2), 401–421. DOI: 10.1177/1475921718757405

Baris, K., & Kucukali, S. (2012). Availibility of renewable energy sources in Turkey: Current situation, potential, government policies and the EU perspective. *Energy Policy*, *42*, 377–391. DOI: 10.1016/j.enpol.2011.12.002

Barrs, H. D. (1966). Root pressure and leaf water potential. *Science*, *152*(3726), 1266–1268. DOI: 10.1126/science.152.3726.1266 PMID: 17769543

Başıbüyük, G. Ö., & Ay F. Acar, S. (2017). Küresel ısınma ve küresel ısınmanın yaşlılar üzerindeki etkisi. *Türkiye Sosyal Araştırmalar Dergisi*, *21*(1), 203–218.

Bassi, N., Kumar, M. D., Sharma, A., & Pardha-Saradhi, P. (2014). Status of wetlands in India: A review of extent, ecosystem benefits, threats and management strategies. *Journal of Hydrology. Regional Studies*, *2*, 1–19. DOI: 10.1016/j.ejrh.2014.07.001

Bates, L. S., Waldren, R. P., & Teare, I. D. (1973). Rapid determination of free proline for water studies. *Plant and Soil*, *39*(1), 205–208. DOI: 10.1007/BF00018060

Batlle, I., & Tous, J. (1997). *Carob tree Ceratonia siliqua L. Promoting the conservation and use of underutilized and neglected crops. 17* (Vol. 17). Bioversity International.

Behura, A., Sahu, S., & Kabat, M. R. (2021). Advancement of Machine Learning and Cloud Computing in the Field of Smart Health Care. *Machine Learning Approach for Cloud Data Analytics in IoT*, 273-306.

Bekun, F., Emir, F., & Sarkodie, S. (2019). Another look at the relationship between energy consumption, carbon dioxide emissions, and economic growth in South Africa. *The Science of the Total Environment*, *655*, 759–765. DOI: 10.1016/j.scitotenv.2018.11.271 PMID: 30476856

Bellard, C., Bertelsmeier, C., Leadley, P., Thuiller, W., & Courchamp, F. (2012). Impacts of climate change on the future of biodiversity. *Ecology Letters*, *15*(4), 365–377. DOI: 10.1111/j.1461-0248.2011.01736.x PMID: 22257223

Bello, A. S., & Ahmed, T. (2024). Moringa leaf extract alleviates salt stress in tomato (*Solanum lycopersicum* L.) by activating antioxidant defenses, reducing osmolyte accumulation, improving water status, and enhancing yield. *Plant Stress*, *14*, 100640. DOI: 10.1016/j.stress.2024.100640

Bento, J., & Moutinho, V. (2016). CO2 emissions, non-renewable and renewable electricity production, economic growth, and international trade in Italy. *Renewable & Sustainable Energy Reviews*, *55*, 142–155. DOI: 10.1016/j.rser.2015.10.151

Berberoglu, G., & Tosunoglu, C. (1995). Exploratory and confirmatory factor analyses of an environmental attitude scale (EAS) for Turkish university students. *The Journal of Environmental Education*, *26*(3), 40–43. DOI: 10.1080/00958964.1995.9941444

Bhardwaj, L. K., Rath, P., Jain, H., Purohit, S., Yadav, P., & Singh, V. (2024). The Impact of Climate-Induced Livelihood, Health, and Migration on Women and Girls: A Review. *Global Insights on Women Empowerment and Leadership*, 100-118.

Bhardwaj, L. K., Kumar, D., & Kumar, A. (2023). Phytoremediation potential of Ocimum sanctum: A sustainable approach for remediation of heavy metals. In *Phytoremediation potential of medicinal and aromatic plants* (pp. 46–57). CRC Press. DOI: 10.1201/9781003290780-5

Bhardwaj, L. K., Rath, P., Jain, H., & Choudhury, M. (2025a). *A comprehensive review of wetlands in India and their conservation*. Nature-Based Technologies for Wastewater Treatment and Bioenergy Production., DOI: 10.2166/9781789064100_0043

Bhardwaj, L. K., Upadhyay, D., Jain, H., Rath, P., Kumar, D., & Bajpai, S. (2025b). Recent trends of phytoremediation for the wastewater treatment and resource recovery. In *Biotechnologies for Wastewater Treatment and Resource Recovery* (pp. 135–146). Elsevier. DOI: 10.1016/B978-0-443-27376-6.00008-6

Bhateria, R., & Jain, D. (2016). Water quality assessment of lake water: A review. *Sustainable Water Resources Management, 2*(2), 161–173. DOI: 10.1007/s40899-015-0014-7

Bhattacharya, M., Paramati, S., Ozturk, I., & Bhattacharya, S. (2016). The effect of renewable energy consumption on economic growth: Evidence from top 38 countries. *Applied Energy, 162*, 733–741. DOI: 10.1016/j.apenergy.2015.10.104

Bhattacharyya, D., Stephen Neal Joshua, E., & Thirupathi Rao, N. (2023). Medical Image Analysis of Lung Cancer CT Scans Using Deep Learning with Swarm Optimization Techniques. *Machine Intelligence, Big Data Analytics, and IoT in Image Processing: Practical Applications*, 23-50.

Bid, S., & Siddique, G. (2019). Identification of seasonal variation of water turbidity using NDTI method in Panchet Hill Dam, India. *Modeling Earth Systems and Environment, 5*(4), 1179–1200. DOI: 10.1007/s40808-019-00609-8

Bilgen, S., Keles, S., Kaygusuz, A., Sari, A., & Kaygusuz, K. (2008). Global warming and renewable energy sources for sustainable development: A case study in Turkey. *Renewable & Sustainable Energy Reviews, 12*(2), 372–396. DOI: 10.1016/j.rser.2006.07.016

Bilgili, F., Kocak, E., Bulut, U., & Kuloglu, A. (2017b). The impact of urbanization on energy intensity: Panel data evidence considering cross-sectional dependence and heterogeneity. *Energy, 133*, 242–256. DOI: 10.1016/j.energy.2017.05.121

Bilgili, F., Koçak, E., Bulut, Ü., & Kuşkaya, S. (2017a). Can biomass energy be an efficient policy tool for sustainable development? *Renewable & Sustainable Energy Reviews, 71*, 830–845. DOI: 10.1016/j.rser.2016.12.109

Bilgili, S. (2008). *İlköğretim 7. sınıf fen ve teknoloji dersinde çevre konularının öğretiminde, yapılandırmacı yaklaşıma dayalı işbirlikli öğrenmenin öğrencilerin erişine etkisi*. Yayımlanmış Yüksek Lisans Tezi, Gazi Üniversitesi.

Bloch, H., Rafiq, S., & Salim, R. (2015). Economic growth with coal, oil and renewable energy consumption in China: Prospects for fuel substitution. *Economic Modelling*, *44*, 104–115. DOI: 10.1016/j.econmod.2014.09.017

Bölük, G., & Mert, M. (2014). Fossil & renewable energy consumption, GHGs (greenhouse gases) and economic growth: Evidence from a panel of EU (European Union) countries. *Energy*, *74*, 439–446. DOI: 10.1016/j.energy.2014.07.008

Boonjung, H., & Fukai, S. (1996). Effects of soil water deficit at different growth stages on rice growth and yield under upland conditions. 2. Phenology, biomass production and yield. *Field Crops Research*, *48*(1), 47–55. DOI: 10.1016/0378-4290(96)00039-1

Boutwell, J. L., & Westra, J. V. (2016). The role of wetlands for mitigating economic damage from hurricanes. *Journal of the American Water Resources Association*, *52*(6), 1472–1481. DOI: 10.1111/1752-1688.12473

Brassesco, M. E., Brandao, T. R., Silva, C. L., & Pintado, M. (2021). Carob bean (*Ceratonia siliqua* L.): A new perspective for functional food. *Trends in Food Science & Technology*, *114*, 310–322. DOI: 10.1016/j.tifs.2021.05.037

Bulut, U. (2017). The impacts of non-renewable and renewable energy on CO2 emissions in Turkey. *Environmental Science and Pollution Research International*, *24*(18), 15416–15426. DOI: 10.1007/s11356-017-9175-2 PMID: 28508336

Bulut, U., & Muratoglu, G. (2018). Renewable energy in Turkey: Great potential, low but increasing utilization, and an empirical analysis on renewable energy-growth nexus. *Energy Policy*, *123*, 240–250. DOI: 10.1016/j.enpol.2018.08.057

Butt, M. A., Zafar, M., Ahmed, M., Shaheen, S., Sultana, S., Butt, M. A., ... & Sultana, S. (2021). Types of Wetland and Wetland Plants. *Wetland Plants: A Source of Nutrition and Ethno-medicines*, 35-54.

Buzzanca, C., D'Amico, A., Pistorio, E., Di Stefano, V., & Melilli, M. G. (2025). Carob-based functional beverages: Nutritional value and health properties. *Beverages*, *11*(1), 1. DOI: 10.3390/beverages11010001

Cabello, J. C., Karimipour, H., Jahromi, A. N., Dehghantanha, A., & Parizi, R. M. (2020). Big-data and cyber-physical systems in healthcare: Challenges and opportunities. *Handbook of Big Data Privacy*, 255-283.

Çabuk, B., & Karacaoğlu, C. (2003). Üniversiteli öğrencilerinin çevre duyarlılıklarının incelenmesi. [JFES]. *Ankara University Journal of Faculty of Educational Sciences*, *36*(1), 189–198. DOI: 10.1501/Egifak_0000000079

Cai, Y., Sam, C., & Chang, T. (2018). Nexus between clean energy consumption, economic growth and CO2 emissions. *Journal of Cleaner Production*, *182*, 1001–1011. DOI: 10.1016/j.jclepro.2018.02.035

Calhoun, A. J., Mushet, D. M., Bell, K. P., Boix, D., Fitzsimons, J. A., & Isselin-Nondedeu, F. (2017). Temporary wetlands: Challenges and solutions to conserving a 'disappearing'ecosystem. *Biological Conservation*, *211*, 3–11. DOI: 10.1016/j.biocon.2016.11.024

Can, H., & Korkmaz, O. (2019). The relationship between renewable energy consumption and economic growth: the case of Bulgaria. *International Journal of Energy Sector Management*, 1-18.

Capik, M., Yılmaz, A., & Cavusoglu, I. (2012). Present situation and potential role of renewable energy in Turkey. *Renewable Energy*, *46*, 1–13. DOI: 10.1016/j.renene.2012.02.031

Carbon Brief. (2022). *Africa's Carbon Emissions and Climate Impact.* Retrieved from https://www.carbonbrief.org/

Cardoso, A. A., Brodribb, T. J., Kane, C. N., DaMatta, F. M., & McAdam, S. A. (2020). Osmotic adjustment and hormonal regulation of stomatal responses to vapour pressure deficit in sunflower. *AoB Plants*, *12*(4), 12. DOI: 10.1093/aobpla/plaa025 PMID: 32665827

Cardoso, A. A., Brodribb, T. J., Lucani, C. J., DaMatta, F. M., & McAdam, S. A. (2018). Coordinated plasticity maintains hydraulic safety in sunflower leaves. *Plant, Cell & Environment*, *41*(11), 2567–2576. DOI: 10.1111/pce.13335 PMID: 29748980

Casado, C. M. (1995). Allelopathic effects of *Lantana camara* (Verbenaceae) on morning glory (*Imomoea tricolor*). *Rhodora*, *97*, 264–274.

Casalό, L. V., & Escario, J. J. (2018). Heterogeneity in the association between environmental attitudes and pro-environmental behavior: A multilevel regression approach. *Journal of Cleaner Production*, *175*, 155–163. DOI: 10.1016/j.jclepro.2017.11.237

Castro-Díez, M. P., & Fernández, Á. (2017). Effects of non-native riparian plants in riparian and fluvial ecosystems: A review for the Iberian Peninsula. *Limnetica*, *36*, 525–541.

Çetin, T., Turan, M. E., Aytekin, P., & Yurdusev, M. A. (2020). Üniversite öğrencilerinin çevresel yaklaşımlarını ölçmeye yönelik bir araştırma: Manisa Celal Bayar Üniversitesi örneği. *Doğal Afetler ve Çevre Dergisi*, *6*(1), 12–23. DOI: 10.21324/dacd.546369

Chakraborty, C., & Bhattacharya, S. (2020). Healthcare data monitoring under Internet of Things. In *Green Computing and Predictive Analytics for Healthcare* (pp. 1–18). Chapman and Hall/CRC. DOI: 10.1201/9780429317224-1

Chakraborty, S. K., Sanyal, P., & Ray, R. (2023a). Introductory and Basic Eco-biological Aspects of Wetlands. In *Wetlands Ecology: Eco-biological uniqueness of a Ramsar site (East Kolkata Wetlands, India)* (pp. 1–38). Springer International Publishing. DOI: 10.1007/978-3-031-09253-4_1

Chalise, D., Kumar, L., & Kristiansen, P. (2019). Land degradation by soil erosion in Nepal: A review. *Soil Systems*, *3*(1), 12. DOI: 10.3390/soilsystems3010012

Chandra, K. (2021). *Faunal diversity in Ramsar wetlands of India*. Jointly published by Zoological Survey of India & Wetland Division, Ministry of Environment, Forest and Climate Change, Government of India.

Chang, C. H. (2015). Teaching climate change–a fad or a necessity? *International Research in Geographical and Environmental Education*, *24*(3), 181–183. DOI: 10.1080/10382046.2015.1043763

Chang, T., Gupta, R., Simokengne, B., Smithers, D., & Trembling, A. (2013). The causal relationship between renewable energy consumption and economic growth: Evidence from the G7 countries. *Applied Economics Letters*, *23*(1), 1–9.

Chartzoulakhis, K. S. (2005). Salinity and olive: Growth, salt tolerance, photosynthesis and yield. *Agricultural Water Management*, *78*(1-2), 108–121. DOI: 10.1016/j.agwat.2005.04.025

Cheng, B., & Lai, T. (1997). An investigation of co-integration and causality between energy consumption and economic activity in Taiwan. *Energy Economics*, *19*(4), 435–444. DOI: 10.1016/S0140-9883(97)01023-2

Cheung, Y.-W., & Chinn, M. (1997). Further Investigation of the Uncertain Unit Root in GNP. *Journal of Business & Economic Statistics*, *15*(1), 68–73. DOI: 10.1080/07350015.1997.10524688

Chien, T., & Hu, J. (2007). Renewable energy and macroeconomic efficiency of OECD and non-OECD economies. *Energy Policy*, *35*(7), 3606–3615. DOI: 10.1016/j.enpol.2006.12.033

Choat, B., Brodribb, T. J., Brodersen, C. R., Duursma, R. A., Lopez, ´. R., & Medlyn, B. E. (2018). Triggers of tree mortality under drought. *Nature*, *558*(7711), 531–539. DOI: 10.1038/s41586-018-0240-x PMID: 29950621

Choudhury, M., Dixit, G., & Majumdar, S. (Eds.). (2025). *Sustainable Synergy: Harnessing Ecosystems for Climate Resilience*. Springer., DOI: 10.1007/978-3-031-77957-2

Choudhury, M., Fazli, P., Pramanick, P., Gobato, R., Zaman, S., & Mitra, A. (2019). Sensitivity of the Indian Sundarban mangrove ecosystem to local level climate change. *Science and Education*, *5*(3), 24–28.

Choudhury, M., Sharma, A., Singh, P., & Kumar, D. (2021). Impact of climate change on wetlands, concerning Son Beel, the largest wetland of North East, India. In *Global climate change* (pp. 393–414). Elsevier. DOI: 10.1016/B978-0-12-822928-6.00006-X

Cifre, J., Bota, J., Escalona, J. M., Medrano, H., & Flexas, J. (2005). Physiological tools for irrigation scheduling in grapevine (Vitis vinifera L.): An open gate to improve water-use efficiency? *Agriculture, Ecosystems & Environment*, *106*(2-3), 159–170. DOI: 10.1016/j.agee.2004.10.005

Cleveland, C., Costanza, R., Hall, C., & Kaufmann, R. (1997). Energy and the US economy: A biophysical perspective. *International Library of Critical Writings in Economics*, *75*, 295–302.

Clout, M. N., & Veitch, C. R. (2002). *Turning the tide of biological invasion: the potential for eradicating invasive species. Turning the tide: the eradication of invasive species*. IUCN SSC Invasive Species Specialist Group.

Coertjens, L., Boeve-de Pauw, J., De Maeyer, S., & Van Petegem, P. (2010). Do Schools Make a difference in their students' environmental attitudes and awareness? evidence from pisa 2006. *International Journal of Science and Mathematics Education*, *8*(3), 497–522. DOI: 10.1007/s10763-010-9200-0

Correia, P. J., Gama, F., Pestana, M., & Martins-Loução, M. A. (2010). Tolerance of young (*Ceratonia siliqua* L.) carob rootstock to NaCl. *Agricultural Water Management*, *97*(6), 910–916. DOI: 10.1016/j.agwat.2010.01.022

Costanza, R., Anderson, S. J., Sutton, P., Mulder, K., Mulder, O., Kubiszewski, I., Wang, X., Liu, X., Pérez-Maqueo, O., Luisa Martinez, M., Jarvis, D., & Dee, G. (2021). The global value of coastal wetlands for storm protection. *Global Environmental Change*, *70*, 102328. DOI: 10.1016/j.gloenvcha.2021.102328

Creek, D., Lamarque, L. J., Torres-Ruiz, J. M., Parise, C., Burlett, R., Tissue, D. T., & Delzon, S. (2020). Xylem embolism in leaves does not occur with open stomata: Evidence from direct observations using the optical visualization technique. *Journal of Experimental Botany*, *71*(3), 1151–1159. DOI: 10.1093/jxb/erz474 PMID: 31641746

Cronk, Q. C. B., & Fuller, J. L. (1995). *Plant Invaders*. Chapman and Hall.

da Silva Brito, N. D., dos Santos Medeiros, M. J., de Souza, E. S., & de Lima, A. L. A. (2022). Drought response strategies for deciduous species in the semiarid Caatinga derived from the interdependence of anatomical, phenological and bio-hydraulic attributes. *Flora (Jena)*, *288*, 152009. DOI: 10.1016/j.flora.2022.152009

Dadach, M., Bhatt, A., Gairola, S., Elnaggar, A., & El-Keblawy, A. (2024). Germination responses of carob (*Ceratonia siliqua* L.) seeds to salinity and water stresses: Implications for the restoration of Mediterranean marginal lands. *Botany*, *103*, 1–13. DOI: 10.1139/cjb-2024-0070

Dadach, M., Gairola, S., Hameed, A., & El-Keblawy, A. (2023). Effects of different chloride salts on seed germination and early seedling performance of Ceratonia siliqua, a multipurpose tree of the Mediterranean region. *South African Journal of Botany*, *158*, 8–17. DOI: 10.1016/j.sajb.2023.04.044

Dagar, J. C., Sharma, P. C., Chaudhari, S. K., Jat, H. S., & Ahamad, S. (2016). Climate Change vis-a-vis Saline Agriculture: Impact and Adaptation Strategies. In Dagar, J., Sharma, P., Sharma, D., & Singh, A. (Eds.), *Innovative Saline Agriculture*. Springer., DOI: 10.1007/978-81-322-2770-0_2

Daryanto, S., Wang, L., & Jacinthe, P.-A. (2016). Global synthesis of drought effects on maize and wheat production. *PLoS One*, *11*(5), e0156362. DOI: 10.1371/journal.pone.0156362 PMID: 27223810

Dash, S., Shakyawar, S. K., Sharma, M., & Kaushik, S. (2019). Big data in healthcare: Management, analysis and future prospects. *Journal of Big Data*, *6*(1), 1–25. DOI: 10.1186/s40537-019-0217-0

Davidson, D. J., & Freudenburg, W. R. (1996). Gender and environmental risk concerns: A review and analysis of available research. *Environment and Behavior*, *28*(3), 302–339. DOI: 10.1177/0013916596283003

Davis, M. A., Grime, J. P., & Thompson, K. (2000). Fluctuating resources in plant communities: A general theory of invasibility. *Journal of Ecology*, *88*(3), 528–534. DOI: 10.1046/j.1365-2745.2000.00473.x

Day, P. K., LeDuc, H. G., Mazin, B. A., Vayonakis, A., & Zmuidzinas, J. (2003). A broadband superconducting detector suitable for use in large arrays. *Nature*, *425*(6960), 817–821. DOI: 10.1038/nature02037 PMID: 14574407

De Leeuw, A., Valois, P., Ajzen, I., & Schmidt, P. (2015). Using the theory of planned behavior to identify key beliefs underlying pro-environmental behavior in high-school students: Implications for educational interventions. *Journal of Environmental Psychology*, *42*, 128–138. DOI: 10.1016/j.jenvp.2015.03.005

Deloitte. (2023). *Healthcare in India: Trends and Opportunities.* Retrieved from https://www2.deloitte.com/in/en/pages/life-sciences-and-healthcare/articles/shaping-healthcare-for-all-in-india.html

Demarquet, Q., Rapinel, S., Dufour, S., & Hubert-Moy, L. (2023). Long-term wetland monitoring using the landsat archive: A review. *Remote Sensing (Basel)*, *15*(3), 820. DOI: 10.3390/rs15030820

Demirel, M., Gürbüz, B., & Karaküçük, S. (2009). Rekreasyonel aktivitelere katilimin çevreye yönelik tutum üzerindeki etkisi ve yeni ekolojik paradigma ölçeği'nin geçerliği ve güvenirliği. *Spormetre Beden Eğitimi ve Spor Bilimleri Dergisi*, *7*(2), 47–50.

Derouiche, M., Mzabri, I., Ouahhoud, S., Dehmani, I., Benabess, R., Addi, M., Hano, C., Boukroute, A., Berrichi, A., & Kouddane, N. (2023). The effect of salt stress on the growth and development of three *Aloe* species in eastern Morocco. *Plant Stress*, *9*, 100187. DOI: 10.1016/j.stress.2023.100187

Destek, M., & Aslan, A. (2017). Renewable and non-renewable energy consumption and economic growth in emerging economies: Evidence from bootstrap panel causality. *Renewable Energy*, *111*, 757–763. DOI: 10.1016/j.renene.2017.05.008

Devall, B. (2006). Conservation of biodiversity: Opportunities and challenges. *Human Ecology Review*, •••, 60–75. https://www.jstor.org/stable/24707557

Devika, G., & Karegowda, A. G. (2021). Deep Learning in IoT: Introduction, Applications, and Perspective in the Big Data Era. In *Deep Learning Applications and Intelligent Decision Making in Engineering* (pp. 1-54). IGI Global.

Dietz, T., Stern, P. C., & Guagnano, G. A. (1998). Social structural and social psychological bases of environmental concern. *Environment and Behavior*, *30*(4), 450–471. DOI: 10.1177/001391659803000402

Dillon, P. J., & Gayford, C. G. (1997). A psychometric approach to investigating the environmental beliefs, intentions and behaviours of pre-service teachers. *Environmental Education Research*, *3*(3), 283–297. DOI: 10.1080/1350462970030303

Directive, E. U. (2023). Revised Directive EU/2023/2413. *Official Journal of the European Union*, 1-77.

Dixit, G., Majumdar, S., & Choudhury, M. (2025). Introduction to Ecosystem Services for Climate Resilience: A Theoretical Framework. In *Sustainable Synergy: Harnessing Ecosystems for Climate Resilience* (pp. 3-16). Cham: Springer Nature Switzerland. https://doi.org/DOI: 10.1007/978-3-031-77957-2_1

Dlamini, P. J. (2021). Drought stress tolerance mechanisms and breeding effort in sugarcane: A review of progress and constraints in South Africa. *Plant Stress*, *2*, 100027. DOI: 10.1016/j.stress.2021.100027

Dogan, E. (2015). The relationship between economic growth and electricity consumption from renewable and non-renewable sources: A study of Turkey. *Renewable & Sustainable Energy Reviews*, *52*, 534–546. DOI: 10.1016/j.rser.2015.07.130

Dogan, E. (2016). Analyzing the linkage between renewable and non-renewable energy consumption and economic growth by considering structural break in time-series data. *Renewable Energy*, *99*, 1126–1136. DOI: 10.1016/j.renene.2016.07.078

Dono, J., Webb, J., & Richardson, B. (2010). The relationship between environmental activism, pro- environmental behaviour and social identity. *Journal of Environmental Psychology*, *30*(2), 178–186. DOI: 10.1016/j.jenvp.2009.11.006

Dos Santos, G. M., Meléndez-Pastor, I., Navarro-Pedreño, J., & Lucas, I. G. (2019). A Review of Landsat TM/ETM based Vegetation Indices as Applied to Wetland Ecosystems. *Journal of Geographical Research/Ma alla Al-bu ū Al- uġrāfiyya* , *2*(1).

Dubey, D., Kumar, S., & Dutta, V. (2022). Algae and Macrophytes as Bioindicators of Freshwater Ecosystem. In *Environmental Studies and Climate Change* (pp. 441–467). CRC Press. DOI: 10.1201/9781003220824-33

Dubois, M. K., Gils, J. K., Hanniton, P. A., & Smith, F. (1956). Use of phenol reagent for the determination of total sugar. *Analytical Chemistry*, *28*, 350–356. DOI: 10.1021/ac60111a017

Duursma, R. A., Blackman, C. J., Lop'ez, R., Martin-StPaul, N. K., Cochard, H., & Medlyn, B. E. (2019). On the minimum leaf conductance: Its role in models of plant water use, and ecological and environmental controls. *The New Phytologist*, *221*(2), 693–705. DOI: 10.1111/nph.15395 PMID: 30144393

Eagles, P. F., & Demare, R. (1999). Factors influencing children's environmental attitudes. *The Journal of Environmental Education*, *30*(4), 33–37. DOI: 10.1080/00958969909601882

Eden, S., & Jin, J. (1992). Cointegration tests of energy consumption, income, and employment. *Resources and Energy*, *14*(3), 259–266. DOI: 10.1016/0165-0572(92)90010-E

Ediger, V., & Kentel, E. (1999). Renewable energy potential as an alternative to fossil fuels in Turkey. *Energy Conversion and Management*, *40*(7), 743–755. DOI: 10.1016/S0196-8904(98)00122-8

Edsand, H. E., & Broich, T. (2020). The impact of environmental education on environmental and renewable energy technology awareness: Empirical evidence from Colombia. *International Journal of Science and Mathematics Education*, *18*(4), 611–634. DOI: 10.1007/s10763-019-09988-x

Ek, H. N., Kılıç, N., Öğdüm, P., Düzgün, G., & Peker, S. (2009). Adnan Menderes Üniversitesinin farklı akademik alanlarında öğrenim gören ilk ve son sınıf öğrencilerinin çevre sorunlarına yönelik tutumları ve duyarlılıkları. *Kastamonu Eğitim Dergisi*, *17*(1), 125–136.

Elbadawi, M., Gaisford, S., & Basit, A. W. (2021). Advanced machine-learning techniques in drug discovery. *Drug Discovery Today*, *26*(3), 769–777. DOI: 10.1016/j.drudis.2020.12.003 PMID: 33290820

Elloumi, O., Amar, F. B., Trigui, A., Gargouri, K., & Ghrab, M. (2024). Early evaluation of salt-stress tolerance of new released olive cultivars based on physiological and biomass allocation indicators. *Journal of Agriculture and Food Research*, *18*, 101503. DOI: 10.1016/j.jafr.2024.101503

El-Ramady, H., Prokisch, J., Mansour, H., Bayoumi, Y. A., Shalaby, T. A., Veres, S., & Brevik, E. C. (2024). Review of crop response to soil salinity stress: Possible approaches from leaching to nano-management. *Soil Systems*, *8*(1), 11. DOI: 10.3390/soilsystems8010011

Energy Institute database. (2024). 2024 Statistical Review of World Energy. <https://www.energyinst.org/statistical-review> (last visited on 19.03.2025).

Erkal, S., Kilic, I., & Sahin, H. (2012). Comparison of environmental attitudes of university students determined via the new environmental paradigm scale according to the students' personal characteristics. *Eurasian Journal of Educational Research*, *49*, 21–39.

Fang, Y. (2011). Economic welfare impacts from renewable energy consumption: The China experience. *Renewable & Sustainable Energy Reviews*, *15*(9), 5120–5128. DOI: 10.1016/j.rser.2011.07.044

Farooq, K., Shah, F. A., Qayoom, I., Qureshi, K., & Fayaz, I. (2024). Freshwater Floral Diversities as Pollution Indicators. In *Aquatic Pollution* (pp. 135–144). CRC Press. DOI: 10.1201/9781003503705-6

Farooq, M., Gogoi, N., Barthakur, S., Baroowa, B., Bharadwaj, N., Alghamdi, S. S., & Siddique, K. (2017). Drought stress in grain legumes during reproduction and grain filling. *Journal Agronomy & Crop Science*, *203*(2), 81–102. DOI: 10.1111/jac.12169

Fisher, S., Bellinger, D. C., Cropper, M. L., Kumar, P., Binagwaho, A., Koudenoukpo, J. B., Park, Y., Taghian, G., & Landrigan, P. J. (2021). Air pollution and development in Africa: Impacts on health, the economy, and human capital. *The Lancet. Planetary Health, 5*(10), e681–e688. https://www.thelancet.com/journals/lanplh/article/PIIS2542-5196(21)00201-1/fulltext. DOI: 10.1016/S2542-5196(21)00201-1 PMID: 34627472

Fluet-Chouinard, E., Stocker, B. D., Zhang, Z., Malhotra, A., Melton, J. R., Poulter, B., Kaplan, J. O., Goldewijk, K. K., Siebert, S., Minayeva, T., Hugelius, G., Joosten, H., Barthelmes, A., Prigent, C., Aires, F., Hoyt, A. M., Davidson, N., Finlayson, C. M., Lehner, B., & McIntyre, P. B. (2023). Extensive global wetland loss over the past three centuries. *Nature, 614*(7947), 281–286. DOI: 10.1038/s41586-022-05572-6 PMID: 36755174

Food and Agriculture Organization. (2007). *Coping with water scarcity*. FAO Rome.

Forcey, G. M. (2006). *Landscape-level influences on wetland birds in the Prairie Pothole Region of the United States and Canada*. North Dakota State University.

Ford, H., Garbutt, A., Ladd, C., Malarkey, J., & Skov, M. W. (2016). Soil stabilization linked to plant diversity and environmental context in coastal wetlands. *Journal of Vegetation Science, 27*(2), 259–268. DOI: 10.1111/jvs.12367 PMID: 27867297

Freudenburg, W. R. (1991). Rural-urban differences in environmental concern: A closer look. *Sociological Inquiry, 61*(2), 167–198. DOI: 10.1111/j.1475-682X.1991.tb00274.x

Freudenburg, W. R., & Mcginn, B. (1989). Rural Urban Differences. In *Environmental Concern: A Closer Look*. Eric Document Reproduction Service N.

Fullerton, D., & Metcalf, G. (1997). *Environmental taxes and the double-dividend hypothesis: did you really expect something for nothing?* Cambridge: NBER Working Paper No 6199.

Furman, A. (1998). A note on environmental concern in a developing country: Results from an Istanbul survey. *Environment and Behavior, 30*(4), 520–534. DOI: 10.1177/001391659803000406

Gadoum, A., Adda, A., Sahnoune, M., & Aid, F. (2019). Physiological and biochemical responses of three ecotypes of carob (Ceratonia siliqua L.) against drought stress in Algeria. *Applied Ecology and Environmental Research, 17*(2), 1929–1945. DOI: 10.15666/aeer/1702_19291945

Gambro, J. S., & Switzky, H. N. (1999). Variables associated with American high school students' knowledge of environmental issues related to energy and pollution. *The Journal of Environmental Education*, *30*(2), 15–22. DOI: 10.1080/00958969909601866

Gardos, V. T., & Dodd, D. K. (1995). An immediate response to environmentally disturbing news and the environmental attitudes of college students. *Psychological Reports*, *77*(3), 1121–1122. DOI: 10.2466/pr0.1995.77.3f.1121

Gaupp, D. (2008). Turkey's new law on renewable energy sources within the context of the accession negotiations with the EU. *German Law Journal*, *08*(04), 413–416. DOI: 10.1017/S2071832200005666

Gentle, C. B., & Duggin, J. A. (1998). Interference of shape *Choricarpia leptopetala* by shape *Lantana camara* with nutrient enrichment in mesic forests on the Central Coast of NSW. *Plant Ecology*, *136*(2), 205–211. DOI: 10.1023/A:1009784706714

George, D. L., & Southwell, P. L. (1986). Opinion on the Diablo Canyon nuclear power plant: The effects of situation and socialization. *Social Science Quarterly*, *67*(4), 722.

Gharnit, N., & Ennabili, A. (2016). Categories of Carob Tree (*Ceratonia siliqua* L.) from Morocco. *International Journal of Fruit Science*, *16*(3), 259–274. DOI: 10.1080/15538362.2015.1102674

Gill, S. S., & Tuteja, N. (2010). Reactive oxygen species and antioxidant machinery in abiotic stress tolerance in crop plants. *Plant Physiology and Biochemistry*, *48*(12), 909–930. DOI: 10.1016/j.plaphy.2010.08.016 PMID: 20870416

Gladwin, D., Horst, R., James, K., & Sameshima, P. (2022). Imagining futures literacies: A collaborative practice. *Journal of Higher Education Theory and Practice*, *22*(7), 27–40.

Glynn, J., Perera, N., & Verma, R. (2007). Unit root tests and structural breaks: A survey with applications. *Journal of Quantitative Methods for Economics and Business Administration*, *3*(1), 63–79.

Gnansounou, E. (2008). Assessing the energy vulnerability: Case of industrial countries. *Energy Policy*, *36*(10), 3734–3744. DOI: 10.1016/j.enpol.2008.07.004

Gogoi, N., Choudhury, M., Ul Hasan, M. S., Changmai, B., Baruah, D., & Samanta, P. (2024). Soil organic carbon pool in diverse land utilization patterns in North-East India: An implication for carbon sequestration. *Environment, Development and Sustainability*, •••, 1–27. DOI: 10.1007/s10668-024-05037-y

Gökdere, M. (2005). A study on environmental knowledge level of primary students in Turkey. In *Asia-Pacific Forum on Science Learning and Teaching,* 6(2), 1-13.

Gokmenoglu, K., & Kaakeh, M. (2018). Causal relationship between nuclear energy consumption and economic growth: Case of Spain. *Strategic Planning for Energy and the Environment, 37*(3), 58–76. DOI: 10.1080/10485236.2018.11958660

Goldratt, E. M. (1990). *The Goal: A Process of Ongoing Improvement.* North River Press.

Gooden, B., French, K., & Turner, P. J. (2009). Invasion and management of woody plant, *Lantana camara* L., alters vegetation diversity within wet sclerophyll forest in south eastern Australia. *Forest Ecology and Management, 257*(3), 960–967. DOI: 10.1016/j.foreco.2008.10.040

Goodland, R. (1995). The concept of environmental sustainability. *Annual Review of Ecology and Systematics, 26*(1), 1–24. https://www.jstor.org/stable/2097196?seq=1. DOI: 10.1146/annurev.es.26.110195.000245

Gopal, B. (2013). Future of wetlands in tropical and subtropical Asia, especially in the face of climate change. *Aquatic Sciences, 75*(1), 39–61. DOI: 10.1007/s00027-011-0247-y

Gopal, B., & Ghosh, D. (2008). Natural Wetlands. In *Applications in Ecological Engineering*. Elsevier BV Amsterdam. DOI: 10.1016/B978-008045405-4.00067-7

Gozgor, G., Lau, C., & Lu, Z. (2018). Energy consumption and economic growth: New evidence from the OECD countries. *Energy, 153*, 27–34. DOI: 10.1016/j.energy.2018.03.158

Granger, C. (1969). Investigating causal relations by econometric models and cross-spectral methods. *Econometrica, 37*(3), 424–438. DOI: 10.2307/1912791

Granger, C., & Newbold, P. (1974). Spurious regressions in econometrics. *Journal of Econometrics, 2*(2), 111–120. DOI: 10.1016/0304-4076(74)90034-7

Grossiord, C., Buckley, T. N., Cernusak, L. A., Novick, K. A., Poulter, B., Siegwolf, R. T., & McDowell, N. G. (2020). Plant responses to rising vapor pressure deficit. *The New Phytologist, 226*(6), 1550–1566. DOI: 10.1111/nph.16485 PMID: 32064613

Gumiero, B., Mant, J., Hein, T., Elso, J., & Boz, B. (2013). Linking the restoration of rivers and riparian zones/wetlands in Europe: Sharing knowledge through case studies. *Ecological Engineering, 56*, 36–50. DOI: 10.1016/j.ecoleng.2012.12.103

Günden C. Miran B. (2008). Yeni çevresel paradigma ölçeğiyle çiftçilerin çevre tutumunun belirlenmesi: İzmir ili Torbalı ilçesi örneği. *Ekoloji dergisi*, 18(69),41-50.

Gupta, A., Rico-Medina, A., & Cano-Delgado, ˜. A. I. (2020). The physiology of plant responses to drought. *Science, 368*(6488), 266–269. DOI: 10.1126/science.aaz7614 PMID: 32299946

Gupta, K. J., Igamberdiev, A. U., Manjunatha, G., Segu, S., Moran, J. F., Neelawarne, B., Bauwe, H., & Kaiser, W. M. (2011). The emerging roles of nitric oxide (NO) in plant mitochondria. *Plant Science, 181*(5), 520–526. DOI: 10.1016/j.plantsci.2011.03.018 PMID: 21893247

Gupta, K. J., Kaladhar, V. C., Fitzpatrick, T. B., Fernie, A. R., Møller, I. M., & Loake, G. J. (2022). Nitric oxide regulation of plant metabolism. *Molecular Plant, 15*(2), 228–242. DOI: 10.1016/j.molp.2021.12.012 PMID: 34971792

Gürsoy, T. (2008). *Öğretmen adaylarının öğrenme stillerinin çeşitli değişkenler açısından incelenmesi.* Master's thesis, Adnan Menderes Üniversitesi, Sosyal Bilimler Enstitüsü, Manisa.

Gürsoy, M., Balkan, A., & Ulukan, H. (2012). Ecophysiological responses to stresses in plants: A general approach. *Pakistan Journal of Biological Sciences, 15*(11), 506–516. DOI: 10.3923/pjbs.2012.506.516 PMID: 24191624

Gürsoy, S., Hocaoğlu Bahadır, N., & Yalçın, E. C. (2022). Çevre bilinci dinamikleri çevresel tutumda bir farklılık yaratır mı? Kırklareli üniversitesi örneği. *Ankara Üniversitesi SBF Dergisi, 77*(4), 779–799. DOI: 10.33630/ausbf.1081113

Gyimah, J., Yao, X., Tachega, M., Hayford, I., & Opoku-Mensah, E. (2022). Renewable energy consumption and economic growth: New evidence from Ghana. *Energy, 248*, 1–11. DOI: 10.1016/j.energy.2022.123559

Hacker, R., & Hatemi-J, A. (2006). Tests for causality between integrated variables using asymptotic and bootstrap distributions: Theory and application. *Applied Economics, 38*(13), 1489–1500. DOI: 10.1080/00036840500405763

Häggström, M., & Schmidt, C. (2021). Futures literacy–to belong, participate and act!: An educational perspective. *Futures, 132*, 102813. DOI: 10.1016/j.futures.2021.102813

Ham, M., Mrčela, D., & Horvat, M. (2016). Insights for measuring environmental awareness. *Ekonomski vjesnik. Review of Contemporary Entrepreneurship, Business, and Economic Issues, 29*(1), 159–176.

Hamouda, M. M., Badr, A., Ali, S. S., Adham, A. M., Ahmed, H. I. S., & Saad-Allah, K. M. (2023). Growth, physiological, and molecular responses of three phaeophyte extracts on salt-stressed pea (*Pisum sativum* L.) seedlings. *Journal of Genetic Engineering and Biotechnology*, *21*(1), 32. DOI: 10.1186/s43141-023-00483-z PMID: 36929363

Hancock, J. T., & Veal, D. (2021). Nitric oxide, other reactive signalling compounds, redox, and reductive stress. *Journal of Experimental Botany*, *72*(3), 819–829. DOI: 10.1093/jxb/eraa331 PMID: 32687173

Han, J., Kim, M., Mammadov, Z., Lee, S., Elzinga, E. J., Mammadov, G., Hwang, W., & Ro, H. M. (2024). Synergistic effect of climate change and water management: Historical and future soil salinity in the Kur-Araz lowland, Azerbaijan. *The Science of the Total Environment*, *907*, 167720. DOI: 10.1016/j.scitotenv.2023.167720 PMID: 37827308

Harris, C. (1991). Environmental education in England and Wales: A brief review. *European Journal of Education*, *26*(4), 287–290. DOI: 10.2307/1503129

Harvey, D., & Mills, T. (2002). Unit roots and double smooth transitions. *Journal of Applied Statistics*, *29*(5), 675–683. DOI: 10.1080/02664760120098739

Haryono A. Soemarno Djati, M. S., & Setyoleksono, A. (2014). Learning attitude and awareness against students in cultured environmental success in Probolinggo. *Journal of Environment and Earth Science*, *4*(16), 72–81.

Haseeb, M., Abidin, I., Hye, Q., & Hartani, N. (2018). The impact of renewable energy on economic well-being of Malaysia: Fresh evidence from auto regressive distributed lag bound testing approach. *International Journal of Energy Economics and Policy*, *9*(1), 269–275.

Hayta, A. B. (2006). Çevre kirliliğinin önlenmesinde ailenin yeri ve önemi. *Ahi Evran Üniversitesi Kırşehir Eğitim Fakültesi Dergisi*, *7*(2), 359–376.

Hedenus, F., Azar, C., & Johansson, D. (2010). Energy security policies in EU-25: The expected cost of oil supply disruptions. *Energy Policy*, *38*(3), 1241–1250. DOI: 10.1016/j.enpol.2009.01.030

Hemayatkhah Jahromi, M., Ershad F. Danesh, P., & Ghorbani, M. (2017). Explore environmental behavior based on the model of TPB. [Previously Human Development]. *Quarterly Journal of Social Development*, *11*(3), 31–66.

Hepsag, A. (2021). A unit root test based on smooth transitions and nonlinear adjustment. *Communications in Statistics. Simulation and Computation*, *50*(3), 625–632. DOI: 10.1080/03610918.2018.1563154

Herzog, F., Farah, Z., & Amado, R. (1994). Composition and consumption of gathered wild fruits in the V-Baoule, Cote d'Ivoire. *Ecology of Food and Nutrition, 32*(3–4), 181–196. DOI: 10.1080/03670244.1994.9991399

Hes-Quimbita, G., & Pavel, M. (1996) *Assessing an Environmental Attitude Development Model: Factors Influencing the Environmental Attitudes of College Students. Eric Document. American Education Research Association Conference*, April 8-12, New York.

Holmes, P. M., Richardson, D. M., Van Wilgen, B. W., & Gelderblom, C. (2000). Recovery of South African fynbos vegetation following alien woody plant clearing and fire: Implications for restoration. *Austral Ecology*, *25*(6), 631–939. DOI: 10.1111/j.1442-9993.2000.tb00069.x

Holm, L., Doll, J., Holm, E., Pancho, J., & Herberger, J. (1997). *Word weeds: Natural Histories and Distribution*. John Wiley & Sons, Inc.

Holt, R. M. (1983). Dynamic Dallas: A Building Critique. *Public Library Quarterly*, *4*(4), 67–84. DOI: 10.1300/J118v04n04_12

Hopkinson, C. S., Wolanski, E., Cahoon, D. R., Perillo, G. M., & Brinson, M. M. (2019). Coastal wetlands: a synthesis. In *Coastal wetlands* (pp. 1–75). Elsevier. DOI: 10.1016/B978-0-444-63893-9.00001-0

Hoşgör, H., Hoşgör, D. G., & Tosun, N. K. (2015). Sağlık bilimleri fakültesi öğrencilerinin çevreye yönelik tutumlarının belirlenmesi: Kıyaslamalı bir analiz. *Journal of Health Sciences and Professions*, *2*(2), 198–208.

Inglesi-Lotz, R. (2016). The impact of renewable energy consumption to economic growth: A panel data application. *Energy Economics*, *53*, 58–63. DOI: 10.1016/j.eneco.2015.01.003

Intergovernmental Panel on Climate Change (IPCC). (2022). *Climate Change and Africa: Impacts, Adaptation, and Vulnerabilities.* Retrieved from https://www.ipcc.ch/report/ar6/wg2/

International Energy Agency (IEA). (2023). *Africa Energy Outlook.* Retrieved from https://www.iea.org/reports/africa-energy-outlook

Işildar, G. Y. (2008). Meslek yüksekokulları boyutunda" çevre eğitimi" nin çevreci yaklaşımlar ve davranışlar üzerindeki etkilerinin değerlendirilmesi. *Türk Eğitim Bilimleri Dergisi*, *6*(4), 759–778.

Isman, M. B., & Grieneisen, M. L. (2014). Botanical insecticide research: Many publications, limited useful data. *Trends in Plant Science*, *19*(3), 140–145. DOI: 10.1016/j.tplants.2013.11.005 PMID: 24332226

Jacobsen, L., & Engström-Öst, J. (2018). Coping with environments; vegetation, turbidity and abiotics. In *Biology and ecology of pike* (pp. 32–61). CRC press. DOI: 10.1201/9781315119076-4

Jaforullah, M., & King, A. (2015). Does the use of renewable energy sources mitigate CO2 emissions? A reassessment of the US evidence. *Energy Economics*, *49*, 711–717. DOI: 10.1016/j.eneco.2015.04.006

Jain, S. K., Sarkar, A., & Garg, V. (2008). Impact of declining trend of flow on Harike Wetland, India. *Water Resources Management*, *22*(4), 409–421. DOI: 10.1007/s11269-007-9169-9

Jang, W., Kim, J., Kim, J. H., Shin, J. K., Chon, K., Kang, E. T., Park, Y., & Kim, S. (2024). Evaluation of LandsatBased Chlorophyll-a Estimation in a Small-Scale Reservoir: Assessing Accuracy and Availability. *Remote Sensing (Basel)*, *16*(2), 315. DOI: 10.3390/rs16020315

Jaus, H. H. (1982). The effect of environmental education ınstruction on children's attitudes toward the environment. *Science Education*, *66*(5), 689–692. DOI: 10.1002/sce.3730660504

Jeronen, E., & Kaikkonen, M. (2002). Thoughts of children and adults about the environment and environmental education. *International Research in Geographical and Environmental Education*, *11*(4), 341–353. DOI: 10.1080/10382040208667501

Jónasson, J. T. (2016). Educational change, inertia and potential futures: Why is it difficult to change the content of education? *European Journal of Futures Research*, *4*(1), 1–14. DOI: 10.1007/s40309-016-0087-z

Joshi, N., Baumann, M., Ehammer, A., Fensholt, R., Grogan, K., Hostert, P., Jepsen, M., Kuemmerle, T., Meyfroidt, P., Mitchard, E., Reiche, J., Ryan, C., & Waske, B. (2016). A review of the application of optical and radar remote sensing data fusion to land use mapping and monitoring. *Remote Sensing (Basel)*, *8*(1), 70. DOI: 10.3390/rs8010070

Jozefczak, M., Remans, T., Vangronsveld, J., & Cuypers, A. (2012). Glutathione is a key player in metal-induced oxidative stress defenses. *International Journal of Molecular Sciences*, *13*(3), 3145–3175. DOI: 10.3390/ijms13033145 PMID: 22489146

Kamran, M., Imran, Q. M., Ahmed, M. B., Falak, N., Khatoon, A., & Yun, B. W. (2022). Endophyte-mediated stress tolerance in plants: A sustainable strategy to enhance resilience and assist crop improvement. *Cells, 11*(20), 3292. DOI: 10.3390/cells11203292 PMID: 36291157

Kang, J., Wang, X.-L., Yan, S.-J., Guo, H., & Cui, Y.-N. (2024). Comparative physiological and gene expression analyses provide insights into ion transports and osmotic adjustment of sweet sorghum under salt stress. *Agronomy (Basel), 14*(8), 1849. DOI: 10.3390/agronomy14081849

Kapetanios, G., Shin, Y., & Snell, A. (2003). Testing for a unit root in the nonlinear STAR framework. *Journal of Econometrics, 112*(2), 359–379. DOI: 10.1016/S0304-4076(02)00202-6

Kapetanios, G., Shin, Y., & Snell, A. (2006). Testing for cointegration in nonlinear smooth transition error correction models. *Econometric Theory, 22*(2), 279–303. DOI: 10.1017/S0266466606060129

Kar, R. K. (2011). Plant responses to water stress: Role of reactive oxygen species. *Plant Signaling & Behavior, 6*(11), 1741–1745. DOI: 10.4161/psb.6.11.17729 PMID: 22057331

Kathiresan, K., & Bingham, B. L. (2001). Biology of mangroves and mangrove ecosystems.

Kaygusuz, K. (2002). Environmental impacts of energy utilisation and renewable energy policies in Turkey. *Energy Policy, 30*(8), 689–698. DOI: 10.1016/S0301-4215(02)00032-0

Kaygusuz, K., & Sari, A. (2003). Renewable energy potential and utilization in Turkey. *Energy Conversion and Management, 44*(3), 459–478. DOI: 10.1016/S0196-8904(02)00061-4

Keikhosrokiani, P. (Ed.). (2022). *Big data analytics for healthcare: datasets, techniques, life cycles, management, and applications*. Academic Press.

Kharbech, O., Sakouhi, L., Massoud, M. B., Mur, L. A. J., Corpas, F. J., Djebali, W., & Chaoui, A. (2020). Nitric oxide and hydrogen sulfide protect plasma membrane integrity and mitigate chromium-induced methylglyoxal toxicity in maize seedlings. *Plant Physiology and Biochemistry, 157*, 244–255. DOI: 10.1016/j.plaphy.2020.10.017 PMID: 33152643

Kim, K.-H., Kabir, E., & Jahan, S. A. (2017). Exposure to pesticides and the associated human health effects. *The Science of the Total Environment, 575*, 525–535. DOI: 10.1016/j.scitotenv.2016.09.009 PMID: 27614863

Koçak B. Akış A. (2021). Sosyal bilgiler öğretmen adaylarının çevre eğitimi hakkında görüşleri. *Turkish Studies-Educational Sciences*, 16(1).

Koçak, E., & Şarkgüneşi, A. (2017). The renewable energy and economic growth nexus in Black Sea and Balkan countries. *Energy Policy*, *100*, 51–57. DOI: 10.1016/j.enpol.2016.10.007

Kohut A. Shriver J. (1989). *Environment regaining a foothold on the national agenda*. Gallup Report, no. 285.

Kotze, I., Beukes, H., Van Den Berg, E., & Newby, T. (2010). National invasive alien plant survey. Report No. GW/A/2010/21 to the Department of Water and Environmental Affairs, Working for Water Programme. Agricultural Research Council-Institute for Soil, Climate and Water. Pretoria, South Africa.

KPMG. (2023). *Indian Healthcare Sector: Navigating the Future.* Retrieved from https://kpmg.com/in/en/insights/2023/11/indian-healthcare-sector-investment-trends-to-look-for.html

Kraft, J., & Kraft, A. (1978). On the Relationship Between Energy and GNP. *The Journal of Energy and Development*, *3*(2), 401–403.

Kruse, R. (2011). A new unit root test against ESTAR based on a class of modified statistics. *Statistische Hefte*, *52*(1), 71–85.

Kuhlemeier, H., Van Den Bergh, H., & Lagerweij, N. (1999). Environmental knowledge, attitudes, and behavior in Dutch secondary education. *The Journal of Environmental Education*, *30*(2), 4–14. DOI: 10.1080/00958969909601864

Kulawardhana, R. W., Thenkabail, P. S., Vithanage, J. A. G. A. T. H., Biradar, C., Islam, A., Gunasinghe, S., & Alankara, R. (2007). Evaluation of the Wetland Mapping Methods using Landsat ETM+ and SRTM Data. *Journal of Spatial Hydrology*, *7*(2).

Kumar, D., & Choudhury, M. (2021). Recognizing economic values of wetland ecosystem services: A study of emerging role of monetary evaluation of Chandubi ecosystem and biodiversity. Wetlands Conservation: Current Challenges and Future Strategies, 87-110. https://doi.org/DOI: 10.1002/9781119692621.ch5

Kumar, A., Upadhyay, P., & Singh, U. (2023). *Multi-Sensor and Multi-Temporal Remote Sensing: Specific Single Class Mapping*. CRC Press. DOI: 10.1201/9781003373216

Kumar, D., Choudhury, M., & Rathore, A. (2020). Valuation of ecosystem services benefits of Son Beel wetland in Assam, India: a case study of natural solutions to climate change water. *Climate Change Conference*, Hamburg University, Germany.

Kwembeya, A., Rugare, J. T., & Mabasa, S. (2013). Allelopathic effects of lantana (lantana camara) on blackjack (bidens pilosa) and pearl millet (pennisetum glaucum). *Asian Journal of Agriculture and Rural Development*, *3*(8), 543–553.

Laaraj, S., Salmaoui, S., Addi, M., El-rhouttais, C., Tikent, A., Elbouzidi, A., Taibi, M., Hano, C., Noutfia, Y., & Elfazazi, K. (2023). Carob (*Ceratonia siliqua* L.) Seed Constituents: A Comprehensive Review of Composition, Chemical Profile, and Diverse Applications. *Journal of Food Quality*, *2023*, 1–14. DOI: 10.1155/2023/3438179

Lal, L. (1987). Studies on natural repellents against potato tuber moth (*Phthorimaea operculella* Zeller) in country stores. *Potato Research*, *30*(2), 329–334. DOI: 10.1007/BF02357672

Lau, L., Lee, K., & Mohamed, A. (2012). Global warming mitigation and renewable energy policy development from the Kyoto Protocol to the Copenhagen Accord-A comment. *Renewable & Sustainable Energy Reviews*, *16*(7), 5280–5284. DOI: 10.1016/j.rser.2012.04.006

Law, J. W.-F., Ser, H.-L., Khan, T. M., Chuah, L.-H., Pusparajah, P., Chan, K.-G., Goh, B.-H., & Lee, L.-H. (2017). The potential of Streptomyces as biocontrol agents against the rice blast fungus, Magnaporthe oryzae (Pyricularia oryzae). *Frontiers in Microbiology*, *8*, 3. DOI: 10.3389/fmicb.2017.00003 PMID: 28144236

Lee, S., & Jung, Y. (2018). Causal dynamics between renewable energy consumption and economic growth in South Korea: Empirical analysis and policy implications. *Energy & Environment*, *29*(7), 1298–1315. DOI: 10.1177/0958305X18776546

Leonard, B., & Regan, S. (2019). Legal and institutional barriers to establishing non-use rights to natural resources. *Natural Resources Journal*, *59*(1), 135–180. https://digitalrepository.unm.edu/cgi/viewcontent.cgi?article=4037&context=nrj

Leung, P. H., Chui, K. T., Lo, K., & de Pablos, P. O. (2021). A support vector machine–based voice disorders detection using human voice signal. In *Artificial Intelligence and Big Data Analytics for Smart Healthcare* (pp. 197–208). Academic Press. DOI: 10.1016/B978-0-12-822060-3.00014-0

Leybourne, S., Newbold, P., & Vougas, D. (1998). Unit roots and smooth transitions. *Journal of Time Series Analysis*, *19*(1), 83–97. DOI: 10.1111/1467-9892.00078

Lin, E., & Shi, Q. (2014). Exploring individual and school-related factors and environmental literacy: Comparing U.S. and Canada using PISA 2006. *International Journal of Science and Mathematics Education*, *12*(1), 73–97. DOI: 10.1007/s10763-012-9396-2

Liogier, H. A. 1990. Plantas medicinales de Puerto Rico y del Caribe. Ibero americana de Ediciones, Inc., San Juan, PR 563p.

Liu, H. S., Li, F. M., & Xu, H. (2004). Deficiency of water can enhance root respiration rate of drought-sensitive but not drought-tolerant spring wheat. *Agricultural Water Management*, *64*(1), 41–48. DOI: 10.1016/S0378-3774(03)00143-4

Liu, Y., & Hao, Y. (2018). The dynamic links between CO2 emissions, energy consumption and economic development in the countries along "the Belt and Road". *The Science of the Total Environment*, *645*, 674–683. DOI: 10.1016/j.scitotenv.2018.07.062 PMID: 30031325

Li, W., Chai, Y., Khan, F., Jan, S. R. U., Verma, S., Menon, V. G., & Li, X. (2021). A comprehensive survey on machine learning-based big data analytics for IoT-enabled smart healthcare system. *Mobile Networks and Applications*, *26*(1), 234–252. DOI: 10.1007/s11036-020-01700-6

Li, Y., Shi, Y., Zhu, X., Cao, H., & Yu, T. (2014). Coastal wetland loss and environmental change due to rapid urban expansion in Lianyungang, Jiangsu, China. *Regional Environmental Change*, *14*(3), 1175–1188. DOI: 10.1007/s10113-013-0552-1

Lopez, ´R., Ramírez-Valiente, J.A., Pita, P., (2022). How plants cope with heatwaves in a drier environment. Flora 295, 152148. https://doi.org/. flora.2022.152148.DOI: 10.1016/j

Lowe, G. D., & Pinhey, T. K. (1982). Rural-urban differences in support for environmental protection. *Rural Sociology*, *47*(1), 114.

Luqman, M., Ahmad, N., & Bakhsh, K. (2019). Nuclear energy, renewable energy and economic growth in Pakistan: Evidence from non-linear autoregressive distributed lag model. *Renewable Energy*, *139*, 1299–1309. DOI: 10.1016/j.renene.2019.03.008

Mabe, F. N., Talabi, K., & Danso-Abbeam, G. (2017). Awareness of health implications of agrochemical use: Effects on maize production in Ejura-Sekyedumase municipality, Ghana. *Advances in Agriculture*, *2017*, 2017. DOI: 10.1155/2017/7960964

Mahmood, Q., Bilal, M., & Jan, S. (2014). Herbicides, pesticides, and plant tolerance: an overview. In *Emerging technologies and management of crop stress tolerance* (pp. 423–448). Elsevier. DOI: 10.1016/B978-0-12-800876-8.00017-5

Mainieri, T., Barnett, E. G., Valdero, T. R., Unipan, J. B., & Oskamp, S. (1997). Green buying: The influence of environmental concern on consumer behavior. *The Journal of Social Psychology*, *137*(2), 189–204. DOI: 10.1080/00224549709595430

Malaviya, P., & Singh, A. (2012). Constructed wetlands for management of urban stormwater runoff. *Critical Reviews in Environmental Science and Technology, 42*(20), 2153–2214. DOI: 10.1080/10643389.2011.574107

Malik, S., & Tyagi, A. K. (Eds.). (2022). *Intelligent Interactive Multimedia Systems for E-Healthcare Applications*. CRC Press. DOI: 10.1201/9781003282112

Manafi, H., Baninasab, B., Gholami, M., & Talebi, M. (2021). Nitric oxide induced thermotolerance in strawberry plants by activation of antioxidant systems and transcriptional regulation of heat shock proteins. *The Journal of Horticultural Science & Biotechnology, 96*(6), 783–796. DOI: 10.1080/14620316.2021.1927206

Mangnus, A. C., Oomen, J., Vervoort, J. M., & Hajer, M. A. (2021). Futures literacy and the diversity of the future. *Futures, 132*, 102793. DOI: 10.1016/j.futures.2021.102793

Marr, R. S. (1964). Get rid of *lantana.Farming in South Africa*, (February), 42–45.

Martins-Loução, M. A., Correia, P. J., & Romano, A. (2024). Carob: A Mediterranean resource for the future. *Plants, 13*(9), 1188. DOI: 10.3390/plants13091188 PMID: 38732403

Mashila, T. (2014). Spatial planning for climate change adaptation: Developing a climate change local area adaptation plan for Khayelitsha. https://open.uct.ac.za/server/api/core/bitstreams/4ab3b5fe-2f63-4d05-bbc2-aefa27671370/content

Maslin, M., Ramnath, R. D., Welsh, G. I., & Sisodiya, S. M. (2025). Climate health emergency: Understanding the health impacts of the climate crisis. *Future Healthcare Journal, 12*(1), 100240. DOI: 10.1016/j.fhj.2025.100240 PMID: 40236934

Mason, T. J., & French, K. (2007). Management regimes for a plant invader differentially impact resident communities. Institute for Conservation Biology, University of Wollonong, NSW 2522, Australia. DOI: 10.1016/j.biocon.2006.11.023

Mattah, M. M., Mattah, P. A., & Futagbi, G. (2015). Pesticide application among farmers in the catchment of Ashaiman irrigation scheme of Ghana: Health implications. *Journal of Environmental and Public Health, 2015*, 2015. DOI: 10.1155/2015/547272 PMID: 26798369

Ma, X., & Bateson, D. J. (1999). A multivariate analysis of the relationship between attitude toward science and attitude toward the environment. *The Journal of Environmental Education, 31*(1), 27–32. DOI: 10.1080/00958969909598629

Mbarek, M., Saidi, K., & Amamri, M. (2017). The relationship between pollutant emissions, renewable energy, nuclear energy and GDP: Empirical evidence from 18 developed and developing countries. *International Journal of Sustainable Energy*, *37*(6), 597–615. DOI: 10.1080/14786451.2017.1332060

McAdam, S. A., & Cardoso, A. A. (2019). The recurrent evolution of extremely resistant xylem. *Annals of Forest Science*, *76*(1), 1–4. DOI: 10.1007/s13595-018-0786-7

McClain, M. E. (2013). Balancing water resources development and environmental sustainability in Africa: A review of recent research findings and applications. *Ambio*, *42*(5), 549–565. https://link.springer.com/article/10.1007/s13280-012-0359-1. DOI: 10.1007/s13280-012-0359-1 PMID: 23238866

McDowell, N. G., Michaletz, S. T., Bennett, K. E., Solander, K. C., Xu, C., Maxwell, R. M., & Middleton, R. S. (2018). Predicting chronic climate-driven disturbances and their mitigation. *Trends in Ecology & Evolution*, *33*(1), 15–27. DOI: 10.1016/j.tree.2017.10.002 PMID: 29146414

McDowell, N. G., Sapes, G., Pivovaroff, A., Adams, H. D., Allen, C. D., Anderegg, W. R., & Xu, C. (2022). Mechanisms of woody-plant mortality under rising drought, CO2 and vapour pressure deficit. *Nature Reviews. Earth & Environment*, *3*(5), 294–308. DOI: 10.1038/s43017-022-00272-1

McDowell, N., Pockman, W. T., Allen, C. D., Breshears, D. D., Cobb, N., Kolb, T., & Yepez, E. A. (2008). Mechanisms of plant survival and mortality during drought: Why do some plants survive while others succumb to drought? *The New Phytologist*, *178*(4), 719–739. DOI: 10.1111/j.1469-8137.2008.02436.x PMID: 18422905

McEvoy J. (1972). The American concern with the environment. *Social behavior, natural resources and the environment*, 214-236.

Mehta, D., & Vyas, S. (2023). Comparative bio-accumulation of osmoprotectants in saline stress tolerating plants: A review. *Plant Stress*, *9*, 100177. DOI: 10.1016/j.stress.2023.100177

Mello, F. B., Jacobus, D., Carvalho, K., & Mello, J. R. (2005). Effects of *Lantana camara* (Verbenaceae) on general reproductive performance and teratology in rats. *Toxicon*, *45*(4), 459–466. DOI: 10.1016/j.toxicon.2004.12.004 PMID: 15733567

Menyah, K., & Wolde-Rufael, Y. (2010). CO2 emissions, nuclear energy, renewable energy and economic growth in the US. *Energy Policy*, *38*(6), 2911–2915. DOI: 10.1016/j.enpol.2010.01.024

Mezghani, I., & Haddad, H. (2017). Energy consumption and economic growth: An empirical study of the electricity consumption in Saudi Arabia. *Renewable & Sustainable Energy Reviews*, *75*, 145–156. DOI: 10.1016/j.rser.2016.10.058

Middleton, B., Grootjans, A. B., Jensen, K., Venterink, H. O., & Margóczi, K. (2006). Fen management and research perspectives: an overview. *Wetlands: functioning, biodiversity conservation, and restoration*, 247-268.

Miller, D. L. (2007). The seeds of learning: Young children develop important skills through their gardening activities at a midwestern early education program. *Applied Environmental Education and Communication*, *6*(1), 49–66. DOI: 10.1080/15330150701318828

Ministry of Environment. Forests and Climate Change Annual report. (2012). MoEFCC, Government of India, New Delhi.

Mishra, A. (2015). Allelopathic properties of Lantana camara. *International Research Journal of Basic and Clinical Studies*, *3*(1), 13–28.

Mishra, S., & Mishra, D. R. (2012). Normalized difference chlorophyll index: A novel model for remote estimation of chlorophyll-a concentration in turbid productive waters. *Remote Sensing of Environment*, *117*, 394–406. DOI: 10.1016/j.rse.2011.10.016

Mitsch, W. J., Gosselink, J. G., Zhang, L., & Anderson, C. J. (2009). *Wetland ecosystems*. John Wiley & Sons.

Mohai, P., & Twight, B. W. (1987). Age and environmentalism: An elaboration of the Buttel model using national survey evidence. *Social Science Quarterly*, *68*(4), 798.

Mohamed, N. M., Makboul, M. A., Farag, S. F., Jain, S., Jacob, M. R., Tekwani, B. L., & Ross, S. A. (2016). Triterpenes from the roots of *Lantana montevidensis* with antiprotozoal activity. *Phytochemistry Letters*, *15*, 30–36. DOI: 10.1016/j.phytol.2015.11.008

Mohanta, T.K., Bashir, T., & Hashem, A. and Abd_Allah, E.F. (. (2017). Systems biology approach in plant abiotic stresses. *Plant Physiology and Biochemistry*, *121*, 58–73. DOI: 10.1016/j.plaphy.2017.10.019 PMID: 29096174

Møller, I. M., & Sweetlove, L. J. (2010). ROS signalling–specificity is required. *Trends in Plant Science*, *15*(7), 370–374. DOI: 10.1016/j.tplants.2010.04.008 PMID: 20605736

Money, R. P., Wheeler, B. D., Baird, A. J., & Heathwaite, A. L. (2009). Replumbing Wetlands–managing water for the restoration of bogs and fens. *The wetlands handbook*, 755-779.

Monteiro, A. C. B., França, R. P., Arthur, R., & Iano, Y. (2021). An overview of medical Internet of Things, artificial intelligence, and cloud computing employed in health care from a modern panorama. *The Fusion of Internet of Things, Artificial Intelligence, and Cloud Computing in Health Care*, 3-23.

Morris, T. L., Witkowski, E. T. F., & Coetzee, J. A. (2008). Initial response of riparian plant community structure to clearing of invasive alien plants in Kruger National Park, South Africa. *South African Journal of Botany*, *74*(3), 485–494. DOI: 10.1016/j.sajb.2008.01.177

Morton, J. F. (1994). *Lantana*, or red sage (*Lantana camara* L. Verbenaceae), notorious weed and popular garden flower; some cases of poisoning in Florida. *Economic Botany*, *48*(3), 259–270. DOI: 10.1007/BF02862327

Mouhouche, B., Ruget, F., & Delécolle, R. (1998). Effects of water stress applied at different phenological phases on yield components of dwarf bean (Phaseolus vulgaris L.). *Agronomie*, *18*(3), 197–205. DOI: 10.1051/agro:19980303

Mouniane, Y., Chriqui, A., El-Khadir, I., Hbyaj, K., Rochd, A., Aitouhanni, I., Kouighat, M., & Hmouni, D. (2024b). Assessing Morphological Trait Variability in Moroccan Carob (*Ceratonia siliqua* L.) Ecotypes for Adaptive Breeding in Response to Climate Change and Sustainable Food Security. In BIO Web of Conferences. 109:01040. https://doi.org/DOI: 10.1051/bioconf/202410901040

Mouniane, Y., Chriqui, A., El-Khadir, I., Bensaid, A., Arabi, R., Zoubir, M. H., Mabrouki, J., & Hmouni, D. (2024a). Investigation on the Phytosociology and Soil Characteristics of Carob Tree (*Ceratonia siliqua* L.) in Northeast Morocco: Implications for Biodiversity Conservation and Ecosystem Restoration. In Mabrouki, J., & Azrour, M. (Eds.), *Integrated Solutions for Smart and Sustainable Environmental Conservation. Studies in Systems, Decision and Control* (Vol. 527). Springer., DOI: 10.1007/978-3-031-55787-3_10

Mpofu, F. Y. (2022). Green Taxes in Africa: Opportunities and challenges for environmental protection, sustainability, and the attainment of sustainable development goals. *Sustainability (Basel)*, *14*(16), 10239. https://www.mdpi.com/2071-1050/14/16/10239. DOI: 10.3390/su141610239

Müderrisoğlu, H., & Altanlar, A. (2011). Attitudes and behaviors of undergraduate students toward environmental ıssues. *International Journal of Environmental Science and Technology*, *8*(1), 159–168. DOI: 10.1007/BF03326205

Muhumuza, M., & Balkwill, K. (2013). Factors affecting the success of conserving biodiversity in national parks: A review of case studies from Africa. *International Journal of Biodiversity*, *2013*(1), 798101. https://onlinelibrary.wiley.com/doi/full/10.1155/2013/798101. DOI: 10.1155/2013/798101

Munir, A. A., & Adelaide, J. 1996. A taxonomic review of *Lantana camara* L. and *Lantana montevidensis* (Spreng) Briq. (Verbenaceae) in Australia. Board of the Botanic Gardens and State Herbarium, Adelaide, South Australia. 1976–2015 (Vol. 1, 29).

Munyua, S. J. M., Nienga, M. J., Karitu, T. P., Kimoro, T. P., Kiptoon, J. E., & Buoro, I. B. J. (1990). A note on clinical-pathological findings and serum enzyme activity in sheep, goats and Friesian calves with acute *Lantana camara* poisoning. *Bulletin of Animal Health and Production in Africa*, *38*(3), 275–279.

Mushi, G. V. (2019). *Local communities' perceptions on lantana camara and management responses in East Usambara, Tanzania* (Doctoral dissertation, Sokoine University of Agriculture).

Musser, L. M., & Diamond, K. E. (1999). The children's attitudes toward the environment scale for preschool children. *The Journal of Environmental Education*, *30*(2), 23–30. DOI: 10.1080/00958969909601867

Muthu, B., Sivaparthipan, C. B., Manogaran, G., Sundarasekar, R., Kadry, S., Shanthini, A., & Dasel, A. (2020). IOT based wearable sensor for diseases prediction and symptom analysis in healthcare sector. *Peer-to-Peer Networking and Applications*, *13*(6), 2123–2134. DOI: 10.1007/s12083-019-00823-2

Muzenda, E., Ntuli, F., & Pilusa, T. J. (2012). Waste management, strategies and situation in South Africa: An overview. Johansson, E. L. (2021). Participatory futures thinking in the African context of sustainability challenges and socio-environmental change. *Ecology and Society*, *26*(4). https://pdfs.semanticscholar.org/6970/c6a60b8a7c2236f6c36290e8eb6b0ac06f9e.pdf

Myers, N., Mittermeier, R. A., Mittermeier, C. G., Da Fonseca, G. A. B., & Kent, J. (2000). Biodiversity hotspots for conservation priorities. *Nature*, *403*(6772), 853–858. DOI: 10.1038/35002501 PMID: 10706275

Nakawuka, P., Langan, S., Schmitter, P., & Barron, J. (2018). A review of trends, constraints and opportunities of smallholder irrigation in East Africa. *Global Food Security*, *17*, 196–212. https://oaktrust.library.tamu.edu/server/api/core/bitstreams/4ab0ea7f-027e-428e-a3d4-6b6c8481b16f/content. DOI: 10.1016/j.gfs.2017.10.003

Namdech, P., Rotjanawasuthorn, S., Singh, B., & Kaunert, C. (2025). Harnessing artificial intelligence for lifelong education: opportunities and challenges. *Asian Education and Learning Review*, *3*(2), 4–4.

Nardini, A. (2022). Hard and tough: The coordination between leaf mechanical resistance and drought tolerance. *Flora (Jena)*, *288*, 152023. DOI: 10.1016/j.flora.2022.152023

Naser, H. (2015). Can Nuclear Energy Stimulates Economic Growth? Evidence from Highly Industrialised Countries. *International Journal of Energy Economics and Policy*, *5*(1), 164–173.

Nathaniel, S. P., & Iheonu, C. O. (2019). Carbon dioxide abatement in Africa: The role of renewable and non-renewable energy consumption. *The Science of the Total Environment*, *679*, 337–345. https://www.sciencedirect.com/science/article/abs/pii/S0048969719320224. DOI: 10.1016/j.scitotenv.2019.05.011 PMID: 31085413

National Health Profile, India. (2023). *Annual Reports.* Ministry of Health and Family Welfare, Government of India. Retrieved from https://cbhidghs.mohfw.gov.in

Nazlioglu, S., Lebe, F., & Kayhan, S. (2011). Nuclear energy consumption and economic growth in OECD countries: Cross-sectionally dependent heterogeneous panel causality analysis. *Energy Policy*, *39*(10), 6615–6621. DOI: 10.1016/j.enpol.2011.08.007

Nejat, P., Jomehzadeh, F., Taheri, M., Gohari, M., & Majid, M. (2015). A global review of energy consumption, CO2 emissions and policy in the residential sector (with an overview of the top ten CO2 emitting countries). *Renewable & Sustainable Energy Reviews*, *43*, 843–862. DOI: 10.1016/j.rser.2014.11.066

Nguyen, T. T., Grote, U., Neubacher, F., Do, M. H., & Paudel, G. P. (2023). Security risks from climate change and environmental degradation: Implications for sustainable land use transformation in the Global South. *Current Opinion in Environmental Sustainability*, *63*, 101322. DOI: 10.1016/j.cosust.2023.101322

Nichanian, D. (2016). Seizing a Seat at the Table: Participatory Politics in the Face of Disqualification. https://knowledge.uchicago.edu/record/1733?v=pdf

Nisari, A. R., & Sujatha, C. H. (2021). Assessment of trace metal contamination in the Kol wetland, a Ramsar site, Southwest coast of India. *Regional Studies in Marine Science*, *47*, 101953. DOI: 10.1016/j.rsma.2021.101953

Noor Ouarzazate Solar Complex. (2023). *Renewable Energy in North Africa.* Retrieved from https://www.masen.ma/en/our-projects/noor-ouarzazate

Nyiwul, L. M. (2019). *Climate change mitigation and adaptation in Africa: Strategies, synergies, and constraints*. Springer International Publishing., https://link.springer.com/chapter/10.1007/978-3-030-02662-2_11

O'Donoghue, R., & Rosenberg, E. A review of the staging and enactment of environment and sustainability education in South Africa: An illustrative case study. *World Review*, 22-39. https://www.taylorfrancis.com/chapters/edit/10.1201/9781003145202-3/review-staging-enactment-environment-sustainability-education-south-africa-rob-donoghue-eureta-rosenberg

Ocak, M., Ocak, Z., Bilgen, S., Keles, S., & Kaygusuz, K. (2004). Energy utilization, environmental pollution and renewable energy sources in Turkey. *Energy Conversion and Management*, *45*(6), 845–864. DOI: 10.1016/S0196-8904(03)00192-4

Ocal, O., & Aslan, A. (2013). Renewable energy consumption–economic growth nexus in Turkey. *Renewable & Sustainable Energy Reviews*, *28*, 494–499. DOI: 10.1016/j.rser.2013.08.036

Ogeto, G. M. (2017). *Role of Media in Political Conflict: A Comparative Study of Kenya and Zimbabwe*. https://erepository.uonbi.ac.ke/bitstream/handle/11295/108075/Ogeto_Role%20Of%20Media%20In%20Political%20Conflict.pdf?sequence=1&isAllowed=y

Omokhua, A. G., Madikizela, B., Aro, A., Uyi, O. O., Van Staden, J., & McGaw, L. J. (2018). Noxious to ecosystems, but relevant to pharmacology: Four South African alien invasive plants with pharmacological potential. *South African Journal of Botany*, *117*, 41–49. https://repository.up.ac.za/bitstream/handle/2263/69205/Omokhua_Noxious_2018.pdf?sequence=1&isAllowed=y. DOI: 10.1016/j.sajb.2018.04.015

Omri, A., Mabrouk, N., & Sassi-Tmar, A. (2015). Modeling the causal linkages between nuclear energy, renewable energy and economic growth in developed and developing countries. *Renewable & Sustainable Energy Reviews*, *42*, 1012–1022. DOI: 10.1016/j.rser.2014.10.046

Oyedapo, O. O., Sab, F. C., & Olagunju, J. A. (1999). Bioactivity of fresh leaves of *Lantana camara.Biometrical Letters*, *59*, 179–183.

Ozcan, B., & Ari, A. (2015). Nuclear Energy Consumption-Economic Growth Nexus in OECD: A Bootstrap Causality Test. *Procedia Economics and Finance*, *30*, 586–597. DOI: 10.1016/S2212-5671(15)01271-X

Ozcan, B., & Ozturk, I. (2019). Renewable energy consumption-economic growth nexus in emerging countries: A bootstrap panel causality test. *Renewable & Sustainable Energy Reviews*, *104*, 30–37. DOI: 10.1016/j.rser.2019.01.020

Ozcan, M. (2018). The role of renewables in increasing Turkey's self-sufficiency in electrical energy. *Renewable & Sustainable Energy Reviews*, *82*, 2629–2639. DOI: 10.1016/j.rser.2017.09.111

Ozgur, M. (2008). Review of Turkey's renewable energy potential. *Renewable Energy*, *33*(11), 2345–2356. DOI: 10.1016/j.renene.2008.02.003

Özkivrak, Ö. (2005). Electricity restructuring in Turkey. *Energy Policy*, *33*(10), 1339–1350. DOI: 10.1016/j.enpol.2003.12.010

Ozturk, I. (2010). A literature survey on energy–growth nexus. *Energy Policy*, *38*(1), 340–349. DOI: 10.1016/j.enpol.2009.09.024

Ozturk, I., & Acaravci, A. (2010). CO2 emissions, energy consumption and economic growth in Turkey. *Renewable & Sustainable Energy Reviews*, *14*(9), 3220–3225. DOI: 10.1016/j.rser.2010.07.005

Padmaja, M., Shitharth, S., Prasuna, K., Chaturvedi, A., Kshirsagar, P. R., & Vani, A. (2022). Grow of artificial intelligence to challenge security in IoT application. *Wireless Personal Communications*, *127*(3), 1829–1845. DOI: 10.1007/s11277-021-08725-4

Palaiogianni, A., Stylianou, M., Sarris, D., & Agapiou, A. (2022). Carob-Agro-Industrial Waste and Potential Uses in the Circular Economy. In Ramadan, M. F., & Farag, M. A. (Eds.), *Mediterranean Fruits Bio-wastes*. Springer., DOI: 10.1007/978-3-030-84436-3_33

Palm, F., Smeekes, S., & Urbain, J. (2006). *Bootstrap unit root tests: comparison and extensions.* Maastricht: METEOR, Maastricht University School of Business and Economics. METEOR Research Memorandum No. 015.

Panesar, A. (2019). *Machine learning and AI for healthcare*. Apress. DOI: 10.1007/978-1-4842-3799-1

Pao, H., & Fu, H. (2013). Renewable energy, non-renewable energy and economic growth in Brazil. *Renewable & Sustainable Energy Reviews*, *25*, 381–392. DOI: 10.1016/j.rser.2013.05.004

Parsons, W. T., Parsons, W. T., & Cuthbertson, E. G. (2001). *Noxious weeds of Australia*. CSIRO publishing.

Payne, J. (2009). On the dynamics of energy consumption and output in the US. *Applied Energy*, *86*(4), 575–577. DOI: 10.1016/j.apenergy.2008.07.003

Payne, J. (2011). On biomass energy consumption and real Output in the U.S. *Energy Sources. Part B, Economics, Planning, and Policy*, *6*(1), 47–52. DOI: 10.1080/15567240903160906

Pazokinejad, Z., & Salehi, S. (2014). An analysis of social factors influencing studentsâ environmental attitudes and performance. *Journal of Applied Sociology*, *25*(3), 71–88.

Pegkas, P. (2020). The impact of renewable and non-renewable energy consumption on economic growth: The case of Greece. *International Journal of Sustainable Energy*, *39*(4), 380–395. DOI: 10.1080/14786451.2019.1700261

Pennsylvania Department of Conservation and Natural Resources, Invasive species Management Plan., 2005.

Perron, P. (1989). The great crash, the oil price shock, and the unit root hypothesis. *Econometrica*, *57*(6), 1361–1401. DOI: 10.2307/1913712

Perron, P. (1997). Further Evidence on Breaking Trend Functions in Macroeconomic Variables. *Journal of Econometrics*, *80*(2), 355–385. DOI: 10.1016/S0304-4076(97)00049-3

Phukan, P., & Saikia, R. (2014). Wetland degradation and its conservation: A case study of some selected wetlands of Golaghat District, Assam, India. *Research Journal of Recent Sciences ISSN*, *2277*, 2502.

Piłatowska, M., Geise, A., & Włodarczyk, A. (2020). The effect of renewable and nuclear energy consumption on decoupling economic growth from CO2 emissions in Spain. *Energies*, *13*(9), 1–18. DOI: 10.3390/en13092124

Pinheiro, C., David, T. S., Baptista, P., & Guerra-Guimarães, L. (2022). Mediterranean woody agroecosystems in a warming and drier climate: The importance of sustainable knowledge-based management. *Flora (Jena)*, *291*, 152070. DOI: 10.1016/j.flora.2022.152070

Pisani, O., Gao, M., Maie, N., Miyoshi, T., Childers, D. L., & Jaffé, R. (2018). Compositional aspects of herbaceous litter decomposition in the freshwater marshes of the Florida Everglades. *Plant and Soil*, *423*(1-2), 87–98. DOI: 10.1007/s11104-017-3495-3

Platt, R. V., Ogra, M. V., Badola, R., & Hussain, S. A. (2016). Conservation-induced resettlement as a driver of land cover change in India: An object-based trend analysis. *Applied Geography (Sevenoaks, England)*, *69*, 75–86. DOI: 10.1016/j.apgeog.2016.02.006

Porter, M. E. (1980). *Competitive Strategy: Techniques for Analyzing Industries and Competitors*. Free Press.

Poveda, J. (2022). Effect of volatile and non-volatile metabolites from Leptosphaeria maculans on tomato calli under abiotic stresses. *Plant Stress*, *3*, 100054. DOI: 10.1016/j.stress.2021.100054

Pramanik, P. K. D., Pal, S., & Mukhopadhyay, M. (2022). Healthcare big data: A comprehensive overview. *Research anthology on big data analytics, architectures, and applications*, 119-147.

Prasad, S. N., Ramachandra, T. V., Ahalya, N., Sengupta, T., Kumar, A., Tiwari, A. K., & Vijayan, L. (2002). Conservation of wetlands of India-a review. *Tropical Ecology*, *43*(1), 173–186.

Puglielli, G., Laanisto, L., Gori, A., & Cardoso, A. A. (2023). Woody plant adaptations to multiple abiotic stressors: Where are we? *Flora (Jena)*, *299*, 152221. DOI: 10.1016/j.flora.2023.152221

Pyšek, P., Hulme, P. E., Simberloff, D., Bacher, S., Blackburn, T. M., Carlton, J. T., Dawson, W., Essl, F., Foxcroft, L. C., Genovesi, P., Jeschke, J. M., Kühn, I., Liebhold, A. M., Mandrak, N. E., Meyerson, L. A., Pauchard, A., Pergl, J., Roy, H. E., Seebens, H., & Richardson, D. M. (2020). Scientists' warning on invasive alien species. *Biological Reviews of the Cambridge Philosophical Society*, *95*(6), 1511–1534. DOI: 10.1111/brv.12627 PMID: 32588508

Pyšek, P., & Pyšek, A. (1995). Invasion by *Heracleum mantegazzianum* in different habitats in the Czech Republic. *Journal of Vegetation Science*, *6*(5), 711–718. DOI: 10.2307/3236442

Qureshi, R., Irfan, M., Ali, H., Khan, A., Nittala, A. S., Ali, S., & Alam, T. (2023). Artificial Intelligence and Biosensors in Healthcare and its Clinical Relevance: A Review. *IEEE Access : Practical Innovations, Open Solutions*, *11*, 61600–61620. DOI: 10.1109/ACCESS.2023.3285596

Rafiq, S., & Salim, R. (2009). Temporal causality between energy consumption and income in six Asian emerging countries. *Applied Economics Quarterly*, *55*(4), 335–350. DOI: 10.3790/aeq.55.4.335

Rafiq, S., & Salim, R. (2011). The linkage between energy consumption and income in six emerging economies of Asia: An empirical analysis. *International Journal of Emerging Markets*, *6*(1), 50–73. DOI: 10.1108/17468801111104377

Rakotomamonjy, S. N., Jones, J. P. G., Razafimanahaka, J. H., Ramamonjisoa, B., & Williams, S. J. (2015). The effects of environmental education on children's and parents' knowledge and attitudes towards lemurs in rural Madagascar. *Animal Conservation*, *18*(2), 157–166. DOI: 10.1111/acv.12153

Ramachandra, T. V. (2001). Restoration and management strategies of wetlands in developing countries. *Electronic Green Journal*, *1*(15). Advance online publication. DOI: 10.5070/G311510435

Ramon-Laca, L., & Mabberley, D. J. (2004). The ecological status of the carob-tree (*Ceratonia siliqua* Leguminosae) in the Mediterranean. *Botanical Journal of the Linnean Society*, *144*(4), 431–436. DOI: 10.1111/j.1095-8339.2003.00254.x

Rebelo, L. M., Finlayson, C. M., & Nagabhatla, N. (2009). Remote sensing and GIS for wetland inventory, mapping and change analysis. *Journal of Environmental Management*, *90*(7), 2144–2153. DOI: 10.1016/j.jenvman.2007.06.027 PMID: 18367311

Reddy, B. M. (2023). Amalgamation of Internet of Things and Machine Learning for Smart Healthcare Applications–A Review. *Int. J Comp. Eng. Sci. Res, 5*, 08-36.

Reşat Cica, H., & Karabulut, T. (2024). Çalışanların örgütsel bağlılık ile örgütsel özdeşleşme davranışının yeşil örgütsel davranış üzerindeki etkisinde çevre tutkusunun aracı rolü. *Gaziantep Üniversitesi Sosyal Bilimler Dergisi*, *23*(1), 304–326. DOI: 10.21547/jss.1312993

Richane, A., Rim, B. M., Wided, M., Riadh, K., Khaoula, A., Nizar, M., & Hanen, B. I. (2022). Variability of phenolic compounds and antioxidant activities of ten *Ceratonia siliqua* L. provenances. *Biochemical Systematics and Ecology*, *104*, 104486. DOI: 10.1016/j.bse.2022.104486

Richardson, D. M., & Van Wilgen, B. W. (2004). Invasive alien plants in South Africa: How well do we understand the ecological impacts? *South African Journal of Science*, *100*, 45–52.

Rideout, B. E., Hushen, K., McGinty, D., Perkins, S., & Tate, J. (2005). Endorsement of the new ecological paradigm in systematic and e-mail samples of college students. *The Journal of Environmental Education*, *36*(2), 15–23. DOI: 10.3200/JOEE.36.2.15-23

Rodrigues, J. F.Jr, Florea, L., de Oliveira, M. C., Diamond, D., & Oliveira, O. N.Jr. (2021). Big data and machine learning for materials science. *Discover Materials*, *1*(1), 1–27. DOI: 10.1007/s43939-021-00012-0 PMID: 33899049

Rojas, R., Flexas, J., & Coopman, R. E. (2022). The highest elevation treeline in the world: Polylepis tarapacana Phil. as a unique opportunity to study ecophysiological adaptations to extreme conditions. *Flora : the Journal of Infectious Diseases and Clinical Microbiology = Infeksiyon Hastaliklari ve Klinik Mikrobiyoloji Dergisi*, *292*, 152076. DOI: 10.1016/j.flora.2022.152076

Ronald, P. C., Albano, B., Tabien, R., Abenes, L., Wu, K., McCouch, S., & Tanksley, S. D. (1992). Genetic and physical analysis of the rice bacterial blight disease resistance locus, Xa21. *Molecular & General Genetics*, *236*(1), 113–1. DOI: 10.1007/BF00279649 PMID: 1362973

Rosacia, W.Z., Achivar, A.N. and Avanzado, M.B., (2004). *Lantana* and Hagonoy: Poisonous weeds prominent in rangeland and grassland areas. *Research Information Series on Ecosystems (Department of the Environment and Natural Resources, Republic of the Philippines),* 16(2).

Ross, I. A. (1999). *Medicinal plants of the world. Chemical constituents, traditional and modern medical uses*. Human Press.

Rtibi, K., Selmi, S., Grami, D., Amri, M., Eto, B., El-Benna, J., & Marzouki, L. (2017). Chemical constituents and pharmacological actions of carob pods and leaves (*Ceratonia siliqua* L.) on the gastrointestinal tract: A review. *Biomedicine and Pharmacotherapy*, *93*, 522–528. DOI: 10.1016/j.biopha.2017.06.088 PMID: 28686965

Rudin, C. (2019). Stop explaining black box machine learning models for high stakes decisions and use interpretable models instead. *Nature Machine Intelligence*, *1*(5), 206–215. DOI: 10.1038/s42256-019-0048-x PMID: 35603010

Ruehr, N. K., Grote, R., Mayr, S., & Arneth, A. (2019). Beyond the extreme: Recovery of carbon and water relations in woody plants following heat and drought stress. *Tree Physiology*, *39*(8), 1285–1299. DOI: 10.1093/treephys/tpz032 PMID: 30924906

Saadat, M. N., & Shuaib, M. (2020). Advancements in deep learning theory and applications: Perspective in 2020 and beyond. *Advances and Applications in Deep Learning, 3*.

Saatci, M., & Dumrul, Y. (2012). The relationship between energy consumption and economic growth: Evidence from a structural break analysis for Turkey. *International Journal of Energy Economics and Policy*, *3*(1), 20–29.

Sadorsky, P. (2009). Renewable energy consumption, CO2 emissions and oil prices in the G7 countries. *Energy Economics*, *31*(3), 456–462. DOI: 10.1016/j.eneco.2008.12.010

Saidi, K., & Mbarek, M. (2016). Nuclear energy, renewable energy, CO2 emissions, and economic growth for nine developed countries: Evidence from panel Granger causality tests. *Progress in Nuclear Energy*, *88*, 364–374. DOI: 10.1016/j.pnucene.2016.01.018

Salehi, S., Nejad, Z. P., Mahmoudi, H., & Burkart, S. (2016). Knowledge of global climate change: View of Iranian university students. *International Research in Geographical and Environmental Education*, *25*(3), 226–243. DOI: 10.1080/10382046.2016.1155322

Salim, R., & Rafiq, S. (2012). Why do some emerging economies proactively accelerate the adoption of renewable energy? *Energy Economics*, *34*(4), 1051–1057. DOI: 10.1016/j.eneco.2011.08.015

Sam, N., Gürsakal, S., & Sam, R. (2010). Üniversite öğrencilerinin çevresel risk algısı ve çevresel tutumlarının belirlenmesi. *Akademik Bakış Dergisi*, *20*, 1–16.

Samways, M. J., & Taylor, S. T. (2004). Impacts of invasive plants on red-listed South African dragonflies (Odonata). *South African Journal of Science*, *100*, 78–80.

Sanera, M. (1998). Environmental education: Promise and performance. [CJEE]. *Canadian Journal of Environmental Education*, •••, 9–26.

Sanower, H. (2019). Present scenario of global salt affected soils, its management and importance of salinity research. *International Research Journal of Biological Sciences*, *1*, 1–3.

Sapre, S., Gontia-Mishra, I., & Tiwari, S. (2022). Plant growth-promoting rhizobacteria ameliorates salinity stress in pea (*Pisum sativum*). *Journal of Plant Growth Regulation*, *41*(2), 647–656. DOI: 10.1007/s00344-021-10329-y

Sarkar, A., & Jain, S. K. (2007, October). Using Remote Sensing data to study wetland dynamics–A case study of Harike Wetland. In *Proceedings of the Taal2007: The 12th World Lake Conference, Kusatsu, Japan* (Vol. 680, p. 684).

Sarkar, P., & Das, T. (2016). Wetland ecosystem services and its valuation with special reference to India—A review. *Biodiversity and Environmental Conservation*, 59-76.

Sarvestani, A. A. (2012). Environmental attitude and behavior of students of Gorgan University of Agricultural Sciences and Natural Resources. *Ulum-i Tarvij va Amuzish-i Kishavarzi*, *7*(2), 77–92.

Schrier-Uijl, A. P., Silvius, M., Parish, F., Lim, K. H., Rosediana, S., & Anshari, G. (2013). Environmental and social impacts of oil palm cultivation on tropical peat. *Reports from the Technical Panels of the 2nd Greenhouse Gas Working Group of the Roundtable on Sustainable Palm Oil Kuala Lumpur*. https://d2ouvy59p0dg6k.cloudfront.net/downloads/the_environemental_and_social_impacts_of_oil_palm_on_peat.pdf

Scuderi, L. A., Laudadio, C. K., & Fawcett, P. J. (2010). Monitoring playa lake inundation in the western United States: Modern analogues to late-Holocene lake level change. *Quaternary Research*, *73*(1), 48–58. DOI: 10.1016/j.yqres.2009.04.004

Sengan, S., Kamalam, G. K., Vellingiri, J., Gopal, J., Velayutham, P., & Subramaniyaswamy, V. (2020). Medical information retrieval systems for e-Health care records using fuzzy based machine learning model. *Microprocessors and Microsystems*, •••, 103344.

Seward, P. (2010). Challenges facing environmentally sustainable ground water use in South Africa. *Groundwater*, *48*(2), 239–245. https://ngwa.onlinelibrary.wiley.com/doi/full/10.1111/j.1745-6584.2008.00518. DOI: 10.1111/j.1745-6584.2008.00518.x PMID: 19016897

Shafiei, A., & Maleksaeidi, H. (2020). Pro-environmental behavior of university students: Application of protection motivation theory. *Global Ecology and Conservation*, *22*, e00908. DOI: 10.1016/j.gecco.2020.e00908

Shahbaz, M., Raghutla, C., Chittedi, K., Jiao, Z., & Vo, X. (2020). The effect of renewable energy consumption on economic growth: Evidence from the renewable energy country attractive index. *Energy*, *207*, 1–14. DOI: 10.1016/j.energy.2020.118162

Shahid, Z. (2021). Distributed Machine Learning for Anomalous Human Activity Recognition using IoT Systems.

Shari, B. E., Moumouni, Y., Ohunakin, O. S., Blechinger, P., Madougou, S., & Rabani, A. (2024). Exploring the role of green hydrogen for distributed energy access planning towards net-zero emissions in Nigeria. *Sustainable Energy Research*, *11*(1), 16. DOI: 10.1186/s40807-024-00107-1

Sharma, S., & Singh, P. (Eds.). (2021). *Wetlands conservation: Current challenges and future strategies*. John Wiley & Sons. DOI: 10.1002/9781119692621

Shirley, H. L. (1936). Lethal high temperatures for conifers, and the cooling effect of transpiration. *Journal of Agricultural Research*, *53*, 239–258.

Shobeiri, S. M., Omidvar, B., & Prohallada, N. N. (2007). A comparative study of environmental awareness among secondary school students in Iran and India. *International Journal of Environmental Research*, *1*(1), 28–34.

Silvapulle, P. (1996). Testing for a unit root in a time series with mean shifts. *Applied Economics Letters*, *3*(10), 629–635. DOI: 10.1080/135048596355826

Sims, C., Stock, J., & Watson, M. (1990). Inference in linear time series models with some unit roots. *Econometrica*, *58*(1), 113–144. DOI: 10.2307/2938337

Simsek, H., & Simsek, N. (2013). Recent incentives for renewable energy in Turkey. *Energy Policy*, *63*, 521–530. DOI: 10.1016/j.enpol.2013.08.036

Singh, P., Choudhary, K. K., Chaudhary, N., Gupta, S., Sahu, M., Tejaswini, B., & Sarkar, S. (2022). Salt stress resilience in plants mediated through osmolyte accumulation and its crosstalk mechanism with phytohormones. *Frontiers in Plant Science*, *13*, 1006617. DOI: 10.3389/fpls.2022.1006617 PMID: 36237504

Slack, N., Chambers, S., & Johnston, R. (2023). *Operations Management*. Pearson Education.

Slingo, J. M., & Slingo, M. E. (2024). The science of climate change and the effect of anaesthetic gas emissions. *Anaesthesia*, *79*(3), 252–260. DOI: 10.1111/anae.16189 PMID: 38205585

Soytas, U., & Sari, R. (2003). Energy consumption and GDP: Causality relationship in G-7 countries and emerging markets. *Energy Economics*, *25*(1), 33–37. DOI: 10.1016/S0140-9883(02)00009-9

Squalli, J. (2007). Electricity consumption and economic growth: Bounds and causality analyses of OPEC members. *Energy Economics*, *29*(6), 1192–1205. DOI: 10.1016/j.eneco.2006.10.001

Stavrou, I. J., Christou, A., & Kapnissi-Christodoulou, C. P. (2018). Polyphenols in carobs: A review on their composition, antioxidant capacity and cytotoxic effects, and health impact. *Food Chemistry*, *269*, 355–374. DOI: 10.1016/j.foodchem.2018.06.152 PMID: 30100447

Stefanakis, A. I. (2019). The role of constructed wetlands as green infrastructure for sustainable urban water management. *Sustainability (Basel)*, *11*(24), 6981. DOI: 10.3390/su11246981

Steg, L., & Vlek, C. (2009). Encouraging pro-environmental behaviour: An integrative review and research agenda. *Journal of Environmental Psychology*, *29*(3), 309–317. DOI: 10.1016/j.jenvp.2008.10.004

Stern, D. (1993). Energy and economic growth in the USA: A multivariate approach. *Energy Economics, 15*(2), 137–150. DOI: 10.1016/0140-9883(93)90033-N

Stern, P., Dietz, T., & Kalof, L. (1993). Value orientations, gender, and environmental concern. *Environment and Behavior*, ●●●, 25.

Stirton, C. H. 1977. Some thoughts on the polyploid complex *Lantana camara* L. (Verbenaceae). In *Proceedings of the Second National Weeds Conference of South Africa.* pp. 321–340.

Subcommittee, W. (2013). *Classification of wetlands and deepwater habitats of the United States*. Federal Geographic Data Committee.

Subrahmanyam, K. (2000). Self-reliant defence and Indian industry. *Strategic Analysis, 24*(7), 1221–1234. https://www.tandfonline.com/doi/abs/10.1080/09700160008455283DOI: 10.1080/09700160008455283

Sukeshini, S., P., Ved, M., Chintalapti, J., & Pal, S. N. (2021). Big data analytics and machine learning technologies for HPC applications. In Evolving Technologies for Computing, Communication and Smart World: Proceedings of ETCCS 2020 (pp. 411-424). Springer Singapore.

Sulaiman, J., Azman, A., & Saboori, B. (2013). The potential of renewable energy: Using the environmental Kuznets curve model. *American Journal of Environmental Sciences*, *9*(2), 103–112. DOI: 10.3844/ajessp.2013.103.112

Sullivan, R. C., & Pryor, S. C. (2014). Quantifying spatiotemporal variability of fine particles in an urban environment using combined fixed and mobile measurements. *Atmospheric Environment*, *89*, 664–671. https://www.sciencedirect.com/science/article/abs/pii/S135223101400168X?via%3Dihub. DOI: 10.1016/j.atmosenv.2014.03.007

Sutton, P. (2004). A perspective on environmental sustainability. *Paper on the Victorian Commissioner for Environmental Sustainability*, *1*, 32. https://www.donboscogozo.org/images/pdfs/energy/A-Perspective-on-Environmental-Sustainability.pdf

Takeuchi, K., & Aginam, O. (2011). Sustainability challenges and opportunities in Africa. *Sustainability Science*, *6*(1), 3–5. https://link.springer.com/article/10.1007/s11625-010-0124-y. DOI: 10.1007/s11625-010-0124-y

TEIAS. (2024). Turkish electrical energy 10-year generation capacity projection. 1-57. <https://www.teias.gov.tr/ilgili-raporlar> (last visited on 15.03.2025).

Teikari, P., Najjar, R. P., Schmetterer, L., & Milea, D. (2019). Embedded deep learning in ophthalmology: Making ophthalmic imaging smarter. *Therapeutic Advances in Ophthalmology, 11*, 2515841419827172. DOI: 10.1177/2515841419827172 PMID: 30911733

Thapa, C., & Camtepe, S. (2021). Precision health data: Requirements, challenges and existing techniques for data security and privacy. *Computers in Biology and Medicine, 129*, 104130. DOI: 10.1016/j.compbiomed.2020.104130 PMID: 33271399

The World Bank. (2023). *Energy Access in Sub-Saharan Africa.* Retrieved from https://www.worldbank.org/en/programs/africa-energy

Tian, X., Chang, Z., Wang, R., Li, G., & Qi, Y. (2024). The abiotic stress gene (Asg) family member Asg2 as a modulator of plant responses to salt stress. *Plant Stress, 14*, 100596. DOI: 10.1016/j.stress.2024.100596

Tikka, P. M., Kuitunen, M. T., & Tynys, S. M. (2000). Effect of educational background on students' attitude, activity levels, and knowledge concerning the environment. *The Journal of Environmental Education, 31*(3), 12–19. DOI: 10.1080/00958960009598640

Tiwari, S. K., Kaur, J., Singla, P., & Hrisheekesha, P. N. (2022, September). A Comprehensive Review of Big Data Analysis Techniques in Health-Care. In *International Conference on Emergent Converging Technologies and Biomedical Systems* (pp. 401-420). Singapore: Springer Nature Singapore.

Tiwari, A. (2011). A structural VAR analysis of renewable energy consumption, real GDP and CO2 emissions: Evidence from India. *Economic Bulletin, 31*(2), 1793–1806.

Toda, H., & Yamamoto, T. (1995). Statistical inference in vector autoregressions with possibly integrated processes. *Journal of Econometrics, 66*(1-2), 225–250. DOI: 10.1016/0304-4076(94)01616-8

Toklu, E. (2013). Overview of potential and utilization of renewable energy sources in Turkey. *Renewable Energy, 50*, 456–463. DOI: 10.1016/j.renene.2012.06.035

Toth, F., & Rogner, H. (2006). Oil and nuclear power: Past, present, and future. *Energy Economics, 28*(1), 1–25. DOI: 10.1016/j.eneco.2005.03.004

Trenberth, K. E., Dai, A., Van Der Schrier, G., Jones, P. D., Barichivich, J., Briffa, K. R., & Sheffield, J. (2014). Global warming and changes in drought. *Nature Climate Change, 4*(1), 17–22. DOI: 10.1038/nclimate2067

Tugcu, C., Ozturk, I., & Aslan, A. (2012). Renewable and non-renewable energy consumption and economic growth relationship revisited: Evidence from G7 countries. *Energy Economics*, *34*(6), 1942–1950. DOI: 10.1016/j.eneco.2012.08.021

Tuncer, G., Ertepinar, H., Tekkaya, C., & Sungur, S. (2005). Environmental attitudes of young people in Turkey: Effect of school type and gender. *Environmental Education Research*, *11*(2), 215–233. DOI: 10.1080/1350462042000338379

Turkish Central Bank database. (2025). Electronic Data Delivery System. <https://evds2.tcmb.gov.tr/index.php?> (last visited on 19.03.2025).

Tyagi, S., Garg, N., & Paudel, R. (2014). Environmental degradation: Causes and consequences. *European researcher, 81*(8-2), 1491. file:///C:/Users/dasha/Downloads/1409224435.pdf

Tyree, M. T., & Sperry, J. S. (1989). Vulnerability of xylem to cavitation and embolism. *Annual Review of Plant Physiology and Plant Molecular Biology*, *40*(1), 19–38. DOI: 10.1146/annurev.pp.40.060189.000315

Uddin, M. G., Nash, S., Rahman, A., & Olbert, A. I. (2022). A comprehensive method for improvement of water quality index (WQI) models for coastal water quality assessment. *Water Research, 219*, 118532. DOI: 10.1016/j.watres.2022.118532 PMID: 35533623

Umuhire, M. L., & Fang, Q. (2016). Method and application of ocean environmental awareness measurement: Lessons learnt from university students of China. *Marine Pollution Bulletin*, *102*(2), 289–294. DOI: 10.1016/j.marpolbul.2015.07.067 PMID: 26242972

UNESCO–UNEP. (1991). United Nations environment programme: report of the governing council on the work of its 16th session, 20-31 May 1991. UNEP. Governing Council (16th sess.: 1991: Nairobi)

United Nations Framework Convention on Climate Change (UNFCCC). (2022). *Paris Agreement Commitments and African Climate Action.* Retrieved from https://unfccc.int/process-and-meetings/the-paris-agreement

Uyeki, E. S., & Holland, L. J. (2000). Diffusion of pro-environment attitudes? *The American Behavioral Scientist*, *43*(4), 646–662. DOI: 10.1177/00027640021955478

Uzun, N., & Sağlam, N. (2007). The environmental education in secondary school and teachers' opinions on environmental education programs. *Eurasian Journal of Educational Research*, *26*, 176–187.

Vahdati, K., & Leslie, C. (Eds.). (2013). *Abiotic stress: plant responses and applications in agriculture*. BoD–Books on Demand. DOI: 10.5772/45842

Vaillancourt, K., Labriet, M., Loulou, R., & Waaub, J.-P. (2008). The role of nuclear energy in long-term climate scenarios: An analysis with the World-TIMES model. *Energy Policy*, *36*(7), 2296–2307. DOI: 10.1016/j.enpol.2008.01.015

Van Liere, K. D., & Dunlap, R. E. (1981). Environmental concern: Does it make a difference how it's measured? *Environment and Behavior*, *13*(6), 651–676. DOI: 10.1177/0013916581136001

Varsani, A., Navas-Castillo, J., Moriones, E., Hernández-Zepeda, C., Idris, A., Brown, J. K., Zerbini, F. M., & Martin, D. P. (2014). Establishment of three new genera in the family Geminiviridae: Becurtovirus, Eragrovirus and Turncurtovirus. *Archives of Virology*, *159*(8), 2193–2203. DOI: 10.1007/s00705-014-2050-2 PMID: 24658781

Verma, M., Bakshi, N., & Nair, R. P. (2001). Economic valuation of Bhoj Wetland for sustainable use. *Unpublished project report for World Bank assistance to Government of India, Environmental Management Capacity-Building. Bhopal: Indian Institute of Forest Management, 35*.

Versfeld, D. B., Le Maitre, D. C., & Chapman, R. A. (1998). Alien invading plants and Water Resources in South Africa: A Preliminary Assessment. Water Research Commission Report No. TT99/98.

Vicente-Molina, M. A., Fernández-Sainz, A., & Izagirre-Olaizola, J. (2018). Does gender make a difference in pro-environmental behavior? The case of the Basque Country University students. *Journal of Cleaner Production*, *176*, 89–98. DOI: 10.1016/j.jclepro.2017.12.079

Vining, J. R., & Ebro, A. (1990). What makes a recycler? A comparison of recyclers and non-recyclers. *Environment and Behavior*, *22*(1), 56–73. DOI: 10.1177/0013916590221003

Vlahov, S. J., & Treagust, D. F. (1988). Students' knowledge of energy and attitudes to energy conservation. *School Science and Mathematics*, *88*(6), 452–458. DOI: 10.1111/j.1949-8594.1988.tb11837.x

Wang, S., Li, Q., Fang, C., & Zhou, C. (2016). The relationship between economic growth, energy consumption, and CO2 emissions: Empirical evidence from China. *The Science of the Total Environment*, *542*, 360–371. DOI: 10.1016/j.scitotenv.2015.10.027 PMID: 26520261

Warren, J. K., & Warren, J. K. (2016). Sabkhas, saline mudflats and pans. *Evaporites: A Geological Compendium*, 207-301.

Wellburn, A. R. (1994). The spectral determination of chlorophylls a and b, as well as total carotenoids, using various solvents with spectrophotometers of different resolution. *Journal of Plant Physiology*, *144*(3), 307–313. DOI: 10.1016/S0176-1617(11)81192-2

Weyerstahl, P., Marschall, H., Eckhardt, A., & Christiansen, C. (1999). Constituents of commercial Brazilian *Lantana* oil. *Flavour and Fragrance Journal*, *14*(1), 15–28. DOI: 10.1002/(SICI)1099-1026(199901/02)14:1<15::AID-FFJ777>3.0.CO;2-M

White, E., & Kaplan, D. (2017). Restore or retreat? Saltwater intrusion and water management in coastal wetlands. *Ecosystem Health and Sustainability*, *3*(1), e01258. DOI: 10.1002/ehs2.1258

Wolde-Rufael, Y., & Menyah, K. (2010). Nuclear energy consumption and economic growth in nine developed countries. *Energy Economics*, *32*(3), 550–556. DOI: 10.1016/j.eneco.2010.01.004

World Bank database. (2025). World Development Indicators. <https://databank.worldbank.org/source/world-development-indicators> (last visited on 19.03.2025).

World Economic Forum. (2023). *Africa's Energy Future and Sustainable Development.* Retrieved from https://www.weforum.org/agenda/2023/10/renewable-energy-africa

Worsley, A., & Skrzypiec, G. (1998). Environmental attitudes of senior secondary school students in South Australia. *Global Environmental Change*, *8*(3), 209–225. DOI: 10.1016/S0959-3780(98)00016-8

Wu, C., Zhang, M., Liang, Y., Zhang, L., & Diao, X. (2023). Salt stress responses in foxtail millet: Physiological and molecular regulation. *The Crop Journal*, *11*(4), 1011–1021. DOI: 10.1016/j.cj.2023.06.001

Xiao, C., & Hong, D. (2018). Gender differences in environmental behaviors among the Chinese public: Model of mediation and moderation. *Environment and Behavior*, *50*(9), 975–996. DOI: 10.1177/0013916517723126

Xie, X., He, Z., Chen, N., Tang, Z., Wang, Q., & Cai, Y. (2019). The roles of environmental factors in regulation of oxidative stress in plant. *BioMed Research International*, *2019*, 2019. DOI: 10.1155/2019/9732325 PMID: 31205950

Xu, T., Weng, B., Yan, D., Wang, K., Li, X., Bi, W., Li, M., Cheng, X., & Liu, Y. (2019). Wetlands of international importance: Status, threats, and future protection. *International Journal of Environmental Research and Public Health*, *16*(10), 1818. DOI: 10.3390/ijerph16101818 PMID: 31121932

Yamasaki, H. (2005). The NO world for plants: Achieving balance in an open system. *Plant, Cell & Environment, 28*(1), 78–84. DOI: 10.1111/j.1365-3040.2005.01297.x

Yamian, Z., Yifei, J., Shengwu, J., Qing, Z., Duoduo, F., Yumin, G., & Guangchun, L. (2012). Wuliangsuhai wetlands: A critical habitat for migratory water birds. *Journal of Resources and Ecology, 3*(4), 316–323. DOI: 10.5814/j.issn.1674-764x.2012.04.004

Yang, X., Zhao, S., Qin, X., Zhao, N., & Liang, L. (2017). Mapping of urban surface water bodies from LandsatMSI imagery at 10 m resolution via NDWI-based image sharpening. *Remote Sensing (Basel), 9*(6), 596. DOI: 10.3390/rs9060596

Yan, L., Zhou, J., Li, H., Zhong, R., Zhuang, J., Xu, X., Wang, Y., Liao, X., & Shi, B. (2023). Wearable synthetic leather-based high-performance X-ray shielding materials enabled by the plant polyphenol-and hierarchical structure-facilitated dispersion. *Collagen and Leather, 5*(1), 12. https://jlse.springeropen.com/counter/pdf/10.1186/s42825-023-00119-5.pdf. DOI: 10.1186/s42825-023-00119-5

Yaseen, U., & Chalotra, E. S. C. E. S. (2022). Sustainable Management Strategies Urban Water Bodies and Wetlands in the Indian Context. *International Journal of Innovative Research in Engineering & Management, 9*(5), 257–265.

Yazdi, S., & Shakouri, B. (2017). Renewable energy nonrenewable energy consumption, and economic growth. *Energy Sources. Part B, Economics, Planning, and Policy, 12*(12), 1038–1045. DOI: 10.1080/15567249.2017.1316795

Yildirim, E., Saraç, Ş., & Aslan, A. (2012). Energy consumption and economic growth in the USA: Evidence from renewable energy. *Renewable & Sustainable Energy Reviews, 16*(9), 6770–6774. DOI: 10.1016/j.rser.2012.09.004

Yılmaz S. Ertekin Ö. (2019). Bilimsel Kriterlere Dayalı Çevre Bilinci Ölçümü: Bir Literatür Taraması. *İşletme Bilimi Dergisi*, 7(1), 155-181.

Yılmaz, V., & Arslan, T. (2011). Üniversite öğrencilerinin çevre koruma vaatleri ve çevre dostu tüketim davranışlarının incelenmesi. *Anadolu University Journal of Social Sciences, 11*(3), 1–10.

Yılmaz, V., Çelik, E., & Yağızer, C. (2009). Investigating the effects of environmental sensitivity and environmental behavior on ecological product buying behavior through structural equation model. *Anadolu University Journal of Social Science, 9*(2), 1–14.

Yoon, M.-Y., Cha, B., and Kim, J.-C. (2013). Recent trends in studies on botanical fungicides in agriculture. The plant pathology journal 29, 1.20.

Yuksel, I., & Kaygusuz, K. (2011). Renewable energy sources for clean and sustainable energy policies in Turkey. *Renewable & Sustainable Energy Reviews*, *15*(8), 4132–4144. DOI: 10.1016/j.rser.2011.07.007

Yu, W., Ramanathan, R., & Nath, P. (2017). Environmental pressures and performance: An analysis of the roles of environmental innovation strategy and marketing capability. *Technological Forecasting and Social Change*, *117*, 160–169. DOI: 10.1016/j.techfore.2016.12.005

Zachariades, C., Paterson, I. D., Strathie, L. W., Hill, M. P., & Van Wilgen, B. W. (2017). Assessing the status of biological control as a management tool for suppression of invasive alien plants in South Africa. *Bothalia –. African Biodiversity and Conservation*, *47*(2), 1–19.

Zarat, L., Alfaiz, C., Chiadmi, N., Kallida, R., Gaboun, F., & Ibriz, M. (2020). Plant growth, physiological response and osmotic adjustment of wild *bituminaria bituminosa* (L.) stirton accessions under salt stress. *Plant Cell Biotechnology and Molecular Biology*, *21*(1&2), 59–69.

Zedler, J. B., & Kercher, S. (2005). Wetland resources: Status, trends, ecosystem services, and restorability. *Annual Review of Environment and Resources*, *30*(1), 39–74. DOI: 10.1146/annurev.energy.30.050504.144248

Zhang, M., Ma, C., Qiao, S., Li, H., Zhao, W., & Liu, B. (2025). Compensatory growth and ion balance adaptation mechanisms of Salix matsudana Koidz under heterogeneous salinity stress. *BMC Plant Biology*, *25*(1), 231. DOI: 10.1186/s12870-025-06252-z PMID: 39979843

Zhe, L., Yüksel, S., Dinçer, H., Mukhtarov, S., & Azizov, M. (2021). The Positive Influences of Renewable Energy Consumption on Financial Development and Economic Growth. *SAGE Open*, *11*(3), 1–10. DOI: 10.1177/21582440211040133

Zhou, S. K., Greenspan, H., Davatzikos, C., Duncan, J. S., Van Ginneken, B., Madabhushi, A., Prince, J. L., Rueckert, D., & Summers, R. M. (2021). A review of deep learning in medical imaging: Imaging traits, technology trends, case studies with progress highlights, and future promises. *Proceedings of the IEEE*, *109*(5), 820–838. DOI: 10.1109/JPROC.2021.3054390 PMID: 37786449

Zhou, S., Zhang, R., Chen, D., & Zhu, X. (2021). A novel framework for bringing smart big data to proactive decision making in healthcare. *Health Informatics Journal*, *27*(2), 14604582211024698. DOI: 10.1177/14604582211024698 PMID: 34159834

Zhou, X., Joshi, S., Khare, T., Patil, S., Shang, J., & Kumar, V. (2021). Nitric oxide, crosstalk with stress regulators and plant abiotic stress tolerance. *Plant Cell Reports*, *40*(8), 1395–1414. DOI: 10.1007/s00299-021-02705-5 PMID: 33974111

Zhu-Salzman, K., Salzman, R. A., Ahn, J.-E., & Koiwa, H. (2004). Transcriptional regulation of sorghum defense determinants against a phloem-feeding aphid. *Plant Physiology*, *134*(1), 420–431. DOI: 10.1104/pp.103.028324 PMID: 14701914

Zimmermann, L. K. (1996). Knowledge, affect, and the environment: 15 years of research (1979–1993). *The Journal of Environmental Education*, *27*(3), 41–44. DOI: 10.1080/00958964.1996.9941466

Zivot, E., & Andrews, D. (1992). Further evidence on the great crash, the oil-price shock, and the unit-root hypothesis. *Journal of Business & Economic Statistics*, *10*(3), 25–44. DOI: 10.1080/07350015.1992.10509904

Zörb, C., Geilfus, C. M., & Dietz, K. J. (2019). Salinity and crop yield. *Plant Biology*, *21*(S1), 31–38. DOI: 10.1111/plb.12884 PMID: 30059606

About the Contributors

Alabi Olusoji David is a dedicated doctoral student at the Faculty of Applied Science, Durban University of Technology, where he is pursuing research in the field of phytoremediation and the decontamination of polluted sites. His academic journey has been marked by notable achievements, including the completion of a Master's degree in 2021 at Mangosuthu University of Technology in South Africa. His teaching experience encompasses diverse subjects, as he lectured on Conservation Ecology and Animal Behavior in 2021 and more recently on Project Management and Wildlife Management in the Department of Nature Conservation at Mangosuthu University of Technology. Alabi's contributions extend beyond academia, as he has authored and co-authored papers in accredited journals and served as a reviewer for various scholarly publications. His diverse interests span across climate change, carbon sequestration, integrated pest control, waste recycling, and food security, reflecting a holistic approach to addressing environmental and agricultural challenges. Alabi O D's academic and professional endeavors highlight his dedication to advancing knowledge and contributing to sustainable development initiatives

Melford Mbedzi is a lecturer at Mangosuthu university of Technology. I teach Soil Sciences 1(second year), Vegetation Management 3(Third year), Strategic Management for Environmental Management (Msc), Environmental Education (Advanced Diploma). He has a PhD in Botany, focusing on invasion biology.

Sithembile Nkosi is a lecturer at Mangosuthu University of Technology Department of Nature Conservation, Natural Sciences. She conservation ecology, and administration. Doctorate in view

***

Umut Akduğan has his BA degree in the field of Mathematics Education at Marmara University in 2007. Subsequently, he received his MA degree in the field of Econometrics at Trakya University and PhD degrees in the field of Economics at Istanbul University. The author currently continues his career at Trakya University, Faculty of Economics and Administrative Sciences, Department of Public Finance, since 2017. Dr. Akduğan is interested in Econometrics of Macroeconomic Modelling, Environmental Economics and Fiscal Policy.

Ferda Alper Ay has a PhD in Social Sciences in Strategy and Management. She is working as a professor in the Department of Health Management at Sivas Cumhuriyet University in Turkey. Skills and areas of expertise: Management and organization, organizational psychology, human resources management, work motivation, organizational cynicism, organizational culture, workplace management, favoritism, workplace mobbing, leadership, organizational citizenship behavior, unethical pro-organizational behaviors (UPB), whistleblowing, abusive supervision, counterproductive work behaviors, organizational behavior, health management.

Oumaima Ninich is a prominent researcher based in Morocco, specializing in biosecurity, biosafety, and plant sciences. She is affiliated with notable institutions, including Mohammed V University in Rabat. Her research contributions encompass various topics, particularly in biology and botany, with a focus on the properties of wood tar and its applications in environmental health.

Mikail Pehlivan has his BA degree in the field of Public Administration and Public Finance (double major) at Afyon Kocatepe University in 2015. Subsequently, he received his MA and PhD degrees in the field of Public Finance at Istanbul University. The author currently continues his career at Trakya University, Faculty of Economics and Administrative Sciences, Department of Public Finance, since 2017. Dr. Pehlivan is interested in Public Economics, Environmental Economics and Regulation Economics.

Sebataolo Rahlao (Seb) is an ecologist with a PhD in conservation ecology and over a decade of experience in undertaking and managing research teams in terrestrial and marine science. His expertise lie in generating and synthesizing science to support evidence-based decision-making in biodiversity conservation with focus on invasive alien species. He has been a lead author and coordinating lead author with the Intergovernmental Science-Policy Platform on Biodiversity and Ecosystem Services (IPBES). He is now the Executive Manager at the Oceanographic Research Institute, a research division of the South African Association for Marine Biological Research in Durban, South Africa.

Bhupinder Singh working as Professor at Sharda University, India. Also, Honorary Professor in University of South Wales UK and Santo Tomas University Tunja, Colombia. His areas of publications as Smart Healthcare, Medicines, fuzzy logics, artificial intelligence, robotics, machine learning, deep learning, federated learning, IoT, PV Glasses, metaverse and many more. He has 3 books, 139 paper publications, 163 paper presentations in international/national conferences and seminars, participated in more than 40 workshops/FDP's/QIP's, 25 courses from international universities of repute, organized more than 59 events with international and national academicians and industry people's, editor-in-chief and co-editor in journals, developed new courses. He has given talks at international universities, resource person in international conferences such as in Nanyang Technological University Singapore, Tashkent State University of Law Uzbekistan; KIMEP University Kazakhstan, All'ah meh Tabatabi University Iran, the Iranian Association of International Criminal law, Iran and Hague Center for International Law and Investment, The Netherlands, Northumbria University Newcastle UK,

Nazan Yalçin Erik completed her undergraduate education in the Geological Engineering Department of Sivas Cumhuriyet University, Faculty of Engineering between 1990-1994, her Master's degree in the Institute of Science at the same university in 1998, and her PhD between 1998-2002. She has been continuing her academic activities as a professor in the Geological Engineering Department of Sivas Cumhuriyet University, Faculty of Engineering since 2021. Her fields of study are; Conventional, Unconventional energy (Shale Gas, Coalbed methane) and energy resources research and evaluation, including coal, bituminous shale and oil. In addition to many international and national publications and reports on this subject, she has six scientific books (Global Oil and Gas Industry, Shale Gas, Petroleum and Evaporites, Annotated Dictionary of Energy Terms, Organic Geochemistry) and 4 book chapters.

Index

S

U

W

www.ingramcontent.com/pod-product-compliance
Lightning Source LLC
LaVergne TN
LVHW081333300926
849161LV00008B/1963
9798337315492